C0-AJQ-911

Mathematical Statistics

AN INTRODUCTION BASED
ON THE NORMAL DISTRIBUTION

Mathematical Statistics

AN INTRODUCTION BASED ON THE NORMAL DISTRIBUTION

SIMEON M. BERMAN

Associate Professor of Mathematics
New York University

INTEXT EDUCATIONAL PUBLISHERS

Scranton Toronto London

ISBN 7002 2349 5

 Printed in the United States of America by The Haddon Craftsmen, Inc., Scranton, Pennsylvania. Library of Congress Catalog Card Number: 70-140844.

To Jessica, Migdana, Tehilah

Preface

There is much to study but limited time for it in an introductory course in a mathematical subject—or any subject. The instructor often must choose between two courses of the following types: (a) one consisting of quick visits to many representative, sometimes unrelated areas, and (b) another consisting of a profound study of important special cases.

While courses are often combinations of these types, they are usually more like either one or the other. Each type has good and bad features: (a) furnishes a broad but possibly shallow view of the subject, while (b) gives a penetrating but possibly narrow view. Almost every introductory text in mathematical statistics is meant for a course of the first type. Personally, I prefer a type (b) course, and have written this book as an alternative to the others.

For this introduction to the theory of mathematical statistics, the reader must have a knowledge of differential and integral calculus in one variable, and a superficial knowledge of the two- and three-variable calculus. He should know the concepts of *limit* and *least upper* and *greatest lower bounds*, as presented in most books on calculus. While there is no requirement—formal or informal—of probability, such knowledge would certainly assist the reader.

While there are many excellent introductory books on statistics, there are none at the premeasure theoretic level which are mathematically complete and unified, and cover the classical theorems of inference. This book was motivated by the need for such a text. I have restricted most of this work to the normal distribution because

(i) The theory is complete and unified.
(ii) Many of the results in normal theory are classical, motivating examples of results in general statistical theory.

(iii) The best known and most used statistical procedures are those based on the normal distribution.

(iv) Many distributions arising in applications are approximately normal, or can be transformed into normal distributions.

(v) Statistical procedures based on the assumption of normality are often "robust" in nonnormal situations.

I do not say that normal theory is the *only* worthwhile branch of statistics; indeed, the important subject of nonparametric statistics grew from the need to *avoid* the assumption of normality. At the end of each chapter there is a section which briefly shows how the results for the normal case extend to the general case. These sections are not fully rigorous but are intended to describe the position of the normal distribution in relation to the general theory.

Thorough expositions of many of the results in this book have previously been limited to more advanced works requiring a knowledge of measure theory (e.g., E. Lehmann, *Testing Statistical Hypotheses*, and H. Cramér, *Mathematical Methods of Statistics*). We are able to prove these results using a probability based on the Riemann integral instead of the more general but more abstract Lebesgue integral. Even though the probability measure lacks the power of a countably additive measure on a sigma-field, it has enough to enable us to prove important results.

Here are examples of the theorems proved:

1. The sample mean is a uniformly minimum variance unbiased estimator of the population mean. (Cramér-Rao lower bound.)

2. The one-sided test on the mean of a normal population is uniformly most powerful.

3. The two-sided test is uniformly most powerful unbiased.

4. The t-test is uniformly most powerful invariant.

5. The F-test in the one way analysis of variance has, among all invariant tests, best average power over spheres in the parameter space. (A complete distribution theory is given.)

6. In the linear regression model, the conventional predictor of the next observation is the best unbiased predictor.

Other advanced topics are given complete treatment at the submeasure theory level:

7. The sequential test on the mean of the normal population is described, and the expected number of observations *numerically* estimated without application of Wald's equation.

8. Stein's two-sample confidence interval is defined and justified without resorting to infinite-dimensional sample spaces or conditional distributions.

It is common in the first course in mathematical statistics to use methods from a variety of subjects: matrix algebra, Fourier and Laplace transforms, and integration in higher-dimensional space. In this book just a few basic methods are used, appearing throughout most of the work. Two of these are: *rotation* of the sample space, and *sufficiency*. We give a complete geometric discussion of rotations and their relation to integration. No matrix theory or notation is needed. We do not invoke the general theorem on the transformation of multiple integrals by means of Jacobians but instead give a comprehensive discussion of the most relevant case, the rotation. While the concept of sufficiency has been adequately treated only in the measure-theoretic texts, we are able to present a complete theory in the normal case by using the special properties of the normal sample space. In this space the conditional expectation of a function given the sample mean and variance is a certain spherical average of the function and does not involve the unknown parameters. The theory of such averages is given in Chapter 6; the main theorem is that the integral of a continuous function over the normal sample space is equal to the integral of the conditional expectation. This result is repeatedly applied throughout the book.

One of the ingredients of statistical theory, furnished by measure theory, is a useful definition of a class of subsets of the sample space ("events") and a probability measure—an integral—on it. Its importance is well known; for example, the concept of a rejection region of a statistical test requires that the probability of rejection—the integral of the likelihood function over the region—be well defined. This problem of events and integration has been ignored or dodged at the introductory level; however, I have chosen to meet it. In Chapter 3 an intermediate-level integration theory involving the Riemann integral is developed; the integral is defined over the class of "regular" sets. It is not necessary or desirable to cover all of the proofs on regularity in a first reading or during a brief course; however, the reader must become familiar with the definition and properties of these sets.

It was not my aim to include all topics of current interest; thus Bayes and minimax theory, which are not specifically related to normal theory, have been omitted.

This book has been written in the spirit and in accordance with the standards of other mathematical subjects at the post-calculus level, and is intended not only for those specializing in statistics but also for the many students in the statistics course who specialize in other areas of pure and applied mathematics. I find that the classical theory of inference for normal populations is structurally similar to classical Newtonian potential theory. Both subjects grew directly from applied fields. It is well known that many problems of potential theory are variational problems for multiple integrals; many of the statistical problems presented here are of a similar nature.

The book has this structure. Each chapter is divided into sections. Exercises follow each section. They consist of applications and extensions of results in the text, and of verifications of steps in arguments omitted from the text. Most of the sections in each chapter are about the normal population. Almost every chapter ends with a section showing how the results for the normal case extend to the general case, and also a section containing numerical examples based on the table of normal deviates in the appendix. The numerical exercises are optional, and may be done if a desk calculator or computer is available. There are examples of real statistical data from the fields of biology, finance, education, and agriculture. These are intended to illustrate the usefulness of the mathematical theory.

This text is meant for a first course in mathematical statistics which is usually taught to (i) undergraduates as the second semester of a combined course in probability and statistics, and (ii) graduate students as a semester or year course with a probability prerequisite. In (i) the instructor usually selects a book for the entire year's course combining probability and statistics. The probability content of such a course often consists of those topics which are needed in statistics. Since the present text demands no fixed probability prerequisite, the instructor may choose *any* book on probability for the first semester of such a year course.

This book can serve courses of various lengths and intensities. The syllabus for one semester of moderate intensity includes all the sections not called "optional." (The titles of optional sections are so marked in the table of contents.) In this case some of the material in Chapter 3, as noted there, should be omitted. The more comprehensive syllabus includes all of Chapter 3 and some or all of the unstarred sections.

For permission to reproduce various tables I thank the Rand Corporation, Professor E. S. Pearson and the Biometrika Trustees, The Wistar Institute of Anatomy and Biology, The American Society for Engineering Education, and The Iowa State University Press.

SIMEON M. BERMAN

New York, New York
December, 1970

Contents

CHAPTER 1

Populations, Samples, and Random Variables

1-1. POPULATIONS AND SAMPLES

Statistics is the science of making decisions concerning a numerical characteristic of a population. The population may be not only a group of persons, but also a collection of objects of an arbitrary nature. A biologist tests the effect of a nutrient on the growth of animals: the population consists of the animals, the numerical characteristic is the growth (weight gain), and the decision is whether or not the nutrient should be used. An engineer tests the effect of a chemical additive on the strength of a building material: the population consists of the units of the material, the numerical characteristic is the measured strength of the unit, and the decision is whether or not the additive should be used. Numerous examples arise in economics, psychology, physics, and many other areas.

A fundamental method of statistics is the drawing of a relatively small sample from the population, observing the numerical characteristic in the items in the sample, and basing the decision on the latter observation. The biologist decides whether or not to use the nutrient on the basis of its effect on a sample of experimental animals. The engineer decides on the chemical additive on the basis of experiments with a sample of the material.

The mathematical theory of statistics is concerned with the mathematics of sampling from and making decisions about *abstract* populations; it may be applied to the concrete populations arising in research. In order to assist the intuition, we shall refer to the abstract population as a box of labeled balls. Each label is a real number x. In this chapter we introduce those elements of "finite" probability that are needed as a logical prerequisite for the study of normal populations and their sample spaces.

A finite set of balls in a box is called a *population*. A subset of these balls, endowed with a *complete order relation*, is called a *sample*. We recall that a relation on a set is a relation $<$ for pairs of its elements: $A < B$ expresses the relation between two distinct elements A and B. An *order* relation is a *transitive* relation: if A, B, and C are three elements, and $A < B$ and $B < C$, then $A < C$. A *complete* order relation on a set is one for which for every pair of distinct elements A and B exactly one of the relations $A < B$ or $B < A$ holds. A sample is denoted as the collection of the elements in the subset in the appropriate order; for example, if $A, B, C, \ldots$ belong to the subset, and their order is $A < B < C < \cdots$, then the sample is represented as $[A\,B\,C \cdots]$.

Example 1-1. A box contains four balls labeled A, B, C, and D, respectively. The subset consisting of those labeled A, C, and D can be furnished with six distinct complete order relations:

$$A < C < D, \quad A < D < C, \quad C < A < D,$$

$$C < D < A, \quad D < C < A, \quad D < A < C.$$

The corresponding samples are denoted by

$$[ACD], [ADC], [CAD], [CDA], [DCA], [DAC],$$

respectively. ■

Our first result is an elementary combinatorial lemma on the number of samples of specified size from a population of a given size.

LEMMA 1-1

The number of samples of n balls that can be formed from a population of N is

$$N(N-1)\cdots(N-n+1), \tag{1-1}$$

or, equivalently, $N!/(N-n)!$.

Proof: In the formation of a sample from a population of N balls, there are N balls eligible to fill the first position in the sample. After it has been filled there are $N - 1$ balls remaining to fill the second position. Let $A_1, \ldots, A_N$ be the labels on the balls in the population; then the number of ways of filling the first two positions simultaneously is equal to the number of pairs:

$$\begin{array}{c} A_1A_2, \ldots, A_1A_N \\ A_2A_1, \ldots, A_2A_N \\ \vdots \qquad \vdots \\ A_NA_1, \ldots, A_NA_{N-1}. \end{array}$$

There are $N(N - 1)$ of these; thus, there are $N(N - 1)$ ways of filling the first two positions.

The third position in the sample is filled from among the remaining $N - 2$ balls by combining each of the $N(N - 1)$ pairs for the first two positions with each of the $N - 2$ for the third; we find $N(N - 1)(N - 2)$ ways of simultaneously filling the first three positions. Continuing in this way until n positions have been filled, we find the number of samples to be that in (1-1). ■

Example 1-2. The number of samples of five balls that can be formed from a population of ten is $10 \cdot 9 \cdot 8 \cdot 7 \cdot 6 = 30{,}240$. ■

Suppose that the population is decomposed into k mutually exclusive and exhaustive subsets $E_1, \ldots, E_k$. Let N_j be the number of balls in the subset E_j, $j = 1, \ldots, k$, so that

$$N_1 + \cdots + N_k = N.$$

We may imagine that the balls in a particular subset have a distinguishing color; thus we shall refer to the corresponding symbols $E_1, \ldots, E_k$ as the colors. Now we consider the numbers of samples in which the *colors* occur in specified orders.

Example 1-3. A population consists of balls of colors E_j, $j = 1, \ldots, k$; let N_j be the number of those of color E_j, $j = 1, \ldots, k$. Suppose $n = 3$. The number of samples in which the first, second, and third balls have colors E_1, E_2, and E_3, respectively, is $N_1 \cdot N_2 \cdot N_3$. The reasoning is similar to that in the proof of Lemma 1-1: there are N_1 ways of selecting the first member of the sample, and N_2 and N_3 ways of choosing the second and third, respectively. The number of samples in which the first and second members are of a common color E_i and the third of a different color E_j is $N_i(N_i - 1) \cdot N_j$; indeed, the first two positions can be filled in $N_i(N_i - 1)$ ways and the third in N_j ways. (This exemplifies the proof of Lemma 1-2 below.) ∎

In describing the number of samples in which balls of prescribed color occur in prescribed positions, we introduce a double subscript on the symbol E of the color. Since the single subscript on E stands for the index of the color, we cannot without confusion refer to the ball of color E_j appearing in the jth position of the sample; thus we shall instead let E_{i_j} be the color of the ball in that position.

LEMMA 1-2

A population consists of $N_1, \ldots, N_k$ balls of colors $E_1, \ldots, E_k$, respectively. Consider samples of n balls, where

$$n < N_j, \qquad j = 1, \ldots, k,$$

in which the ball of color E_{i_j} appears in the jth position, for $j = 1, \ldots, n$. The colors E_{i_j}, $j = 1, \ldots, n$, need not be distinct. The number of such samples is at least equal to

$$N_{i_1}(N_{i_2} - 1) \cdots (N_{i_n} - n + 1)$$

and at most equal to

$$N_{i_1} \cdots N_{i_n}.$$

Proof: The number of ways of filling the first position is N_{i_1}. The second can be filled in either N_{i_2} ways (if $i_1 \neq i_2$) or $N_{i_2} - 1$ ways (if $i_1 = i_2$); thus in either case the first two positions can be simultaneously filled in at least

$$N_{i_1}(N_{i_2} - 1)$$

and at most

$$N_{i_1} \cdot N_{i_2}$$

ways. The number of ways of filling the third position (with a ball of color E_{i_3}) is

N_{i_3}	if $i_1 \neq i_3$ and $i_2 \neq i_3$,
$N_{i_3} - 1$	if exactly *one* of the equations $i_1 = i_3$, $i_2 = i_3$ holds,
$N_{i_3} - 2$	if $i_1 = i_2 = i_3$.

In any case, it is at least $N_{i_3} - 2$ and at most N_{i_3}; thus the number of ways of filling the first three positions is at least

$$N_{i_1}(N_{i_2} - 1)(N_{i_3} - 2)$$

and at most

$$N_{i_1} \cdot N_{i_2} \cdot N_{i_3}.$$

This reasoning is extended up to the index i_n. ■

Example 1-4. A population consists of 25 red, 75 white, and 40 blue balls. The number of samples in which the first two balls are red and the third blue is $25 \cdot 24 \cdot 40$. The number of samples in which the first, second, and third balls are white, blue and red, respectively, is $75 \cdot 40 \cdot 25$. ■

EXERCISES

1. Write all the complete order relations for a set of four balls labeled A, B, C, D.

2. How many samples of five balls can be formed from a population of nine?

3. A box contains five white and four red balls. How many samples of size four have red in the first and third positions and white in the second and fourth? List the various color combinations in samples of three, and find the corresponding numbers of samples.

4. A population consists of 100 red, 150 white, and 200 blue balls. Find the number of samples of four balls in which red appears in the first two positions, and white and blue in the third and fourth, respectively. Compare this number to the bounds furnished by Lemma 1-2.

5. Suppose that for fixed n and k the number N_j tends to infinity. Show that the ratio of the lower to the upper bound in Lemma 1-2 converges to 1.

6. *Prove:* The number of samples of k balls consisting of exactly one of each of the colors $E_1, \ldots, E_k$ is $k!N_1 \cdots N_k$.

1-2. THE RANDOM SAMPLE

In the practice of statistics, the small sample of items is drawn "at random" from the population. The intuitive meaning of drawing at random is that every member of the population has the same chance, or likelihood, of being selected for the sample. We shall precisely define this concept for the abstract population and sample in terms of the drawing of balls from the box. Suppose that the balls are indistinguishable to the touch. A person blindly puts his hand in the box, takes a sample of n balls "at random," and draws it out. The identity of the sample actually drawn cannot be predicted with certainty before the drawing: it is a variable which may turn out to be any of the samples; the one drawn is a "random sample." We call it a "Sample" or a "Sample of n from the population"—with a capital letter S. To each Sample that can be formed from the population we attach a number called a *probability* —a measure of the likelihood that *it* is the one selected as the Sample. The precise meaning of the Sample being drawn at *random* is that every one of the samples that can be formed is assigned the same probability. By convention, probabilities are positive numbers whose sum is 1. By Lemma 1-1, there are $N!/(N - n)!$ samples of n from a population of N; thus to each sample we assign a probability equal to the reciprocal of the total number of samples, $(N - n)!/N!$. We summarize: The Sample is the sample actually selected when each of the latter is assigned the same probability.

This definition prompts two immediate comments:

(i) The definition of the Sample—the sample that is actually selected—does not depend on the particular choice of probabilities attached to the system of samples. In the more general theory of probability, the assigned probabilities may be arbitrary; however, the particular assignment of equal probabilities is the only one of interest to us in the present work.

(ii) The assumption of equally probable samples is not provable or disprovable. It is justified by the fact that its consequences furnish a very useful theory.

Example 1-5. In random sampling of three balls from a box of six each sample has probability $1/6 \cdot 5 \cdot 4 = 1/120$. ■

Assume again that the population consists of balls of colors $E_1, \ldots, E_k$, and that there are N_j balls of color E_j, $j = 1, \ldots, n$; put $N = N_1 + \cdots + N_k$. If one ball is drawn at random (a Sample of one) the probability that the ball selected belongs to E_j is defined as the fraction N_j/N of balls in the subset. If a Sample of n balls is drawn, the probability that its members belong to $E_{i_1}, \ldots, E_{i_n}$ in that order is defined as the quotient

$$\frac{\text{Number of samples of } n \text{ balls in which the first}, \ldots, n\text{th members belong to } E_{i_1}, \ldots, E_{i_n}, \text{ respectively}}{\text{Number of samples of } n \text{ balls that can be formed from the population}} \tag{1-2}$$

Example 1-6. A box contains balls numbered 1 through 12: the first four are red, the second five white, and the last three blue. A Sample of two balls is selected at random. In accordance with Lemma 1-1, there are 132 samples ($N = 12, n = 2$). Let us compute the probabilities of various kinds of Samples: first, the probability that the first and second members of the Sample are red and blue, respectively. In accordance with the principle of enumeration in the proof of Lemma 1-1, there are $4 \cdot 3$ ways of selecting such a sample; thus, by the definition (1-2) of the probability, the latter is equal to

$$4 \cdot 3/132 = 1/11.$$

The probabilities of other compositions of the Sample are similarly found; for example, the probability that both members are white is 5/33. ■

In statistical problems we analyze Samples from large populations. For this purpose we prove the following result on the limit of the probability (1-2) for a sequence of populations of sizes increasing to infinity:

LEMMA 1-3

Let the number k of colors and the number n of members of the Sample be fixed. If $N_1, \ldots, N_k$ all tend to infinity in such a way that the ratios N_j/N converge to limits:

$$\lim_{N \to \infty} N_j/N = p_j, \qquad j = 1, \ldots, k, \tag{1-3}$$

then the corresponding sequence of probabilities (1-2) converges to the product

$$p_{i_1} \cdots p_{i_n}. \tag{1-4}$$

Proof: By Lemmas 1-1 and 1-2, the ratio (1-2) is at least equal to

$$N_{i_1}(N_{i_2} - 1) \cdots (N_{i_n} - n + 1)/N(N - 1) \cdots (N - n + 1), \qquad (1\text{-}5)$$

and at most equal to

$$N_{i_1}N_{i_2} \cdots N_{i_n}/N(N - 1) \cdots (N - n + 1). \qquad (1\text{-}6)$$

By the assumption (1-3) each of the quotients

$$N_{i_1}/N, (N_{i_2} - 1)/(N - 1), \ldots, (N_{i_n} - n + 1)/(N - n + 1)$$

converges to the corresponding limit p_{i_j}, $j = 1, \ldots, n$, because n is fixed; thus the ratio (1-5) converges to the limit (1-4). For the same reason the ratio (1-6) has the same limit. ■

The limit p_j is the limiting proportion of balls of color E_j. Since the quantities N_j/N have the sum 1, so do their limits p_j.

EXERCISES

1. A box contains three white and four red balls. Find the probabilities that a Sample of three balls consists of balls of various colors in various orders, e.g., the probability that the first and third balls are white and the second red.
2. Find the probabilities of all other color compositions in Example 1-6.
3. Explain why the probability of a Sample of a given color composition (specified colors in specified orders) is independent of the specified order and depends only on the numbers of balls of the various colors.

1-3. RANDOM VARIABLES, PROBABILITY DISTRIBUTIONS, AND SAMPLE SPACES

Let each ball in the population be labeled by one of the numbers $x_1, \ldots, x_k$; suppose that each of the latter is the label of at least one ball. The labels on different balls are not necessarily distinct: several balls may bear the same number x. A ball is selected at random and its number is observed. Since the *ball* actually drawn cannot be predicted with certainty before the drawing, the same holds for the *number* observed on the ball actually drawn; the latter is a number selected at random. We call it a *random variable* and denote it by the capital letter X. Let $E_1, \ldots, E_k$ be the sets of balls with the labels $x_1, \ldots, x_k$, respectively; these can be imagined to be distinguished by colors. The set of labels $x_1, \ldots, x_k$ is called the set of *values* of X. The probability that "X assumes the value x_j" is defined as the probability that the ball drawn is a member of the set E_j; thus it is equal to N_j/N, and is denoted $\Pr(X = x_j)$, $j = 1, \ldots, n$:

$$\Pr(X = x_j) = N_j/N, \qquad j = 1, \ldots, k. \tag{1-7}$$

This set of equations is called the *probability distribution* of X.

Example 1-7. A box contains four balls bearing the number -1, two bearing 0, and four bearing $+1$. Let X be the random variable representing the label of a ball drawn at random; its probability distribution is

$$\Pr(X = -1) = 2/5, \quad \Pr(X = 0) = 1/5, \quad \Pr(X = +1) = 2/5. \quad \blacksquare$$

Suppose that a Sample of n balls is drawn from the population. Let $X_1, \ldots, X_n$ be the random variables representing the labels on the balls in the Sample. We call the n-component vector $(X_1, \ldots, X_n)$ a "Sample of n observations from the population with the probability distribution (1-7)." It is also called a *Sample point* because it is a random point in n-dimensional space. The probability that the Sample point has the coordinates $x_{i_1}, \ldots, x_{i_n}$ is defined as the probability that the members of the Sample belong to the subsets $E_{i_1}, \ldots, E_{i_n}$ in the given order; the latter probability is the quotient (1-2). The values assumed by the Sample point constitute the set of n-component vectors whose coordinates take on any of the values $x_1, \ldots, x_k$. (In the language of set theory, this is the n-fold Cartesian product of the set $(x_1, \ldots, x_k)$; it contains k^n elements.) The probability that the Sample point has the coordinates $x_{i_1}, \ldots, x_{i_n}$ is denoted by

$$\Pr(X_1 = x_{i_1}, \ldots, X_n = x_{i_n}).$$

Example 1-8. Consider the population of Example 1-7, and a Sample of two observations. The values assumed by the coordinates of the Sample point, and the corresponding probabilities are

(X_1, X_2)	$(-1,-1)$	$(-1,0)$	$(-1,1)$	$(0,-1)$	$(0,0)$
Probability	6/45	4/45	8/45	4/45	1/45

$(0,1)$	$(1,-1)$	$(1,0)$	$(1,1)$
4/45	8/45	4/45	6/45

■

In statistical problems we suppose that the numbers $N_1, \ldots, N_k$ of balls with labels $x_1, \ldots, x_k$, respectively, are all very large, and that the ratios N_j/N are approximately equal to p_j, $j = 1, \ldots, k$. Lemma 1-3 implies that the following set of equations approximately holds:

$$\Pr(X_1 = x_{i_1}, \ldots, X_n = x_{i_n}) = p_{i_1} \cdots p_{i_n}. \tag{1-8}$$

This system defines a function whose domain is the collection of n-component vectors $(x_{i_1}, \ldots, x_{i_n})$ and whose range is the corresponding collection of products on the right-hand side of (1-8). This function is called the *likelihood function of the Sample* because it describes the probabilities of the various sets of coordinates that can be assumed by the Sample point.

Example 1-9. A large population consists of balls bearing the numbers -1, 0, and $+1$ in the proportions 2/5, 1/5, and 2/5, respectively (cf. Example 1-7). A Sample of two is drawn. The nine equations defined by (1-8) are

$$\Pr(X_1 = -1, X_2 = -1) = .16$$
$$\Pr(X_1 = -1, X_2 = 0) = .08$$
$$\Pr(X_1 = -1, X_2 = +1) = .16$$
$$\Pr(X_1 = 0, X_2 = -1) = .08$$
$$\Pr(X_1 = 0, X_2 = 0) = .04$$
$$\Pr(X_1 = 0, X_2 = 1) = .08$$
$$\Pr(X_1 = 1, X_2 = -1) = .16$$
$$\Pr(X_1 = 1, X_2 = 0) = .08$$
$$\Pr(X_1 = 1, X_2 = 1) = .16$$

■

Let A be a set of real numbers. We say that "X falls in A" if X takes on a value belonging to A; for example, if A is a closed interval on the real line with end points b and c, $b < c$, then X is said to fall in A if X assumes some value between b and c, inclusive. Now we define the *probability that X falls in A*:

DEFINITION 1-1

The probability that X falls in A is the sum of all probabilities p_j corresponding to values x_j belonging to A; in mathematical notation it is

$$\sum_{x_j \in A} p_j. \tag{1-9}$$

Example 1-10. Let X be the random variable in Example 1-7. If A is the set of all positive real numbers, then the probability that X falls in A is just the probability that X assumes the value $+1$. If A is the interval from -1.5 to 0.5, inclusive, the probability is .6. ■

Let $(X_1, \ldots, X_n)$ be a Sample of n observations from a large population whose probability distribution is given approximately by the limiting proportions p_j, $j = 1, \ldots, k$. Let A be an arbitrary subset of n-dimensional real space. The Sample point is said to fall in A if it takes on a (vector) value belonging to A—that is, $X_j = x_{i_j}$, $j = 1, \ldots, n$, and $(x_{i_1}, \ldots, x_{i_n})$ belongs to A. Generalizing Definition 1-1, we define the probability that the Sample point falls in A:

DEFINITION 1-2

The probability that the Sample point falls in A, denoted $P(A)$, is the sum of the probabilities (1-8) corresponding to vectors $(x_{i_1}, \ldots, x_{i_n})$ belonging to A:

$$P(A) = \sum_{(x_{i_1}, \ldots, x_{i_n}) \in A} p_{i_1} \cdots p_{i_n}. \tag{1-10}$$

The correspondence between subsets A of R^n (n-dimensional space) and the numbers $P(A)$ defines a function whose domain is the class of all subsets of R^n and whose range is a set of numbers in the closed unit interval. Endowed with this system of probabilities, R^n is called a "Sample space corresponding to a Sample of n observations from a population with the probability distribution (p_j)." To be more brief, we shall usually refer to it simply as a Sample space.

Example 1-11. Let (X_1, X_2) be a Sample of two as in Example 1-9. The Sample space is the x_1x_2-plane, furnished with the probabilities determined in Example 1-9. If A is the set in the plane consisting of the x_1-axis, then the probability that the Sample point falls in A is the sum of the probabilities attached to the three points $(-1,0)$, $(0,0)$, and $(1,0)$—namely, $.08 + .04 + .08 = .20$. If A is the x_2-axis, then the probability is also .20. If A is the positive quarter-plane (excluding the axes), then $P(A) = .16$. Finally, if A is the closed disk centered at the origin and of radius 1, then $P(A) = .68$. ■

The function $P(A)$ has a particularly simple form over the class of *product* sets. Let $A_1, \ldots, A_n$ be subsets of the real line; their product set A, denoted by

$$A = A_1 \times \cdots \times A_n, \tag{1-11}$$

is the subset of n-dimensional space consisting of all points whose jth coordinate belongs to the (one-dimensional) set A_j, $j = 1, \ldots, n$; in other words, the vector $(x_{i_1}, \ldots, x_{i_n})$ belongs to A if and only if x_{i_j} belongs to A_j, $j = 1, \ldots, n$. The set in n-space consisting of the single vector $(x_{i_1}, \ldots, x_{i_n})$ is a product set—the product of the sets consisting of the single points $x_{i_1}, \ldots, x_{i_n}$, respectively. The probability of this product set is the product of the probabilities of the component points $x_{i_1}, \ldots, x_{i_n}$; indeed, this is the meaning of Eq. 1-8. This property extends to all product sets.

LEMMA 1-4

The probability of the product set (1-11) is the product of the probabilities of $A_1, \ldots, A_n$, respectively, where the latter probabilities are in the sense of Definition 1-1.

Proof: We shall give the proof just in the case $n = 2$; the general case entails the same ideas but requires more notation. The probability that (X_1, X_2) falls in $A_1 \times A_2$ is the sum of the products p_ip_j corresponding to values x_i in A_1 and x_j in A_2:

$$P(A_1 \times A_2) = \sum_{x_i \in A_1} \sum_{x_j \in A_2} p_i p_j.$$

The double sum factors into a product of two sums, namely,

$$\left(\sum_{x_i \in A_1} p_i\right)\left(\sum_{x_j \in A_2} p_j\right).$$

By Definition 1-1 this is the product of the probabilities of A_1 and A_2, respectively. ■

Example 1-12. Let (X_1, X_2) be a Sample of two observations as in Example 1-9. Let A_1 be the set consisting of -1 and 1, and A_2 the set consisting of 0 and 1. The product set $A_1 \times A_2$ is the set of the four pairs

$$(-1,0),\ (-1,1),\ (1,0),\ (1,1).$$

It is clear from the given probability distribution that

$$P(A_1 \times A_2) = (4/5)\cdot(3/5) = 12/25,$$

which is the product of the probabilities of A_1 and A_2. ■

EXERCISES

1. The balls in a large population are labeled by one of three numbers 1, 2, and 3 in proportions .3, .5, and .2, respectively. Determine the likelihood function for $n = 3$.

2. Prove that the sum of the probabilities (1-8) over all indices $i_1, \ldots, i_n$ is equal to 1.

3. Let X be a random variable with the probability distribution in Exercise 1. Find the probabilities that X falls in the following closed intervals: [0,1], [1,2], [.5,10], [−1,2.5], [−2,0].

4. For the Sample space in Exercise 1 ($n = 3$) find the probability that the Sample point falls in the set of points (x_1, x_2, x_3) for which the sum of the coordinates, $x_1 + x_2 + x_3$, is less than or equal to 6.

5. In Exercise 1 find the probability that the Sample point falls within the sphere centered at the origin and of radius 4; in other words, it falls in the set of points whose sum of squares of coordinates is less than 16.

6. Let A_1, A_2, and A_3 be the closed intervals [0,1.5], [.5,3], and [2,3], respectively. Verify the conclusion of Lemma 1-4 for the product set $A_1 \times A_2 \times A_3$ in the Sample space of Exercise 1.

7. Repeat Exercise 1 for balls labeled -1 and $+1$ in proportions .4 and .6, respectively; determine the likelihood function for $n = 4$.

8. Extend the proof of Lemma 1-4 from the case $n = 2$ to all n by mathematical induction: show that the validity of Lemma 1-4 for all m-dimensional product sets, $m \leq n$, implies its validity for all $(n + 1)$-dimensional product sets. (*Hint:* The main idea is contained in the proof for the case $n = 2$: fix the last factor $p_{i_{n+1}}$ in the product and sum over the first n factors.)

1-4. RANDOM DIGITS

Table V, Appendix consists of "random digits." These can be interpreted as actual random samplings of balls from a huge population in which each ball is labeled by one of the ten digits, and in which each digit appears in the same proportion 1/10. By furnishing numerical examples of random Samples, the table spares us the work of mechanically drawing samples from populations, or of generating fresh random numbers on a computer.

The table is not limited to the ten digits. Pairs of numbers in the table may also be interpreted as random numbers from 0 through 99; for example, the first twenty pairs of digits may be considered as a Sample of twenty from a large population with labels 0 through 99 in equal proportions. In the same way triples of numbers may be considered as random integers from 0 through 999.

One of the important properties of random digits is that the actual proportions of digits appearing in a large Sample tend to be close to the proportion 1/10 in the underlying population; thus the Sample is "representative" of the population. This phenomenon

TABLE 1-1
Cumulative Proportions of Random Digits.

Digit	First 100	Percent	Second 100	Percent	Additional	Percent
0	卌 卌 ||||	14	(16 strokes)	15.0	(18)	12.0
1	卌 卌 ||	12	(6)	9.0	(22)	10.0
2	卌 |||	8	(9)	8.5	(22)	9.8
3	卌 卌 ||	12	(8)	10.0	(19)	9.8
4	卌 卌 |	11	(8)	9.5	(20)	8.5
5	卌 ||	7	(8)	7.5	(24)	9.8
6	卌 卌 |||	13	(15)	14.0	(20)	12.0
7	卌 |	6	(9)	7.5	(15)	7.5
8	卌 |||	8	(12)	10.0	(21)	10.2
9	卌 ||||	9	(9)	9.0	(19)	9.2

is based on the "Law of Large Numbers" in the theory of probability (see Bibliography).

This principle is illustrated in Table 1-1, for the first 400 digits appearing in Table V. The digits are recorded by single strokes, and are grouped in clusters of five. The cumulative proportions are recorded after the first 100, 200, and 400 digits.

Although not *all* the proportions have moved close to 1/10, most have done so.

EXERCISES

(*Optional: These may be done if a calculator or computer is available.*)

1. Extend Table 1-1 to 600 digits.

2. From the data in Table 1-1 compute the cumulative proportions of odd digits.

3. Find the cumulative proportions of digits between 0 and 4, inclusive; compare to the result in Exercise 2.

BIBLIOGRAPHY

The best-known introductory book on probability at the postcalculus level is William Feller, *An Introduction to Probability Theory and Its Applications*. Vol. 1. 3rd ed. (New York: Wiley, 1968).

Books on the precalculus level are:

B. V. Gnedenko and A. Ya. Khinchin, *An Elementary Introduction to the Theory of Probability* (Translated from Russian.) (New York: Dover, 1962).

Simeon M. Berman, *The Elements of Probability* (Reading, Mass.: Addison-Wesley, 1969).

Tables and information about random digits are in: Rand Corporation, *A Million Random Digits with 100,000 Normal Deviates* (New York: Free Press, 1955).

CHAPTER 2

The Normal Distribution

2-1. THE PRINCIPLE OF DUHAMEL AND THE STANDARD NORMAL DISTRIBUTION

In the abstract population considered in Chapter 1, the set of labels is a finite set of numbers $x_1, x_2, \ldots,$. Most populations occurring in research are so large that they can be thought, from an applied mathematical point of view, to be of infinite size; furthermore, the labeling set of numbers is not only considered infinite, but even imagined to form a *continuum* of numbers—that is, consist of an *interval* of real numbers. This happens when the numerical characteristic of the population is a continuous variable. Consider the weights of animals in the statistical problem of the biologist, described in Sec. 1-1. The population consists of all the animals of the given species living in the world. There are only finitely many such animals; thus there is only a finite number of their corresponding weight values. Even so, we think of the weight of an animal as a variable which may assume any value in some interval, or continuum, of real numbers. Now the system of probabilities was defined in Chapter 1 only for a finite set of labels. The definition breaks down when the latter set is a continuum; indeed, it can be shown that the sum of positive numbers p_x, where x runs through an interval, is necessarily infinite; so that the sum of the probabilities is infinite. For this reason it is necessary to redefine probabilities for a continuum of values.

Let $x_1, \ldots, x_k$ be the labels on the N balls in the population; assume $x_1 < x_2 < \cdots < x_k$. These are representable as points on the real number axis. Let N be very large, and $p_1, \ldots, p_k$ the corresponding proportions of balls bearing the number $x_1, \ldots, x_k$, respectively. The latter can be thought of as masses of amounts p_j placed at the points $x_j, j = 1, \ldots, k$. (Fig. 2-1.)

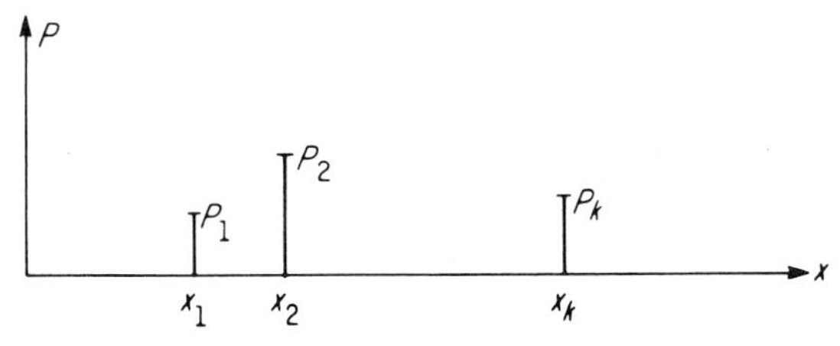

Fig. 2-1. A mass distribution.

Let I be a closed interval on the real line, and $P(I)$ the probability that the random variable X falls in I; then $P(I)$ is equal to the sum of the probabilities or masses that belong to points in I:

$$P(I) = \sum_{x_j \in I} p_j.$$

Now suppose that the masses are individually small but that there are so many of them that they are very dense in every interval on the real line. The "density" may vary from interval to interval. In the analysis of such mass distributions, as in physics, the discrete mass points are replaced by a continuous mass distribution, characterized by a density function $f(x)$; for every interval $I = [a, b]$, the total mass contained in it is expressed as the integral

$$P(I) = \int_a^b f(x)\,dx. \tag{2-1}$$

This is made possible by the application of the Principle of Duhamel; it appears in the standard textbooks on the integral calculus, to which we refer the reader. The density function f represents the local mass per unit interval:

$$f(x) = \lim_{h \to 0} \frac{1}{2h} \int_{x-h}^{x+h} f(y)\,dy;$$

in other words, the mass in a small interval $[x - dx, x + dx]$ is approximately equal to $2f(x)\,dx$.

This principle—the replacement of a dense discrete mass distribution by a continuous one—is applied in the same way to probability distributions over continua. Consider an infinite population which is the "limit" of an increasing sequence of finite populations in the following sense. The successive populations are indexed by the integers $1, 2, \ldots, n, \ldots$. Let $P_n(A)$ be the probability for the nth population that the random variable X falls in A; then we assume that the limit

$$\lim_{n \to \infty} P_n(I) = P(I) \tag{2-2}$$

exists for every closed interval I, and is representable in the form (2-1), as the integral of a continuous function. The latter function is called the *density function*, or *probability density function*. In the general theory of probability the density function may be arbitrary; however, in most of this book we shall restrict the class of density functions to *normal* densities.

We define the standard normal population. It is an infinite population—the limit of an increasing sequence of finite populations with the property (2-2). It is distinguished by the special form of its density function: for every interval $I = [a, b]$, the limit (2-2) is representable as

$$P(I) = \int_a^b \phi(x)\,dx, \tag{2-3}$$

where

$$\phi(x) = (1/\sqrt{2\pi})\,e^{-(1/2)x^2}, \tag{2-4}$$

and where this same function ϕ serves as the integrand in (2-3) for every interval I. (For convenience we shall often write the exponential function e^u as exp (u).) ϕ is called the *standard normal density*; its graph appears in Fig. 2-2.

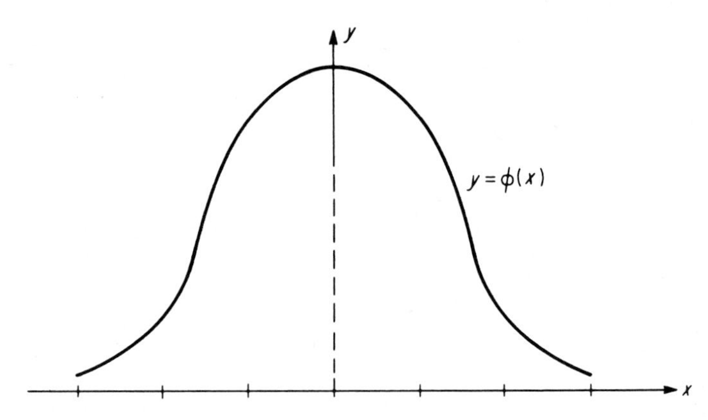

Fig. 2-2. The standard normal density.

This means that the standard normal population is such that the proportion of the population whose labeling numbers are in an interval $[a, b]$ is equal to the integral (2-3). Later we shall actually construct such a population.

The indefinite integral of ϕ,

$$\Phi(x) = \int_{-\infty}^{x} \phi(y)\, dy,$$

is called the *standard normal distribution function*; for each x, $\Phi(x)$ is the limiting proportion of balls whose label value is less than or equal to x. Since ϕ is positive, $\Phi(x)$ increases with x. We shall prove that

$$\Phi(\infty) = \int_{-\infty}^{\infty} \phi(y)\, dy = 1, \tag{2-5}$$

from which also follows

$$\Phi(-\infty) = \lim_{x \to -\infty} \int_{-\infty}^{x} \phi(y)\, dy = 0. \tag{2-6}$$

The relation (2-5) is consistent with the interpretation of $\Phi(x)$ as a limiting proportion: all the balls have labels with values less than infinity. The geometric interpretation of (2-5) is that the area under ϕ from $-\infty$ to $+\infty$ is equal to 1.

For the proof of (2-5), it is sufficient to demonstrate that the *square* of the integral is 1; or, equivalently,

$$(2\pi)^{-1} \int_{-\infty}^{\infty} \int_{-\infty}^{\infty} \exp\left[-(x^2 + y^2)/2\right] dx\, dy = 1.$$

Upon transformation to polar coordinates, the double integral above becomes

$$(2\pi)^{-1} \int_{0}^{2\pi} \int_{0}^{\infty} \exp(-r^2/2)\, r\, dr\, d\theta.$$

The integrand is the derivative of $-\exp(-r^2/2)$ with respect to r; thus the inner integral is equal to

$$-\exp(-r^2/2)\,\Big|_0^{\infty} = 1.$$

Integration over θ yields 1.

Here are several more integral formulas for the standard normal density:

$$\int_{-\infty}^{\infty} x\phi(x)\, dx = 0 \tag{2-7}$$

$$\int_{-\infty}^{\infty} x^2\phi(x)\,dx = 1 \tag{2-8}$$

$$\int_{-\infty}^{\infty} e^{tx}\phi(x)\,dx = \exp(t^2/2). \tag{2-9}$$

(We recall that these improper integrals above are defined as the limits, for $a \to \infty$, $b \to -\infty$, of the corresponding proper integrals from a to b.) The proof of (2-7) is based on the fact that

$$\phi(x) = \phi(-x),$$

so that the integral over the positive axis cancels that over the negative axis (Exercise 2). The formula (2-8) is obtained by integration by parts (Exercise 3). Formula (2-9) depends on the relation

$$\int_{-\infty}^{\infty} \phi(x - t)\,dx = 1 \qquad \text{for all } t. \tag{2-10}$$

(Exercise 5); the latter is obtained from (2-5) by a change of variable of integration (Exercise 4).

EXERCISES

1. Why does Eq. 2-6 follow from Eq. 2-5?

2. Evaluate: $\int_0^{\infty} x\phi(x)\,dx.$

3. Verify (2-8) by integration by parts: put $u = x$, $v = -\phi(x)$.

4. Verify (2-10).

5. Show that Eq. 2-9 follows from Eq. 2-10.

6. Put

$$\mu_{2n} = \int_{-\infty}^{\infty} x^{2n}\,\phi(x)\,dx \qquad n \geq 1.$$

By integration by parts show that

$$\mu_{2n} = (2n - 1)\,\mu_{2n-2};$$

thus, $\mu_{2n} = 1 \cdot 3 \cdots (2n - 1)$.

2-2. THE MEAN AND VARIANCE; GENERAL NORMAL DISTRIBUTION

As in the previous section, consider a population of N balls labeled $x_1, \ldots, x_k$ in proportions $p_1, \ldots, p_k$, respectively. We define the *mean* of the population as the average of the values of the labels on the balls, where the label on *each ball* is counted once in computing the average. This is not the same as the average of the k numbers $x_1, \ldots, x_k$ because the proportions of balls with different labels are not necessarily equal; for example, if a population consists of balls numbered 1, 1, 1, 0, 0, 2, respectively, then the mean is

$$\tfrac{1}{6}(1 + 1 + 1 + 0 + 0 + 2) = \tfrac{5}{6},$$

but the average of the three values 0, 1, and 2 is 1. The general expression for the mean is obtained by weighting each value x_j by the corresponding proportion p_j, and summing over $j, j = 1, \ldots, n$:

$$m = x_1 p_1 + \cdots + x_k p_k;$$

this follows from the definitions of the average and the proportions p_j. The sum m has a physical interpretation: if $p_1, \ldots, p_k$ are masses placed on the real axis at the points $x_1, \ldots, x_k$, respectively, then m is the center of gravity of the mass distribution.

Next we define the *variance* of the population, which is a measure of the spread of the distribution of the label values about m. The variance v is the average of the sum of the squares of the differences between the label values and m, or, equivalently,

$$v = \sum_{i=1}^{k} (x_i - m)^2 p_i.$$

(As in the case of m, each value $(x_i - m)^2$ is weighted by the corresponding proportion p_i, $i = 1, \ldots, k$.) v is an index of the spread of the distribution: if there are many balls whose labels are far from m, then the corresponding terms $(x_i - m)^2 p_i$ add a substantial amount to the sum defining v.

An alternative formula for v is obtained from the identity

$$\sum_{i=1}^{k} (x_i - m)^2 p_i = \sum_{i=1}^{k} x_i^2 p_i - m^2, \tag{2-11}$$

from Exercise 1.

The positive square root of v is called the *standard deviation* of the population.

We may think of the integral

$$\int_{-\infty}^{\infty} x\,\phi(x)\,dx$$

as the limit of the means of an increasing sequence of finite populations whose limit—in the sense of (2-2)—is the standard normal population. The integral is the limiting form of the sequence of sums defining the means: the continuous variable x replaces the discrete variable x_j, and the density times the element of integration, $\phi(x)\,dx$, replaces p_j. Since, by (2-7), the integral is equal to 0, we say that the standard normal population has mean 0.

The integral

$$\int_{-\infty}^{\infty} x^2\,\phi(x)\,dx$$

may similarly be conceived as the limit of the variances of an increasing sequence of finite populations with means 0, which converges to the standard normal population. This is verified by putting $m = 0$ in the sum defining v, replacing x_i and p_i by x and $\phi(x)\,dx$, respectively, and then integrating instead of summing. By (2-8) the above integral is equal to 1; thus we say that the standard normal population has the variance 1. Since the standard deviation is the square root of the variance, the standard deviation of the standard normal population is also equal to 1.

We now introduce a more general normal population, one having mean μ and standard deviation σ, where $\sigma > 0$. An infinite population is said to be such a population if it satisfies the conditions in the definition of the standard normal population with one modification: the density function $\phi(x)$ appearing in the integrand in (2-3) is replaced by the transformed function

$$\frac{1}{\sigma}\,\phi\left(\frac{x-\mu}{\sigma}\right). \tag{2-12}$$

This, by the definition (2-4) of ϕ, is equal to

$$(2\pi\sigma^2)^{-1/2}\exp\left(\frac{-(x-\mu)^2}{2\sigma^2}\right).$$

The latter is called the "normal density function with mean μ and standard deviation σ." The reason for this name will be explained in terms of subsequent integral formulas. The integral

$$\int_{-\infty}^{x} \frac{1}{\sigma} \phi\left(\frac{y - \mu}{\sigma}\right) dy$$

is called the "normal distribution function with mean μ and standard deviation σ." By a change of variable of integration, it is equal to

$$\Phi\left(\frac{x - \mu}{\sigma}\right),$$

where Φ is the standard normal distribution function, defined above. The graph of the normal density,

$$y = \frac{1}{\sigma} \phi\left(\frac{x - \mu}{\sigma}\right),$$

in the xy-plane is obtained from that of the standard normal by a change of coordinates: the center of symmetry is shifted from $x = 0$ to $x = \mu$, and the units on the x- and y-axes are multiplied by the factors σ and $1/\sigma$, respectively. Most of the area under the curve is contained under that portion between $x = \mu - 3\sigma$ and $x = \mu + 3\sigma$; thus if σ is small, the density is concentrated near μ, and if σ is large, the density is well spread about μ. Graphs of the density for various values of μ and σ appear in Fig. 2-3.

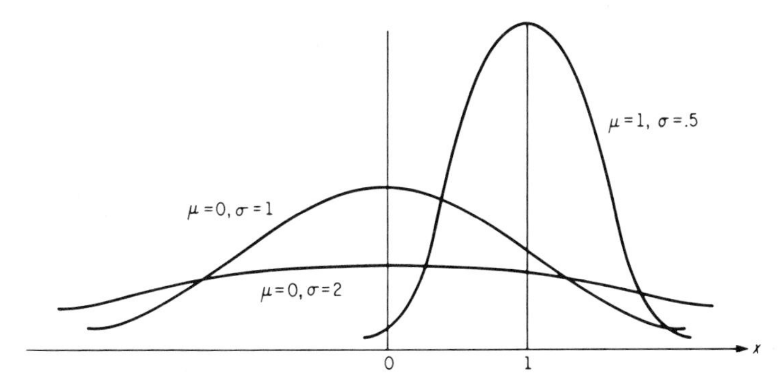

Fig. 2-3. Normal densities with various values of μ and σ.

The area under the normal density is 1 because

$$\int_{-\infty}^{\infty} \frac{1}{\sigma}\,\phi\left(\frac{x-\mu}{\sigma}\right) dx = \lim_{x\to\infty}\left[\Phi\left(\frac{x-\mu}{\sigma}\right) - \Phi\left(\frac{-x-\mu}{\sigma}\right)\right] = 1.$$

The equations corresponding to (2-7) and (2-8) are

$$\int_{-\infty}^{\infty} x\,\frac{1}{\sigma}\,\phi\left(\frac{x-\mu}{\sigma}\right) dx = \mu \tag{2-13}$$

$$\int_{-\infty}^{\infty} (x-\mu)^2\,\frac{1}{\sigma}\,\phi\left(\frac{x-\mu}{\sigma}\right) dx = \sigma^2. \tag{2-14}$$

These follow from (2-7) and (2-8), respectively (Example 2).

The concepts of mean and variance were defined for finite populations as certain weighted sums; now we explain why we call the population with the density (2-12) a normal population with *mean* μ and *standard deviation* σ. First of all the standard normal population is the special case in which $\mu = 0$ and $\sigma = 1$: it is the normal population with mean 0 and standard deviation 1. This terminology is justified by the integral formulas (2-7) and (2-8). The formulas (2-13) and (2-14) justify the name in the general case. The integrals in (2-13) and (2-14) are limiting forms of corresponding sums representing the means and variances of large populations which are "nearly" normal. The reasoning is similar to that in the case of the standard normal population: the density $\phi(x)$ is replaced by $(1/\sigma)\,\phi((x-\mu)/\sigma)$.

EXERCISES

1. Verify the identity (2-11) by expanding the square on the left-hand side and summing.

2. Verify Eqs. 2-13 and 2-14.

3. What is the formula corresponding to (2-9) when $\phi(x)$ is replaced by $(1/\sigma)\,\phi((x-\mu)/\sigma)$?

4. Prove: For every $t > 0$

$$\int_{\mu-\sigma t}^{\mu+\sigma t} \frac{1}{\sigma}\,\phi\left(\frac{x-\mu}{\sigma}\right) dx \geq 1 - t^{-2}.$$

This is *Chebyshev's Inequality*. What does it mean in terms of the relation between the standard deviation and the "spread" of the normal distribution? (For the proof of the inequality, note that the integral (2-14) is diminished by two successive operations: removal of the part of the domain outside $\mu \pm \sigma t$ and the replacement of $(x - \mu)^2$ in the integrand by $\sigma^2 t^2$.)

5. Extend the result of Exercise 4 to an arbitrary finite population. Prove that the sum of the probabilities p_j corresponding to values x_j inside $m \pm \sqrt{v}\ t$ is at least equal to $1 - t^{-2}$.

2-3. TABLE OF THE NORMAL DISTRIBUTION

The standard normal distribution function $\Phi(x)$ is frequently used in statistical practice; however, the defining integral cannot be written in a closed analytic form. It has been computed by numerical methods and tabulated. It is given in Table I, Appendix. For each value $z \geq 0$, the table furnishes the difference $\Phi(z) - \Phi(0)$. Using these, we can find all values of $\Phi(z)$ and of the differences

$$\Phi(b) - \Phi(a) = \int_a^b \phi(x)\,dx. \tag{2-15}$$

The integral above represents the area under the standard normal curve between the points a and b. Here are some of the facts used for the computation of the difference (2-15) for various numbers a and b.

1. $\Phi(0) = 1 - \Phi(0) = \frac{1}{2}$ because ϕ is symmetric about 0 and $\Phi(\infty) = 1$.
2. $1 - \Phi(z)$, $z > 0$, is obtained from the table by writing
$$1 - \Phi(z) = \tfrac{1}{2} - (\Phi(z) - \Phi(0)).$$
3. $\Phi(-z) = 1 - \Phi(z)$, by the symmetry of ϕ.
4. $\Phi(z)$, $z > 0$, is obtained from the table by writing
$$\Phi(z) = \tfrac{1}{2} + (\Phi(z) - \Phi(0)).$$
5. If $0 < a < b$, the difference (2-15) is obtained from the table by means of the equation
$$\Phi(b) - \Phi(a) = \Phi(b) - \Phi(0) - [\Phi(a) - \Phi(0)].$$
6. If $a < 0 < b$, the difference (2-15) is written as $\Phi(b) - 1 + \Phi(-a)$ (by 3.), and is equal to $\Phi(b) - \Phi(0) + \Phi(-a) - \Phi(0)$, the sum of two tabled values.

7. If $a < b < 0$, then $\Phi(b) - \Phi(a) = \Phi(-a) - \Phi(-b)$, and is obtained as in 5.

Example 2-1. The difference (2-15) for $a = 1.71$, $b = +\infty$ is, by 2 (above) equal to

$$\tfrac{1}{2} - (\Phi(1.71) - \Phi(0)) = .5000 - .4564 = .0436.$$

By 3., $\Phi(-1.71)$ is equal to $1 - \Phi(1.71)$, which is equal to the above difference. By 4., $\Phi(1.71)$ is equal to

$$\tfrac{1}{2} + \Phi(1.71) - \Phi(0) = .5000 + .4564 = .9564. \quad \blacksquare$$

Example 2-2. If $a = .94$, $b = 2.25$, then, by 5., the difference (2-15) is $.4878 - .3264 = .1614$. If $a = -.94$, $b = 2.25$, the difference is, by 6., equal to $.4878 + .3264 = .8142$. If $a = -2.25$, $b = .94$, the difference is, by 7., equal to that for $a = .94$, $b = 2.25$, namely, .1614. ■

To find the values of the differences for the normal distribution with mean μ and standard deviation σ, we note that these can be obtained after transformation from the table of the standard normal: if $a < b$, then

$$\int_a^b \frac{1}{\sigma}\,\phi\left(\frac{x-\mu}{\sigma}\right) dx = \Phi\left(\frac{b-\mu}{\sigma}\right) - \Phi\left(\frac{a-\mu}{\sigma}\right). \quad \blacksquare \quad (2\text{-}16)$$

Example 2-3. For a normal distribution with mean $\mu = 1$ and standard deviation $\sigma = 3$, let us find the difference (2-16) for $a = .80$, $b = 2.13$:

$$\frac{b-\mu}{\sigma} = \frac{1.13}{3} = .38; \qquad \frac{a-\mu}{\sigma} = -\frac{.20}{3} = -.07,$$

and

$$\Phi(.38) - \Phi(-.07) = .1480 + .0279 = .1759. \quad \blacksquare$$

EXERCISES

1. Find the values of the difference (2-15) for: $a = 2.41$; $b = 2.81$; $a = .87$, $b = 1.43$; $a = -1.32$, $b = 2.00$; $a = -2.03$, $b = .98$; $a = -2.00$, $b = -.86$.

2. Find the values of (2-16) when $\mu = .50$, $\sigma = 2$ for all values a,b in Exercise 1.

3. Repeat Exercise 2 for $\mu = -.50$, $\sigma = 2.5$.

2-4. CONSTRUCTION OF A NORMAL POPULATION

In Sec. 3-1 we defined the standard normal population as the "limit" of an increasing sequence of finite populations. At that time we did not show that such a sequence actually exists but promised to construct such a one. This will now be done.

For each integer $n > 0$ consider a population of n balls, labeled by the integers $0, 1, \ldots, n-1$. The standard normal distribution function is continuous and strictly increasing on the line; hence, for every real number t, $0 < t < 1$, there is a number $x = x(t)$ such that

$$\Phi(x(t)) = t. \tag{2-17}$$

The relation between t and $x(t)$ is illustrated in terms of the graph of $y = \Phi(x)$ in Fig. 2-4. $x(t)$ is the inverse function of Φ. It is also strictly increasing and continuous on the interval $0 < t < 1$, and

$$\lim_{t \to 0} x(t) = -\infty, \qquad \lim_{t \to 1} x(t) = \infty. \tag{2-18}$$

(Exercise 1)

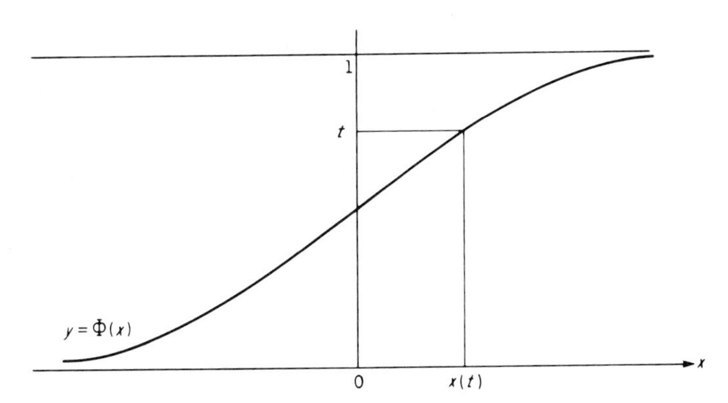

Fig. 2-4. The relation between t and $x(t)$.

Now change the labels on the balls in accordance with the rule: The ball labeled j is given the new label $x(j/n)$, $j = 0, 1, \ldots, n-1$; here, in accordance with (2-18), we put $x(0) = -\infty$. This defines a new labeling system for the nth population, for $n = 1, 2, \ldots$. Now we prove that this sequence of finite populations converges to the standard normal in the sense of the relation (2-2):

THEOREM 2-1

For any two real numbers a and b, $a < b$, the proportion of the nth population whose label values $x(j/n)$ are between a and b, inclusive, converges to the integral (2-3) as $n \to \infty$.

Proof: Since $x(t)$ is an increasing function of t, the numbers $x(j/n)$ are also in increasing order of magnitude:

$$x(0) < \cdots < x\left(\frac{j}{n}\right) < \cdots < x\left(\frac{n-1}{n}\right);$$

furthermore, by (2-18), $x((n-1)/n) \to +\infty$. If n is large enough, at least one of the numbers $x(j/n)$ is less than a. For each such n there are integers $j = j(n)$ and $k = k(n)$ such that

$$x\left(\frac{j}{n}\right) \le a \le x\left(\frac{j+1}{n}\right), \qquad x\left(\frac{k}{n}\right) \le b \le x\left(\frac{k+1}{n}\right).$$

(See Fig. 2-5.)

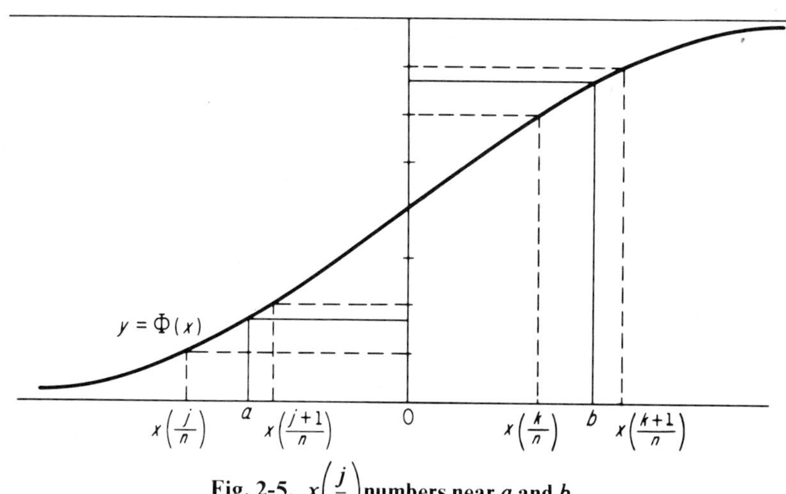

Fig. 2-5. $x\left(\frac{j}{n}\right)$ **numbers near a and b.**

If n is large, the numbers $x(i/n)$, $i = 0, 1, \ldots, n-1$, are very dense in $[a, b]$; thus we may assume that $x((j+1)/n)$ is so close to a and

$x(k/n)$ so close to b that

$$x\left(\frac{j+1}{n}\right) < x\left(\frac{k}{n}\right).$$

The proportion of balls whose labels are between a and b is approximately the same as the proportion whose labels are between $x((j+1)/n)$ and $x(k/n)$, inclusive. The latter are the balls whose original labels were $j+1, \ldots, k$. The proportion of such balls is $(k-j)/n$. In terms of Φ, this proportion may, by (2-17), be written as

$$\Phi\left(x\left(\frac{k}{n}\right)\right) - \Phi\left(x\left(\frac{j}{n}\right)\right).$$

It follows from the definition of $j = j(n)$ and $k = k(n)$, that

$$x\left(\frac{j}{n}\right) \to a, \qquad x\left(\frac{k}{n}\right) \to b, \qquad n \to \infty;$$

indeed, the successive differences between the numbers $x(i/n)$ gets smaller and smaller as $n \to \infty$, and

$$\left|x\left(\frac{j}{n}\right) - a\right| \le \left|x\left(\frac{j+1}{n}\right) - x\left(\frac{j}{n}\right)\right| \to 0$$

$$\left|x(b) - x\left(\frac{k}{n}\right)\right| \le \left|x\left(\frac{k+1}{n}\right) - x\left(\frac{k}{n}\right)\right| \to 0.$$

Since Φ is a continuous function it follows that

$$\Phi\left(x\left(\frac{k}{n}\right)\right) - \Phi\left(x\left(\frac{j}{n}\right)\right) \to \Phi(b) - \Phi(a). \quad \blacksquare$$

By a simple extension of the argument, one can prove an analogous result for the general normal population (Exercise 2).

EXERCISES

1. Show that $x(t)$ is strictly increasing and continuous, and satisfies (2-18).
2. Modify the proof of Theorem 2-1 to cover the case of a normal population with mean μ and standard deviation σ.
3. The result of the previous exercise can be extended even further: Theorem 2-1 is valid for any continuous increasing function Φ assuming values in the unit interval.

2-5. NUMERICAL EXAMPLES OF NORMAL POPULATIONS (OPTIONAL)

Many populations in natural and social science have values with distributions which are approximately normal: for each interval I on the real line, the proportion of the population whose label values fall in I is given approximately by the integral of a normal density over I. If the values of the numbers of the population are measured with a given precision, then the intervals I in the previous statement are restricted to those whose lengths are positive integral multiples of the smallest unit of measurement; for example, if the values are given to two decimal places, then the intervals should have lengths which are multiples of .01.

The observation that many populations are approximately normal has been explained by the mathematical theory of probability in terms of the central limit theorem. This will be briefly described in Sec. 4-6. More complete discussions are given in books on probability; in particular, the book by Cramér (cf. Bibliography) contains a rigorous proof. It must be remarked that not all distributions are normal: many other distributions appear in empirical data.

Two numerical examples of populations with an approximately normal distribution are presented below. These are included for the purpose of illustrating applications of the theory; therefore, we do not emphasize the mechanical numerical procedures used in the computations.

Example 2-4. Data on the physical characteristics of individual major league baseball players is given on small cards bearing a portrait of the corresponding player. Among other information, the height of the player in inches is listed on the card. A deck of 160 cards gave the height distribution in columns 1, 2, and 3 of Table 2-1.

We shall compute the mean and standard deviation of this population, and then show that it is approximately normal. In order to compute the mean, we multiply the heights by the corresponding numbers of players; add these products, and divide the resulting sum by 160. We do *not* multiply the heights by the corresponding proportions because the latter contain rounding errors, which are inflated by the multiplication. A direct numerical calcu-

TABLE 2-1

Distribution of Heights of 160 Players Compared with Normal Distribution, with Same Mean and Standard Deviation.

(1) Height	(2) Number of Players	(3) Proportion	(4) x	(5) $y = \frac{x - 72.88}{2.23}$	(6) $\Phi(y)$	(7) $\Phi\left(y + \frac{.5}{2.23}\right) - \Phi(y)$
66	1	.006	65.5	−3.31	.0005	.0016
67	1	.006	66.5	−2.86	.0021	.0059
68	2	.012	67.5	−2.41	.0080	.0170
69	5	.031	68.5	−1.96	.0250	.0368
70	12	.075	69.5	−1.54	.0618	.0805
71	23	.144	70.5	−1.07	.1423	.1253
72	24	.150	71.5	−.62	.2676	.1649
73	26	.162	72.5	−.17	.4325	.1778
74	30	.187	73.5	+.28	.6013	.1539
75	19	.112	74.5	+.72	.7642	.1148
76	8	.050	75.5	+1.17	.8790	.0750
77	5	.031	76.5	+1.64	.9495	.0411
78	4	.025	77.5	+2.35	.9906	.0035
			78.5	+2.52	.9941	

lation yields the value

$$m = 72.88$$

for the mean. In computing the variance, we use the right-hand side of the identity (2-11) instead of the left because the computed value of m has a rounding error which is exaggerated by the squaring of $(x_i - m)$ and its summation. The first term on the right-hand side of (2-11) is computed as follows: the heights are squared, multiplied by the corresponding numbers of players, and then the products are summed; the sum is then divided by 160. The value of v is then obtained by subtracting m^2; finally, the standard deviation is the square root of v. The results of the calculation are

$$v = 4.98, \qquad \sqrt{v} = 2.23.$$

Now we show that the population is approximately normal with mean $\mu = 72.88$ and standard deviation $\sigma = 2.23$. The population values have been measured in (inch) units; thus in order to show that the distribution is normal we will show that the integral of the normal density over intervals of unit length is approximately equal to the corresponding proportions of heights contained in the intervals. For example, the integral of the normal density from 69.5 to 70.5 will be shown to be close to the proportion .075 contained in that interval. We list the values of $(x - 72.88)/2.23$ for $x = 65.5, \ldots, 78.5$, and the corresponding values of Φ in columns 4 and 5. The differences (2-16) for the intervals of unit length from 65.5 to 78.5, and the corresponding proportions of heights are in column 7.

The approximation above can be geometrically represented by comparing the histogram, or bar graph, of the observed proportions of heights to the graph of the normal density function with $\mu = 72.88$ and $\sigma = 2.23$. This comparison is based on the following: If the interval $[a, b]$ is small relative to σ, then the integral (2-16), by the law of the mean, is approximately equal to

$$\frac{1}{\sigma}\,\phi\left(\frac{\frac{1}{2}(a+b)-\mu}{\sigma}\right)(b-a);$$

therefore, the proportion of heights x is approximately

$$\frac{1}{\sigma}\,\phi\left(\frac{x-\mu}{\sigma}\right)$$

for $x = 66, \ldots, 78$. In Fig. 2-6 this function is compared to the histogram. ■

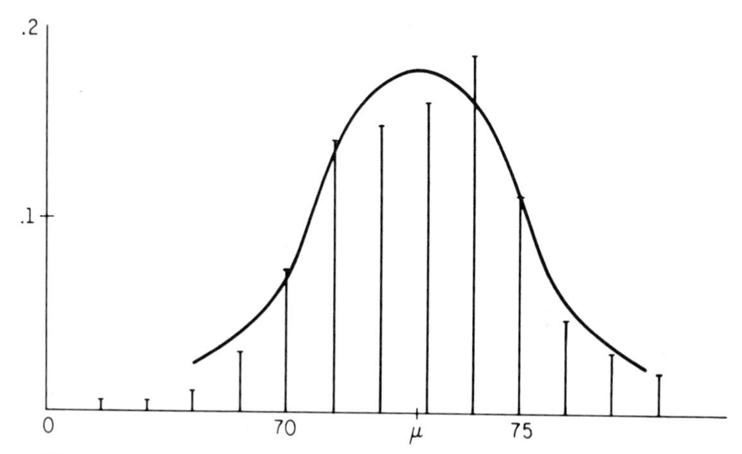

Fig. 2-6. The normal density and the observed proportion of heights.

Example 2-5. The prices of stocks listed on the New York Stock Exchange are given in one-eighths of a dollar. On a certain day the first 225 stocks listed by corporation name in alphabetical order had gains and losses distributed in accordance with the data

TABLE 2-2
Gains and Losses of 225 Stocks on New York Stock Exchange.

Gain	Number of Stocks	Proportion	Gain	Number of Stocks	Proportion
0	32	.142			
1/8	25	.111	−1/8	27	.120
1/4	15	.067	−1/4	20	.089
3/8	10	.044	−3/8	11	.049
1/2	14	.062	−1/2	15	.067
5/8	4	.018	−5/8	7	.031
3/4	7	.031	−3/4	2	.009
7/8	3	.013	−7/8	3	.013
1	3	.013	−1	8	.036
1 1/8			−1 1/8	1	.004
1 1/4	1	.004	−1 1/4	6	.027
1 3/8	1	.004	−1 3/8		
1 1/2			−1 1/2	2	.009
1 5/8	1	.004	−1 5/8		
1 3/4			−1 3/4	1	.004
1 7/8	1	.004	−1 7/8		
2	1	.004	−2		
2 1/8			−2 1/8		
2 1/4			−2 1/4	1	.004
2 3/8	1	.004	−2 3/8		
2 1/2	1	.004	−2 1/2	1	.004

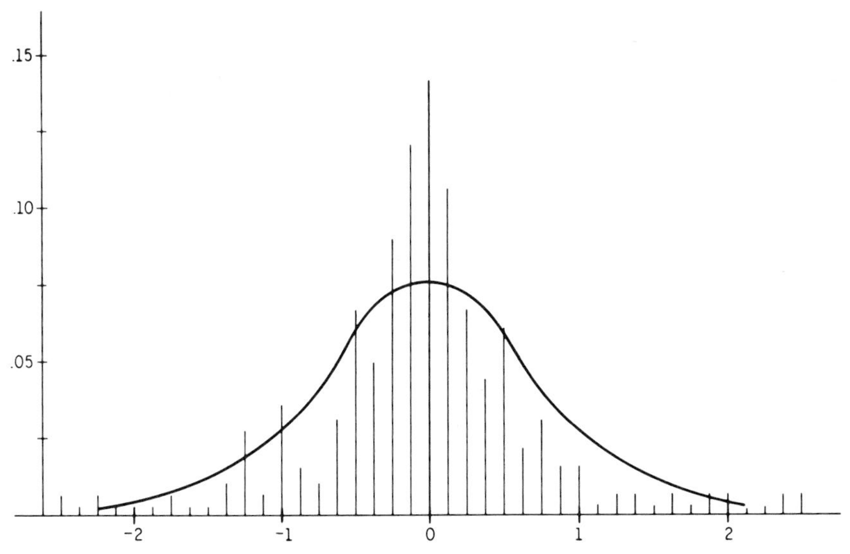

Fig. 2-7. Gains of stocks compared to the normal density with mean −.0161 and standard deviation .667.

in Table 2-2. The mean of the distribution was found to be $-.0161$ and the standard deviation approximately .667. The histogram of stock gains is compared to the normal density with the same mean and standard deviation in Fig. 2-7. The density has been divided by a factor of 8. ■

EXERCISES (Optional)

1. Using the data in Table 2-2 and the computed values of μ and σ given above, find the proportions of stocks whose gains fall in each of the intervals $[\mu + (k - 1)\sigma, \mu + k\sigma]$, for $k = -2, -1, \ldots, 3$.
2. Compare the proportions in Exercise 1 to the corresponding distribution with mean μ and standard deviation σ.

2-6. RANDOM VARIABLES FROM A NORMAL POPULATION

The definition of a random variable does not directly extend from a finite to an infinite population. In Sec. 2-1 we pointed out that if each ball had equal (positive) probability of being selected, the sum of the probabilities would be infinite if there were infinitely many balls. For this reason we have difficulty in the definition of a random variable for a normal population.

We shall meet this problem in the following way: we shall furnish an intuitive definition but not a rigorous one, and then indicate why the latter is not needed for the purposes of this book. The rigorous definition is given in books on the measure-theoretic foundations of probability; for example, we mention the classical *Foundations of the Theory of Probability*, by A. N. Kolmogoroff (New York: Chelsea, 1950).

In its definition for a finite population, a random variable is considered a variable whose value is determined in accordance with a probability distribution. We may think of a random variable from the infinite normal population in the same way—as a variable whose value is determined by the random selection of a real number: the probability that the selected value belongs to the interval I, or the probability that "X falls in I," is the integral of the normal density over I. This is the intuitive definition of X.

It will become apparent as the reader goes on that all our definitions and theorems are really about *probability distributions* of random variables—not random variables themselves. For this reason it is here unnecessary to give a rigorous definition of the latter because the former has already been defined. The theorems could have been stated as results about definite integrals; however, we have used the term "random variable from a normal population" even in the formal analysis in order to exhibit the statistical meaning of the results.

2-7. RANDOM NORMAL DEVIATES

The content of Theorem 2-1 is: If each of a population of n balls is labeled by exactly one of the integers $0, 1, \ldots, n-1$, and if $x(t)$ is the function defined by (2-17), then the labels

$$x(0), x\left(\frac{1}{n}\right), \ldots, x\left(\frac{n-1}{n}\right)$$

have a distribution approaching the standard normal. This implies that if j is a number selected at random from the set $\{0, 1, \ldots, n-1\}$ then the (random) transformed number $x(j/n)$ has a distribution which, for large n, is close to the standard normal. The random number $x(j/n)$ may be considered itself as a random variable from a normal population. Numbers constructed in this way are called "normal deviates" or "Gaussian deviates." Table VI in the Appendix contains an array of such numbers.

TABLE 2-3
Distribution of 100 Normal Deviates Compared to Standard Normal Distribution.

Interval	Numbers	Percentage	Normal Probability
$-\infty$ to -3.0			.0013
-3.0 to -2.5			.0049
-2.5 to -2.0	||	2	.0166
-2.0 to -1.5	|	1	.0440
-1.5 to -1.0	卌 卌	10	.0919
-1.0 to -0.5	卌 卌 ||	12	.1498
-0.5 to 0.0	卌 卌 卌	15	.1915
0.0 to 0.5	卌 卌 卌 卌 卌 ||	27	.1915
0.5 to 1.0	卌 卌 卌 ||||	19	.1498
1.0 to 1.5	卌 ||	7	.0919
1.5 to 2.0	卌 |	6	.0440
2.0 to 2.5			.0166
2.5 to 3.0	|	1	.0049
3.0 to ∞			.0013

In accordance with the principle that the Sample represents the population, the distribution of a large number of normal deviates in the table is close to the standard normal. In Table 2-3 there is given the distribution of the first 100 numbers in Table VI, Appendix. The frequencies are grouped in intervals of half units. Even though the Sample is relatively small, it is still quite representative. Each number is recorded by a single stroke.

2-8. GENERAL CONTINUOUS POPULATIONS

This book is primarily about the normal population; however we shall usually show how concepts developed in the normal case may be extended to a general "continuous" population.

Let $h(x)$ be a nonnegative continuous function such that

$$\int_{-\infty}^{\infty} h(x)\,dx = 1.$$

We define a "population with the density function $h(x)$" in the same way as the standard normal population except that the general function $h(x)$ takes the place of the special function $\phi(x)$: It is an infinite population which is the limit of a sequence of finite populations in the sense that (2-2) holds and the limiting probability is representable [cf. 2-3] as

$$P(I) = \int_{a}^{b} h(x)\,dx.$$

The basic definitions involving the normal population extend to the general one. The integral

$$H(x) = \int_{-\infty}^{x} h(y)\,dy,$$

as a function of x, is called the *distribution function*. It has the same interpretation as does Φ in the normal case: $H(x)$ is the limiting proportion of balls whose labels are less than or equal to x. Since h is nonnegative, $H(x)$ is nondecreasing in x. By analogy with Φ:

$$H(\infty) = 1, \qquad H(-\infty) = 0.$$

If the integral

$$\int_{-\infty}^{\infty} |\,x\,|\,h(x)\,dx$$

is finite, then the integral

$$\mu = \int_{-\infty}^{\infty} xh(x)\,dx \tag{2-19}$$

converges absolutely, and is called the mean of the population (cf. Eq. 2-13). It is interpreted as the center of gravity of the continuous mass distribution with density h. If

$$\int_{-\infty}^{\infty} x^2 h(x)\,dx < \infty \tag{2-20}$$

then the integral

$$\sigma^2 = \int_{-\infty}^{\infty} (x - \mu)^2 h(x)\,dx \tag{2-21}$$

is finite (Exercise 1) and is called the *variance* (cf. Eq. 2-14); its square root, σ, is called the *standard deviation*. The variance may also be expressed as

$$\sigma^2 = \int_{-\infty}^{\infty} x^2 h(x)\,dx - \mu^2 \tag{2-22}$$

(Exercise 2).

While we have assumed that $h(x)$ is continuous, all of the definitions and results above are simply carried over to the case—which frequently occurs—where $h(x)$ consists of continuous "pieces," as in Fig. 2-8. Even this is not essential: the only properties that h must have are nonnegativity and integrability.

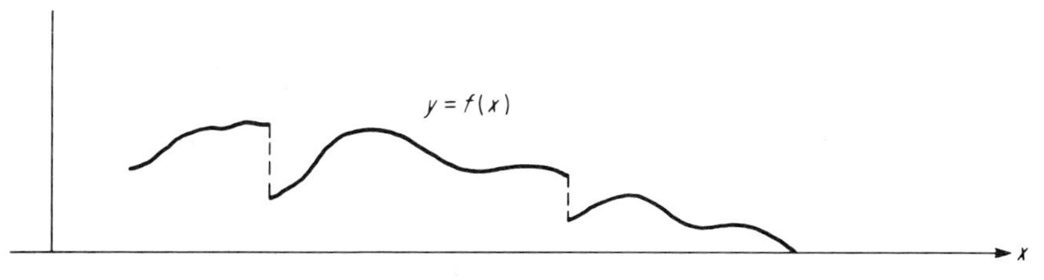

Fig. 2-8. A piecewise continuous function.

Here are examples of densities occurring in theoretical and applied statistics. Several will appear later as "sampling distributions" of certain statistics in sampling from a normal population.

Example 2-6. Let $h(x)$ be defined as

$$h(x) = 1, \quad 0 < x < 1$$
$$= 0, \quad \text{elsewhere;}$$

this is the *uniform density on the unit interval.* The distribution function is

$$H(x) = 0, \quad x \leq 0$$
$$= x, \quad 0 < x < 1$$
$$= 1, \quad x \geq 1.$$

For any $\theta > 0$, the function $(1/\theta)h(x/\theta)$ is also a density, and is called the "uniform density on $[0, \theta]$; furthermore, for $b > 0$

$$\frac{1}{\theta} h \left(\frac{x - b}{\theta}\right)$$

is called the *uniform density* on $[b, b + \theta]$ (see Fig. 2-9). ■

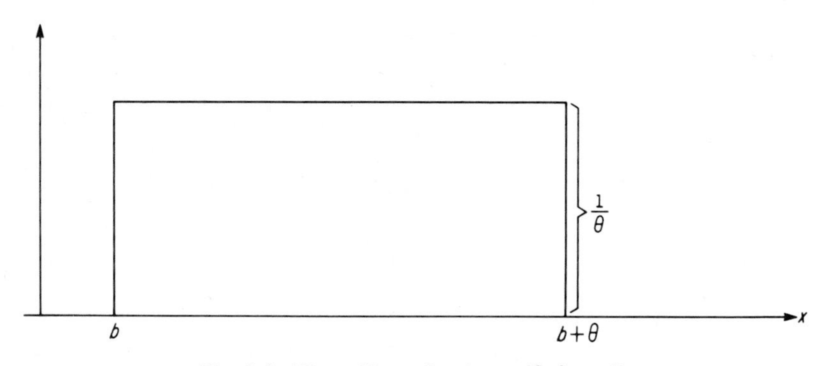

Fig. 2-9. The uniform density on $[b, b + \theta]$.

Example 2-7. Put

$$h(x) = 0, \quad x \leq 0$$
$$= e^{-x}, \quad x > 0;$$

this is the *standard exponential density*. The distribution function is

$$H(x) = 0, \quad x \leq 0$$
$$= 1 - e^{-x}, \quad x > 0.$$

For fixed $\theta > 0$, the density

$$\frac{1}{\theta} h\left(\frac{x}{\theta}\right)$$

is called *exponential with mean* θ because its mean is equal to θ (Exercise 3). (See Fig. 2-10.) ■

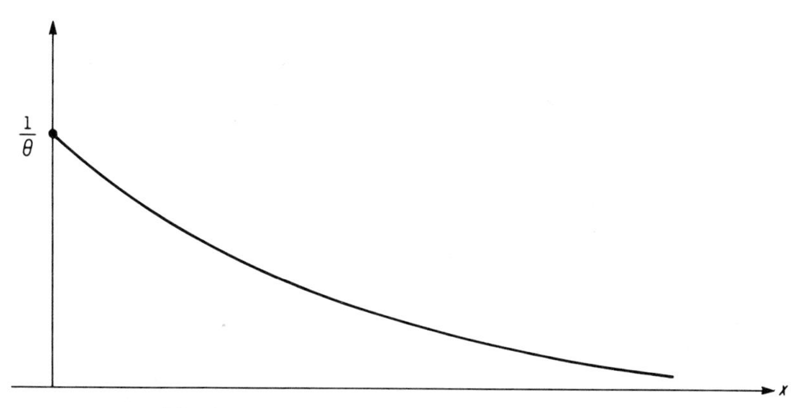

Fig. 2-10. The exponential density with mean θ.

Example 2-8. Put

$$\begin{aligned} h(x) &= 0, && x \leq 0 \\ &= xe^{-(1/2)x^2}, && x > 0; \end{aligned}$$

this is the *Rayleigh density*. (See Fig. 2-11.) The distribution function is 0 for $x \leq 0$, and

$$H(x) = 1 - e^{-(1/2)x^2}, \quad x > 0. \quad ■$$

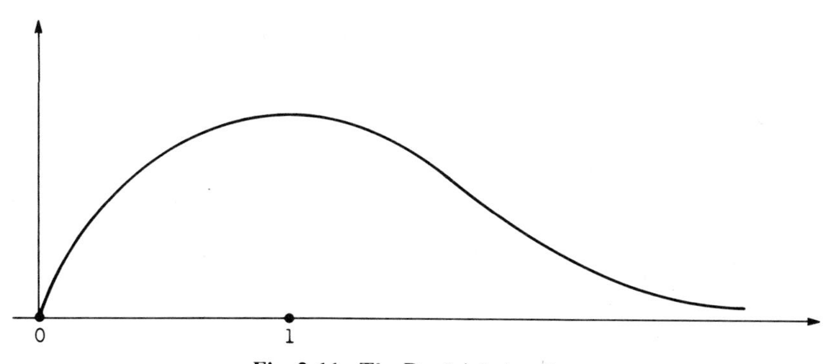

Fig. 2-11. The Rayleigh density.

Example 2-9. The *Laplace density* is

$$h(x) = \tfrac{1}{2}e^{-|x|}, \quad -\infty < x < \infty;$$

the distribution function is

$$\begin{aligned} H(x) &= \tfrac{1}{2}e^{x}, & x \le 0 \\ &= 1 - \tfrac{1}{2}e^{-x}, & x > 0. \end{aligned}$$ ■

Example 2-10. The *Cauchy density* is

$$h(x) = \frac{1}{\pi}\frac{1}{1 + x^2}, \quad -\infty < x < \infty.$$

The distribution function is

$$H(x) = \frac{1}{2} + \frac{1}{\pi}\arctan x,$$

where the arctan function assumes values between $-\pi/2$ and $\pi/2$ as x goes from $-\infty$ to $+\infty$. The Cauchy density does not have a well-defined mean because the integral

$$\int_{-\infty}^{\infty} |x| \, h(x) \, dx$$

is infinite. The integral in (2-20) is, of course, also infinite. ■

By analogy to the normal case in Sec. 2-4, we can construct a population with an arbitrary density $h(x)$. For simplicity, suppose that $H(x)$ is strictly increasing on the set where it is positive, and also continuous. With H in place of Φ, define $x(t)$, $0 < t < 1$, by means of Eq. 2-17 as

$$H(x(t)) = t, \quad 0 < t < 1. \tag{2-23}$$

By the same argument as in the normal case (cf. Sec. 2-4, Exercise 3) it can be shown that the population with the n members with labels $x(j/n)$, $j = 0, 1, \ldots, n-1$, has the limiting distribution $H(x)$.

Example 2-11. If $h(x)$ is the uniform density on the unit interval, then $x(t) = t$. If $h(x)$ is exponential with mean θ, then

$$x(t) = -\theta \log(1 - t), \quad 0 < t < 1;$$

thus if n is very large, the population with the labels

$$-\theta \log\left(1 - \frac{j}{n}\right), j = 0, 1, \ldots, n - 1$$

has an approximate exponential distribution. ■

The table of random digits can be used to construct Samples from arbitrary populations, just as it was used to construct normal Samples (Sec. 7-2). If the number j is drawn at random from among $0, 1, \ldots, n - 1$, and if n is very large, then $x(j/n)$ represents a value drawn at random from the population with the density $h(x)$. As an example, let us interpret the bunches of five digits in Table V, Appendix, as random numbers from 0 through 99,999; then $x(j \cdot 10^{-5})$ is very much like a random variable from a population with the given density. Table 2-4 shows the distribution of the values of 200 random variables from a standard exponential population constructed in this way. In Fig. 2-12 the histogram of observed values is compared to the exponential density.

TABLE 2-4
Distribution of 200 Random Numbers from a Standard Exponential Population Compared to the Theoretical Distribution. (Probability for an Interval $[a,b]$ is $e^{-a} - e^{-b}$.)

Interval	Observed Number	Proportion	Probability of the Interval
0–.5	87	.435	.393
.5–1.0	38	.190	.239
1.0–1.5	29	.145	.145
1.5–2.0	11	.055	.088
2.0–2.5	14	.070	.053
2.5–3.0	13	.065	.032
3.0–3.5	2	.010	.020
3.5–4.0	3	.015	.012
4.0+	3	.015	.018

One of the justifications for our emphasis on the normal population is that it is sometimes possible to transform the labels of a given population so that it becomes normal. If the underlying population has a *known* distribution $H(x)$, then the transformation is explicitly constructable.

THEOREM 2-2
If $\{x\}$ represents the family of labels for the population with the distribution function $H(x)$, then the set of numbers $\{H(x)\}$ represents a uniform distribution on the unit interval.

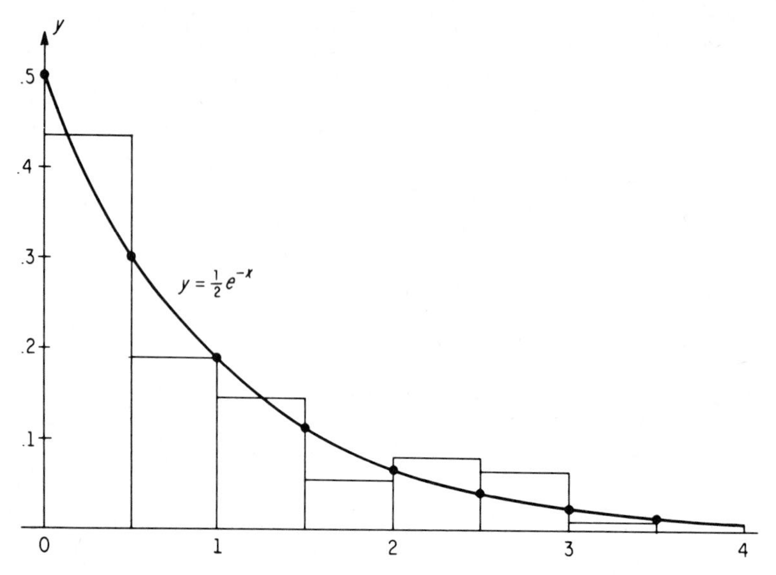

Fig. 2-12. Observed distribution of 200 exponentially distributed random numbers compared to the exponential density.

Proof: Let t be a number such that $0 < t < 1$. The proportion of the population for which the labels x' satisfy

$$H(x') \leq t$$

is the same as the proportion for which $x' \leq x(t)$, because $t = H(x(t))$, and H is nondecreasing:

$$H(x') \leq t \textit{ if and only if } H(x') \leq H(x(t))$$

$$\textit{if and only if } x' \leq x(t).$$

By the definition of H and $x(t)$, the proportion of the population satisfying the last inequality is t; thus $\{H(x)\}$ has a uniform distribution. ■

This theorem states that any population with an increasing continuous distribution can be transformed into a uniform one. On the other hand, the function $x(t)$ in (2-23) transforms the uniform population into one with the distribution H; in particular, the function $x(t)$ defined by Eq. 2-17—with Φ in place of H—transforms it into a standard normal one. It follows that the composition of the two transformations—to the uniform and then to the normal—changes an arbitrary distribution into a normal one. In most problems of interest in statistics the underlying distribution is unknown or only partially known, so that the required transformation is not available; however, it is often found on the basis of previous information that a certain transformation changes a class of partially known distributions into a family of normal distributions. It has been observed that the logarithm function transforms the distributions of certain biological and economic data into normal distributions (cf. Bibliography).

EXERCISES

1. Show that the variance (2-21) is finite if (2-20) holds. (*Hint:* $|x - \mu|^2 \leq 2x^2 + 2\mu^2$.)

2. Verify (2-22).

3. *Prove*: If $h(x)$ is a density function, then so is the function

$$\frac{1}{a} h\left(\frac{x - b}{a}\right)$$

for any $a \neq 0$ and b.

4. Compute the mean and variance of each of the densities in Examples 2-6 through 2-9.

5. Find the function $x(t)$ for the Rayleigh distribution (Example 2-8).

BIBLIOGRAPHY

The Principle of Duhamel is given in Angus E. Taylor, *Calculus with Analytic Geometry* (Englewood Cliffs, N. J.: Prentice-Hall, 1959).

The rigorous definition of random variables from normal or more general populations is contained in the fundamental book, A. N. Kolmogorov, *Foundations of the Theory of Probability* (trans. from the German) (New York: Chelsea, 1950).

A table of normal deviates is given in *A Million Random Digits with 100,000 Normal Deviates*, the Rand Corporation (New York: Free Press, 1955).

Many examples of normal populations appear in George W. Snedecor and William G. Cochran, *Statistical Methods*, 6th ed. (Ames, Iowa: Iowa State U. P., 1967); G. Udny Yule and M. G. Kendall, *An Introduction to the Theory of Statistics*, 14th ed. (New York: Hafner, 1950).

The logarithmic transformation of biological and economic data to normal distributions is explained in Harald Cramér, *Mathematical Methods of Statistics* (Princeton, N. J.; Princeton U. P., 1946).

CHAPTER 3

The Normal Sample Space of *n* Dimensions

This chapter contains many of the mathematical tools necessary for our study of statistics. The *logical* manner of introducing these is to stock them in an early portion of the book and refer back to them whenever necessary. This is not always the best way from the *pedagogical* point of view, because it is hard to study tools without knowing how they are used; as a result, students may be discouraged from ever getting to the main subject.

In order to make the book useful to readers preferring one or the other approach, we have put most of the tools in *this* chapter but have indicated which of these can be studied later without loss of continuity. These subjects will also be of various degrees of difficulty for readers of diverse backgrounds; hence we shall quickly summarize the contents and suggest several alternative paths of study. This will present every reader with an efficient program; in no case should this chapter be a barrier to the subsequent material.

Section 3-1 consists of the extension of analytic plane geometry to the geometry of n-dimensional space. Vector notation is used. The concepts are elementary. The reader who knows this material may skip it, yet he who does not should find it simple.

Rotations in n-dimensions are studied in Sec. 3-2. The reader who knows rotations as orthogonal matrices will recognize the main results, and may omit the proofs; however, he should become familiar with the geometric nature of rotations. We assume that all readers know rotations in the plane from analytic geometry. In a first reading one should study all the details of this material on rotations at least for $n = 3$, and all the results—but not necessarily all proofs—for $n \geq 4$.

Section 3-3 has the usual definition of the multiple integral of a continuous function of n variables over an n-dimensional rectangle, the "Fubini theorem" on the reduction of multiple to iterated integrals, and a definition of the multiple integral over a ball. The important theorem of this section is the invariance of integration under rotation (Theorem 3-4). In a first reading one may cover the details just for the cases $n = 2, 3$; these are easily visualized. The statements of the results for $n \geq 4$ will then have sufficient meaning, so that the proofs do not have to be studied right away.

Section 3-4 contains a description of the class of regular sets in n-space, and its properties. If one is familiar with epsilon-delta methods and least upper bounds, he can go through this section as it is written. A reader not familiar with these analytic methods should think of a regular set as one with a smooth boundary, and note—without proof—the properties of this class of sets as given in the statements of Theorems 3-5–3-9. After this he should study with care the *definition of the regular set* (Definition 3-1).

Section 3-5 is about integration of continuous functions over regular sets. As for the previous section, the reader with a good understanding of real variables can read it through; however, others should read—at first without proofs—the following important results:

(i) Equation 3-26, on the approximation of the integral over a regular set by integrals of upper and lower class functions

(ii) Definition 3-2, of the integral of a continuous function over a regular set

(iii) Theorem 3-10 (positivity) and Theorem 3-11 (additivity) of the indefinite integral

Section 3-6 is essential for the study of the normal Sample space, and Sec. 3-7 for the general space. These are brief but will often be quoted in the rest of the book.

3-1. THE GEOMETRY OF n-DIMENSIONAL SPACE

The Sample space corresponding to a Sample of n observations was introduced in Sec. 1-3 as n-dimensional space endowed with a certain set of probabilities. The latter system consists of the assignment of positive probabilities to certain points in the space: the probability of a set is equal to the sum of the probabilities of such points belonging to the set. This assignment fails in the case of an infinite population—in particular, a normal population. The definition of the Sample space associated with a Sample from a normal population is built upon certain results about the structure of multidimensional space. This chapter is devoted to an exposition of these results. We do not attempt to make it completely self-contained because we would have to review all of analysis from its foundations; however, the only incomplete points are fundamental, well-known theorems which will be stated without proof, and which can be found by the reader in any standard text on calculus containing material on two- and three-variables.

The space R^n, n-dimensional Euclidian space, is the n-fold product of the real line: it is the set of n-tuples of real numbers $(x_1, \ldots, x_n)$. This n-tuple is denoted $\mathbf{x}$ (in boldface type); the numbers $x_1, \ldots, x_n$ are the *coordinates* of $\mathbf{x}$. The origin, denoted $\mathbf{0}$, is the point with all coordinates equal to 0. A familiar example of such a space is R^2—the x_1x_2—plane.

Now we define the concept of a *line* in R^n. Although we can visualize a line in a space of two or three dimensions, we lack such a picture in higher dimensions. A line in R^n is defined as a set of points whose coordinates are of a specified form. If $\mathbf{x}$ and $\mathbf{y}$ are two points, and α and β real numbers, then $\alpha\mathbf{x} + \beta\mathbf{y}$ is defined as the point whose coordinates are $\alpha x_i + \beta y_i$, $i = 1, \ldots, n$. For any two points $\mathbf{x}$ and $\mathbf{y}$, the "line passing through both of them" is the set of points of the form $\alpha\mathbf{x} + (1 - \alpha)\mathbf{y}$, for some real number α. If $\mathbf{z}$ is a third point, and $\mathbf{z}$ is on the line passing through $\mathbf{x}$ and $\mathbf{y}$, then the line passing through $\mathbf{x}$ and $\mathbf{z}$ is identical with the former (Exercise 3); in other words, a line is determined by any two points on it. As an example consider points (x_1, x_2) and (y_1, y_2) in R^2. The line passing through both of them is the set of points whose coordinates are of the form $\alpha x_1 + (1 - \alpha)y_1$, $\alpha x_2 + (1 - \alpha)y_2$, for some α. This definition coincides with the usual one given in analytic geometry (Exercise 4). A line of particular importance in this work is the *diagonal:* it is the line passing through the origin and the point $\mathbf{1} = (1, \ldots, 1)$; equivalently, the diagonal is the set of points with equal coordinates. The line *segment* from $\mathbf{x}$ to $\mathbf{y}$ is the set of points of the form

$$\alpha\mathbf{x} + (1 - \alpha)\mathbf{y}, \quad 0 \le \alpha \le 1. \tag{3-1}$$

Now we define the concept of *distance* in R^n. In the familiar case of the plane, R^2, the distance between two points $\mathbf{x} = (x_1, x_2)$ and $\mathbf{y} = (y_1, y_2)$ is the positive square root of $(x_1 - y_1)^2 + (x_2 - y_2)^2$. Generalizing this to R^n, we define the distance of a point $\mathbf{x}$ from the origin as the positive square root of

$$\|\mathbf{x}\|^2 = \sum_{i=1}^{n} x_i^2; \tag{3-2}$$

more generally, the distance between two points $\mathbf{x}$ and $\mathbf{y}$, denoted $\|\mathbf{x} - \mathbf{y}\|$, is the positive square root of

$$\|\mathbf{x} - \mathbf{y}\|^2 = \sum_{i=1}^{n} (x_i - y_i)^2.$$

The concept of *perpendicular* lines in R^n is defined in terms of distances. Let $\mathbf{x}$, $\mathbf{y}$, and $\mathbf{z}$ be points in R^n; then the lines passing through $\mathbf{x}$ and $\mathbf{y}$ and through $\mathbf{y}$ and $\mathbf{z}$, respectively, are called *perpendicular* if

$$\|\mathbf{x} - \mathbf{z}\|^2 = \|\mathbf{x} - \mathbf{y}\|^2 + \|\mathbf{y} - \mathbf{z}\|^2. \tag{3-3}$$

This is equivalent to the condition that the Pythagorean theorem hold for the three segments from $\mathbf{x}$ to $\mathbf{y}$, $\mathbf{y}$ to $\mathbf{z}$, and $\mathbf{z}$ to $\mathbf{x}$, respectively. The reader can verify that this definition of perpendicularity coincides with the usual one in the familiar case of the plane (Exercise 5). It follows from the relation (3-2) defining the distance that the condition (3-3) for perpendicularity is equivalent to the condition on the coordinates:

$$\sum_{i=1}^{n} (x_i - y_i)(y_i - z_i) = 0 \tag{3-4}$$

(Exercise 6).

The *n axes* are the lines passing through $\mathbf{0}$ and one of the points

$$\mathbf{e}_1 = (1, 0, \ldots, 0),\ \mathbf{e}_2 = (0, 1, 0, \ldots, 0), \ldots, \mathbf{e}_n = (0, \ldots, 0, 1).$$

Every two axes are perpendicular (Exercise 8).

Example 3-1. In R^3, put

$$\mathbf{x} = (1, 1, 1), \quad \mathbf{y} = (1, -2, 1), \quad \mathbf{z} = (1, 0, -1);$$

their mutual distances are

$$\|\mathbf{x} - \mathbf{y}\| = (0^2 + 3^2 + 0^2)^{1/2} = 3$$
$$\|\mathbf{x} - \mathbf{z}\| = (0^2 + 1^2 + 2^2)^{1/2} = \sqrt{5}$$
$$\|\mathbf{y} - \mathbf{z}\| = (0^2 + 2^2 + 2^2)^{1/2} = 2\sqrt{2}.$$

$\mathbf{x}$ is on the diagonal. The lines through $\mathbf{0}$ and $\mathbf{x}$ and through $\mathbf{0}$ and $\mathbf{y}$, respectively, are perpendicular:

$$\|\mathbf{x}\|^2 + \|\mathbf{y}\|^2 = 9 = \|\mathbf{x} - \mathbf{y}\|^2.$$

The same is true for the other pairs of lines through the origin. Every point on the line passing through $\mathbf{x}$ and $\mathbf{z}$ has coordinates of the form $(1, \alpha, 2\alpha - 1)$ for some real number α; for example, the point with coordinates $(1, 5, 9)$ is on this line ($\alpha = 5$). ■

EXERCISES

1. Show that the lines passing through the origin and the points $\mathbf{y}$ and $\mathbf{z}$ (Example 3-1) are perpendicular.

2. What are the general forms of the coordinates of points on the lines through $\mathbf{x}$ and $\mathbf{y}$, and $\mathbf{y}$ and $\mathbf{z}$, respectively, for the points in Example 3-1?

3. Let $\mathbf{z}$ be a point on the line through $\mathbf{x}$ and $\mathbf{y}$, where the three points are distinct. *Prove:* The line through $\mathbf{x}$ and $\mathbf{z}$ is equivalent to the former.

4. In analytic geometry, the line passing through the two points (x_1, x_2) and (y_1, y_2) is defined as the locus of points (t_1, t_2) such that

$$\frac{t_2 - y_2}{t_1 - y_1} = \frac{y_2 - x_2}{y_1 - x_1}.$$

Prove: If $t_1 = \alpha x_1 + (1 - \alpha) y_1, t_2 = \alpha x_2 + (1 - \alpha) y_2$, then the above equation holds; conversely, if t_1 and t_2 satisfy such an equation, then there exists α such that t_1 and t_2 are of the indicated form.

5. Verify that the condition for perpendicularity coincides with the usual one given in analytic geometry: show that the condition (3-3) is equivalent to the condition that the slopes are negative reciprocals of each other.

6. Prove that the conditions (3-3) and (3-4) are equivalent.

7. *Prove:* For any $\mathbf{x}$ and any real number c,

$$\| c\mathbf{x} \| = |c| \cdot \| \mathbf{x} \|$$

8. Show that every two axes are perpendicular.

9. Fill in the details of this outline of the proof of the Cauchy-Schwarz inequality:

$$\left| \sum_{i=1}^{n} x_i y_i \right| \leq \| \mathbf{x} \| \cdot \| \mathbf{y} \| .$$

If either $\mathbf{x} = \mathbf{0}$ or $\mathbf{y} = \mathbf{0}$, the inequality is trivial; therefore, assume $\| \mathbf{y} \| > \mathbf{0}$. Define $\lambda = \sum_{i=1}^{n} x_i y_i / \| \mathbf{y} \|^2$; and note that

$$\| \mathbf{x} - \lambda \mathbf{y} \|^2 \geq \mathbf{0}.$$

10. Prove the "triangle inequality"

$$\| \mathbf{x} + \mathbf{y} \| \leq \| \mathbf{x} \| + \| \mathbf{y} \|$$

by squaring both sides and applying the result of Exercise 9.

3-2. ROTATIONS

A rotation of the plane is a one-to-one transformation of the plane onto itself characterized by the following: There exists a real number ω, $0 \leq \omega < 2\pi$, such that for any point in the plane with polar coordinates (r,θ), the image of the point under the rotation is the point with polar coordinates $(r, \theta + \omega)$. The number ω is called the angle of rotation. The effect of a rotation on a point in the plane is illustrated in Fig. 3-1.

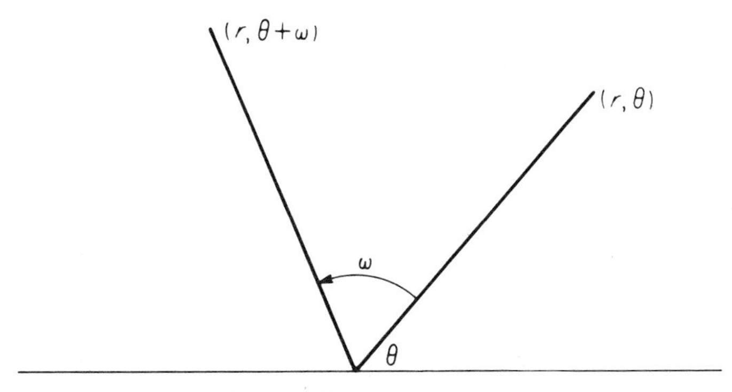

Fig. 3-1. Rotation of angle ω.

The origin is invariant under a rotation. It is clear that a rotation carries geometric figures (lines, triangles, quadrilaterals, circles, etc.) onto congruent ones; furthermore, it preserves circles about the origin. Rotations can be performed in succession, and can be reversed. When the successive application of two rotations is interpreted as "multiplication" of the rotations in the given order, the set of rotations of the plane forms an algebraic *group*: the *product* of two rotations of angles ω_1 and ω_2, respectively, is the rotation of angle $\omega_1 + \omega_2$ (mod 2π); the *identity* rotation is of angle $\omega = 0$; and the *inverse* of the rotation of angle ω is the one of angle $-\omega$. (This group is also commutative.)

A rotation in the plane can be characterized in terms of equations relating the coordinates (x, y) of a point to those of its image (x', y') under the rotation. Let (r, θ) be the polar coordinates of the point (x, y) and $(r, \theta + \omega)$ those of (x', y'); then

$$x = r\cos\theta, \quad y = r\sin\theta.$$

It follows that

$$\begin{aligned} x' = r\cos(\theta + \omega) &= r\cos\theta\cos\omega - r\sin\theta\sin\omega \\ &= x\cos\omega - y\sin\omega, \\ y' = r\sin(\theta + \omega) &= r\sin\theta\cos\omega + r\cos\theta\sin\omega \\ &= y\cos\omega + x\sin\omega; \end{aligned}$$

thus the equations are:

$$\begin{aligned} x' &= x\cos\omega - y\sin\omega \\ y' &= x\sin\omega + y\cos\omega. \end{aligned} \tag{3-5}$$

Rotations in space of dimension greater than two are much more complicated. Consider the rotations in three-dimensional space. In accordance with our intuition, we think of a rotation as a rigid motion of the space about the origin which preserves all mutual distances of pairs of points. The rotation of a globe, such as the earth, is characterized by two angles: one describing the change in latitude and one the change in longitude. In place of the pair of equations (3-5). there is a system of three simultaneous equations relating the coordinates of a point (x_1, x_2, x_3) and its image (x_1', x_2', x_3'). Although the rotations in this space form a group, the latter is not commutative: the order in which the rotations are performed affects the product. (See Example 3-2 below.) Rotations in R^n are conventionally studied in the form of orthogonal $n \times n$ matrices; however, in order to avoid the prerequisite of matrix theory, and for the purposes of our immediate applications, we find it sufficient and preferable to study rotations in higher dimensions from a different point of view.

First we define a *plane* rotation in R^n. For a fixed pair of integers i, j, $1 \leq i, j \leq n$, consider a transformation of R^n acting on a typical point $\mathbf{x}$ in the following manner: The coordinates x_i and x_j are transformed in accordance with the equations

$$\begin{aligned} x_i' &= x_i \cos \omega - x_j \sin \omega \\ x_j' &= x_i \sin \omega + x_j \cos \omega, \end{aligned} \tag{3-6}$$

which, like (3-5), define a rotation in the $x_i x_j$-plane; and all other coordinates are unchanged:

$$x_h' = x_h, \qquad h \neq i, \quad h \neq j. \tag{3-7}$$

This transformation is called a "plane rotation in the ith and jth coordinates." It is a one-to-one transformation of R^n onto itself, and has the following two properties:

I. The origin is transformed into itself.

II. Every line segment is transformed into a line segment of the same length.

The proof of I is trivial. From the equation (3-6) it follows that

$$x_i'^2 + x_j'^2 = x_i^2 + x_j^2 \qquad \text{(Exercise 1);}$$

thus from (3-7) we conclude that the distance of a point $\mathbf{x}$ from the origin is invariant under a plane rotation: if $\mathbf{x}'$ is the image of $\mathbf{x}$, then

$$\|\mathbf{x}'\| = \|\mathbf{x}\|. \tag{3-8}$$

If $\mathbf{x}'$ and $\mathbf{y}'$ are the images of $\mathbf{x}$ and $\mathbf{y}$, respectively, under the plane rotation, then the image of $\alpha\mathbf{x} + \beta\mathbf{y}$ is equal to the corresponding linear combination of the images:

$$(\alpha\mathbf{x} + \beta\mathbf{y})' = \alpha\mathbf{x}' + \beta\mathbf{y}', \qquad \alpha, \beta \textit{ real;} \tag{3-9}$$

(Exercise 2). This implies that the line passing through $\mathbf{x}$ and $\mathbf{y}$ is transformed into the line through $\mathbf{x}'$ and $\mathbf{y}'$; therefore, from (3-8) it follows that *lengths* of line segments are preserved under plane rotations. This completes the proof of property II.

A rotation in R^n is defined as a finite succession of plane rotations in various pairs of coordinates. By this we mean: If $U_1, \ldots, U_k$ are plane rotations, their product $U_1 \cdots U_k$—in the order indicated—is the transformation which acts on a point of R^n by transforming it in accordance with U_1, then transforming the image in accordance with U_2, and so on. Every rotation inherits properties I and II from its component plane rotations because these properties are invariant under composition. The product of two rotations is defined by successive transformation: if U is the product of plane rotations $U_1, \ldots, U_k$, and V is the product of $V_1, \ldots, V_l$, then UV is defined as the product $U_1 \cdots U_k V_1 \cdots V_l$ of plane rotations. The inverse of a rotation is the product of the inverse plane rotations in the inverse order: the inverse of a rotation composed of plane rotations of angles $\omega_1, \ldots, \omega_k$ is composed of plane rotations of angles $-\omega_k, \ldots, -\omega_1$ in the corresponding pairs of coordinates.

Example 3-2. Consider the rotation in R^3 composed of a plane rotation of angle ω_1 in the first two coordinates followed by a plane rotation of angle ω_2 in the second and third coordinates. Let (x_1', x_2', x_3') be the image of (x_1, x_2, x_3) under the first plane rotation, and (x_1'', x_2'', x_3'') the image of the former under the second rotation. The equations describing these are

$$x_1' = x_1 \cos \omega_1 - x_2 \sin \omega_1$$

$$x_2' = x_1 \sin \omega_1 + x_2 \cos \omega_1$$

and

$$x_1'' = x_1'$$

$$x_2'' = x_2' \cos \omega_2 - x_3' \sin \omega_2$$

$$x_3'' = x_2' \sin \omega_2 + x_3' \cos \omega_2;$$

thus the equations relating (x_1, x_2, x_3) and (x_1'', x_2'', x_3'') are

$$x_1'' = x_1 \cos \omega_1 - x_2 \sin \omega_1$$

$$x_2'' = x_1 \sin \omega_1 \cos \omega_2 + x_2 \cos \omega_1 \cos \omega_2 - x_3 \sin \omega_2$$

$$x_3'' = x_1 \sin \omega_1 \sin \omega_2 + x_2 \cos \omega_1 \sin \omega_2 + x_3 \cos \omega_2.$$

The equations are different when the order of the plane rotations is reversed, so that rotations do not form a commutative group (Exercise 3). ■

Let (x_1, x_2) and (y_1, y_2) be points in the plane at a common distance r from the origin. There is a rotation which carries the first point into the second; indeed, if the points have polar angles θ and θ', respectively, then the rotation of angle $\omega = \theta' - \theta$ carries the first point onto the second. In three dimensions, one can also intuitively see that two points at a common distance from the origin can be transformed one into the other by a suitable rotation. We generalize this to rotations in R^n:

THEOREM 3-1

Let $\mathbf{x}$ *and* $\mathbf{y}$ *be points in* R^n *such that*

$$\| \mathbf{x} \| = \| \mathbf{y} \| = r > 0;$$

then there is a rotation carrying $\mathbf{x}$ *onto* $\mathbf{y}$.

Proof: It is sufficient to prove that there is a rotation carrying $\mathbf{x}$ onto the point $(r, 0, \ldots, 0)$; indeed, if the rotations U and V carry $\mathbf{x}$ and $\mathbf{y}$, respectively, onto the latter point, then the product of U and the inverse of V carries $\mathbf{x}$ onto $\mathbf{y}$.

The proof of the existence of a rotation carrying $\mathbf{x}$ onto $(r, 0, \ldots, 0)$ follows from successive application of the (known) result for the plane. Let $x_1, \ldots, x_n$ be the coordinates of $\mathbf{x}$. There is a plane rotation in the first and second coordinates carrying $\mathbf{x}$ onto the point with coordinates

$$(\sqrt{x_1^2 + x_2^2}, 0, x_3, \ldots, x_n).$$

Next apply the plane rotation in the first and third coordinates carrying the latter point onto the one with coordinates

$$(\sqrt{x_1^2 + x_2^2 + x_3^2}, 0, 0, x_4, \ldots, x_n).$$

Continue in this way with the first and kth coordinates, $k = 4, \ldots, n$, finally obtaining the point $(r, 0, \ldots, 0)$. The desired rotation is the product of these plane rotations. ■

This theorem can be extended to a more general form. Let (x_1, x_2) and (y_1, y_2) be two points in the plane such that the lines connecting each of these to the origin are perpendicular. If the plane is rotated so that the point (x_1, x_2) is transformed into the point $(\sqrt{x_1^2 + x_2^2}, 0)$ on the first axis, then the point (y_1, y_2) will necessarily be transformed onto one of the points $(0, \pm\sqrt{y_1^2 + y_2^2})$ on the second axis; indeed, the rotation preserves perpendicularity because it preserves distance. This is now generalized to R^n: (In the following, $\mathbf{x}_1, \ldots, \mathbf{x}_n$ are *n points*, not *n* coordinates.)

THEOREM 3-2
Let $\mathbf{x}_1, \ldots, \mathbf{x}_n$ be n points in R^n at a common distance r from the origin. Let $\mathbf{0x}_1, \ldots, \mathbf{0x}_n$ be the line segments connecting $\mathbf{0}$ to $\mathbf{x}_1, \ldots, \mathbf{x}_n$, respectively. If every two of these are perpendicular, then there is a rotation carrying each of $\mathbf{x}_1, \ldots, \mathbf{x}_n$ onto the corresponding member of the set $r\mathbf{e}_1, \ldots, \pm r\mathbf{e}_n$, where $\mathbf{e}_j$ is defined in Sec. 3-1. The sign of the last member is uniquely determined.

Proof: By Theorem 3-1, there is a rotation carrying $\mathbf{x}_1$ onto $r\mathbf{e}_1$. Let $\mathbf{x}_j'$, $j = 2, \ldots, n$, be the images of $\mathbf{x}_j$, $j = 2, \ldots, n$, respectively. A rotation transforms perpendicular lines into perpendicular lines (Exercise 5); thus, every two of the lines $r\mathbf{e}_1$, $\mathbf{0x}_2', \ldots, \mathbf{0x}_n'$ are perpendicular; therefore, the first coordinate of $\mathbf{x}_j'$ is equal to 0, $j = 2, \ldots, n$ (Exercise 6). By a succession of plane rotations in the coordinates 2 and j, $j = 3, \ldots, n$, as in the proof of Theorem 3-1, the point $\mathbf{x}_2'$ is transformed into a point whose second coordinate is r, and whose other coordinates are 0; this is the point $r\mathbf{e}_2$. The point $r\mathbf{e}_1$ is invariant under rotations in the coordinates $2, \ldots, n$ because it has 0's as the coordinates. It follows that the rotations performed carry $\mathbf{x}_1$ and $\mathbf{x}_2$ onto $r\mathbf{e}_1$ and $r\mathbf{e}_2$, respectively. Repeat similar operations on the images of $\mathbf{x}_3', \ldots, \mathbf{x}_n'$ until each of the points has been properly transformed. After $n - 1$ rotations the first $n - 1$ coordinates of the image of $\mathbf{x}_n$ are 0; therefore the nth is $\pm r$. ■

EXERCISES

1. *Prove:* If (x_i, x_j) and (x_i', x_j') are related by (3-6), then $x_i^2 + x_j^2 = x_i'^2 + x_j'^2$.
2. Show that (3-9) holds for plane rotations.
3. Find the equations relating the coordinates of a point and its image under the product of the plane rotations in the reverse order in Example 3-2.

4. Write the equations describing the inverse of the rotation in Example 3-2.

5. Prove that a rotation transforms perpendicular lines into perpendicular lines.

6. *Prove:* The line through $\mathbf{0}$ and $\mathbf{x}$ is perpendicular to the first axis if and only if $x_1 = 0$.

7. Write the equations of the rotation carrying the point $(1, 2, -1)$ onto $(\sqrt{6}, 0, 0)$.

8. Write the equations of the rotation carrying the three points

$$\left(\frac{1}{\sqrt{3}}, \frac{1}{\sqrt{3}}, \frac{1}{\sqrt{3}}\right), \left(\frac{1}{\sqrt{6}}, \frac{-2}{\sqrt{6}}, \frac{1}{\sqrt{6}}\right), \left(\frac{1}{\sqrt{2}}, 0, \frac{-1}{\sqrt{2}}\right)$$

onto (1,0,0), (0,1,0), (0,0,1), respectively.

9. Repeat Exercise 7 for the points (1,0,4) and $(\sqrt{17},0,0)$.

3-3. INTEGRATION OVER R^n

The theory of integration of functions over multidimensional space is much more complicated than that of functions on the real line. Since this theory is usually not well developed in the standard course of integral calculus, we shall devote several sections to it.

Let $f(x)$ be a real-valued function of a real variable. We recall that f is *continuous at a point* x_0 if

$$\lim_{x \to x_0} f(x) = f(x_0);$$

it is *continuous* if it is so at every point. The continuity of a function means that points in the domain that are close to each other are mapped into points in the range that are also close:

$$\text{“}\,|x_1 - x_2|\text{ small” implies “}\,|f(x_1) - f(x_2)|\text{ small.”}$$

For any interval $[a, b]$, the integral

$$\int_a^b f(x)\,dx$$

is defined in the calculus as the limit, which is shown to exist, of certain approximating sums.

Certain problems complicate the extension of the theory of

integration on the line to higher dimensional space. For any continuous function $f(\mathbf{x})$, $\mathbf{x} \in R^n$, and a subset A of R^n, we formally write the integral of f over A as

$$\int \cdots \int_A f(\mathbf{x})\, d\mathbf{x}, \tag{3-10}$$

where the multiple integral is an n-fold integral, and where $d\mathbf{x}$ is the element of integration, usually denoted $dx_1 \cdots dx_n$. We shall consider the following problems:

I. Determine a class of sets A for which the integral (3-10) is defined; and
II. Derive the properties of the integral as a function on this class of sets.

We recall that a function $f(\mathbf{x})$ on R^n is continuous at a point $\mathbf{x}_0$ if

$$\lim_{\|\mathbf{x}-\mathbf{x}_0\|\to 0} f(\mathbf{x}) = f(x_0);$$

in other words, $f(x)$ approaches $f(\mathbf{x}_0)$ as the distance of the variable point $\mathbf{x}$ to $\mathbf{x}_0$ tends to 0. A function is called *continuous* if it is so at every point of R^n.

Let us begin with some simple sets in R^2. Let A be the finite rectangle

$$A = \{(x_1, x_2): a_1 \le x_1 \le b_1, a_2 \le x_2 \le b_2\},$$

and $f(x_1, x_2)$ a continuous function. The area of A is the product $(b_2 - a_2)(b_1 - a_1)$. The integral

$$\iint_A f(x_1, x_2)\, dx_1\, dx_2$$

is defined as the iterated integral

$$\int_{a_1}^{b_1} \int_{a_2}^{b_2} f(x_1, x_2)\, dx_2\; dx_1,$$

or as

$$\int_{a_2}^{b_2} \int_{a_1}^{b_1} f(x_1, x_2)\, dx_1\; dx_2.$$

The *theorem* justifying this definition is

THEOREM 3-3

Let $f(x_1, x_2)$ be a continuous function. For fixed x_2, $f(x_1, x_2)$

is continuous as a function of x_1 (Exercise 1), so that the integral

$$\int_{a_1}^{b_1} f(x_1, x_2)\, dx_1$$

is well defined. As a function of x_2, this integral is continuous so that

$$\int_{a_2}^{b_2} \left(\int_{a_1}^{b_1} f(x_1, x_2)\, dx_1 \right) dx_2$$

is also well defined; furthermore, the latter is equal to the iterated integral obtained by integrating first with respect to x_2 and then x_1:

$$\int_{a_2}^{b_2} \left(\int_{a_1}^{b_1} f(x_1, x_2)\, dx_1 \right) dx_2 = \int_{a_1}^{b_1} \left(\int_{a_2}^{b_2} f(x_1, x_2)\, dx_2 \right) dx_1$$

The double integral of f over A is defined as the common value of the iterated integrals.

We omit the proof of this fundamental result, which is tacitly applied in every computation of a double integral. The geometric interpretation of the equality of the two iterated integrals is that the same double integral is obtained by the two different procedures:

(i) Cutting the rectangle into "x_1-strips," integrating over x_2, and then summing over the strips; or
(ii) Following a similar procedure, interchanging the roles of x_1 and x_2 (Fig. 3-2).

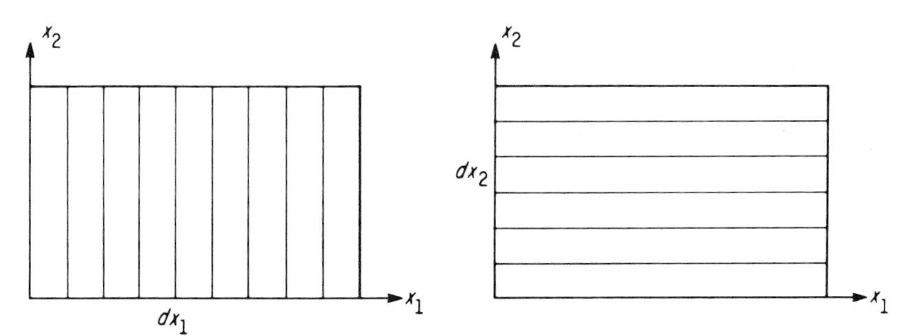

Fig. 3-2. Two methods of iterated integration.

In the above way, the integral (3-10) is defined on the class of rectangles A in R^2. The definition is extended to the class of all finite unions of disjoint rectangles: If $A_1, \ldots, A_k$ are disjoint rectangles, and A is their union, then the integral (3-10) is defined as the sum of the integrals over $A_1, \ldots, A_k$, respectively. The integral is finally extended to the class of all sets which are "approximable" by finite unions of disjoint rectangles. This is the principle underlying the introduction of polar coordinates in the calculation of a double integral over a region in the plane with a smooth boundary (Fig. 3-3).

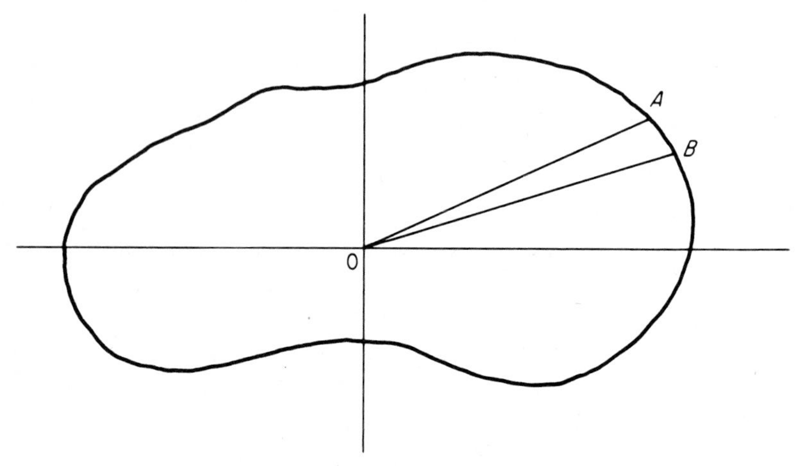

Fig. 3-3. Decomposition of a region for integration in polar coordinates.

The region is decomposed into small sectors; each of the latter is approximable by a narrow triangle of the form AOB (Fig. 3-3). The double integral of a function over such a triangle is approximable by the sum of the double integrals over small rectangles approximating the triangle (Fig. 3-4).

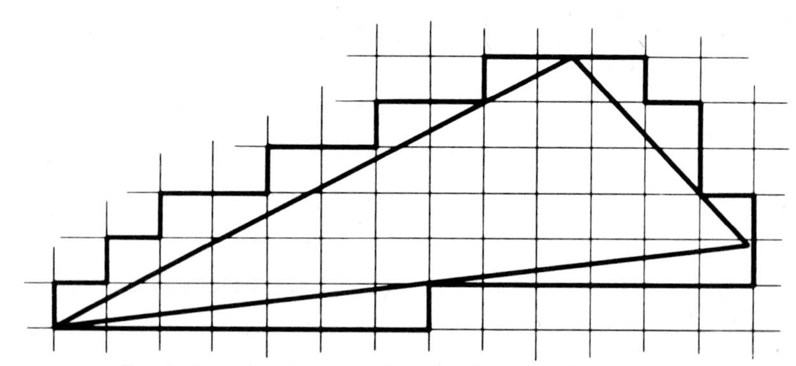

Fig. 3-4. Approximation of a triangle by small rectangles.

The integral over r, times $d\theta$:

$$\int_0^{r(\theta)} f(r,\theta) r\, dr \cdot d\theta,$$

is the approximate value of the double integral over the small triangle; the sum over all triangles is the approximate value of the integral over all θ.

Now consider rectangles in R^n; such sets are of the form

$$A = \{\mathbf{x}: a_i \le x_i \le b_i, \quad k = 1, \ldots, n\}.$$

For a continuous function $f(\mathbf{x})$, the integral (3-10) over the rectangle A is defined as the iterated integral

$$\int_{{}_n a}^{b_n} \cdots \int_{a_1}^{b_1} f(\mathbf{x})\, dx_1 \cdots dx_n, \tag{3-11}$$

where the value of the integral is independent of the order of integration. This is justified by the extended version of Theorem 3-3:

EXTENSION OF THEOREM 3-3

If $f(\mathbf{x})$ is a continuous function on R^n, then it is continuous in a subset of its variables when the others are fixed. The integral with respect to a subset of its variables is a continuous function of the remaining variables, e.g.,

$$\int_{a_n}^{b_n} \cdots \int_{a_{k+1}}^{b_{k+1}} f(\mathbf{x})\, dx_{k+1} \cdots dx_n$$

is a continuous function of $(x_1, \ldots, x_k)$. The iterated integral (3-11) has a value independent of the order of integration.

The integral can be extended from the class of rectangles to the class of finite unions of disjoint rectangles, as in the case $n = 2$; finally, it can be extended to the class of sets which are approximable by finite unions of disjoint rectangles. An important set of this kind is the *ball* in R^n centered at $\boldsymbol{\mu}$ and of radius $r > 0$:

$$\{\mathbf{x}: \|\mathbf{x} - \boldsymbol{\mu}\| \le r\};$$

it is the set of points whose distance from $\boldsymbol{\mu}$ is less than or equal to r. (It is a generalization of the disk in R^2.) Its boundary is the *sphere* centered at $\boldsymbol{\mu}$ and of radius r:

$$\{\mathbf{x}: \|\mathbf{x} - \boldsymbol{\mu}\| = r\}.$$

We shall define the integral (3-10) of a continuous function over a ball in R^n; this will be done by the inductive extension of the integral from $n = 2$ to higher dimensions.

Let $f(x_1, x_2)$ be a continuous function. Its integral over the disk is calculated by changing to polar coordinates. For simplicity let the disk be centered at the origin; the integral is

$$\int_0^{2\pi} \int_0^r f(\rho \cos \theta, \rho \sin \theta)\rho \, d\rho \, d\theta. \tag{3-12}$$

Now let $f(x_1, x_2, x_3)$ be continuous on R^3; we wish to integrate it over the ball centered at the origin and of radius r. Let $(0, 0, z)$ be a fixed point on the x_3-axis such that $-r < z < r$. Pass a plane through this point, parallel to the x_1x_2-plane—that is, perpendicular to the x_3-axis. The plane intersects the ball in a disk centered at $(0, 0, z)$ and of radius $\sqrt{r^2 - z^2}$. (See Fig. 3-5.)

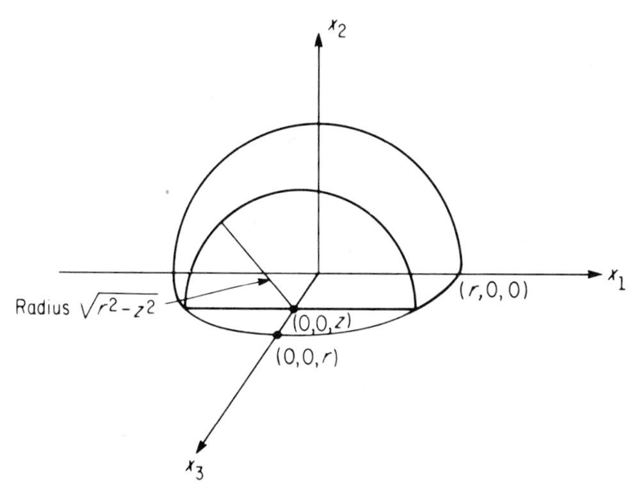

Fig. 3-5. Intersection of plane and ball.

We calculate the integral of f over the ball by slicing it into many thin disks, calculating the integrals over the disks, by means of the double integral (3-12), and then summing over all the disks. For each z, $-r < z < r$, the function $f(x_1, x_2, z)$ is continuous in (x_1, x_2). Its integral over the disk centered at $(0, 0, z)$ and of radius

$\sqrt{r^2 - z^2}$ is, by (3-12), equal to

$$\int_0^{2\pi} \int_0^{\sqrt{r^2 - z^2}} f(\rho \cos \theta, \rho \sin \theta, z) \rho \, d\rho \, d\theta;$$

when multiplied by dz, this is the approximate value of the integral of f over a disk centered at $(0, 0, z)$ and of thickness dz. The integral over the entire ball is the integral of the double integral above over all z, $-r \leq z \leq r$:

$$\int_{-r}^{r} \int_0^{2\pi} \int_0^{\sqrt{r^2 - z^2}} f(\rho \cos \theta, \rho \sin \theta, z) \rho \, d\rho \, d\theta \, dz. \qquad (3\text{-}13)$$

The integral of f over the three-dimensional ball was just obtained by summing integrals over two-dimensional balls. We define the integral over the n-dimensional ball by induction: having defined the integral over the $n - 1$ dimensional ball, we slice the former into many thin $(n - 1)$-dimensional balls (disks), and then sum the integrals over the latter. Let $f(\mathbf{x})$ be a continuous function on R^n, and consider the ball in R^n centered at the origin and of radius r. For a fixed point $(0, \ldots, 0, z)$ on the nth axis, $-r < z < r$, consider the "hyperplane" consisting of all points in R^n whose nth coordinates are equal to z:

$$H(z) = \{\mathbf{x}\colon x_n = z\}.$$

This hyperplane is "perpendicular" to the nth axis in the following sense: If $\mathbf{y}$ is any point in $H(z)$ not equal to $(0, \ldots, 0, z)$, then the line segment connecting both is perpendicular to the segment from $\mathbf{0}$ to the latter (Exercise 3). The intersection of the ball of radius r and $H(z)$ is the set

$$\{\mathbf{x}\colon x_1^2 + \cdots + x_{n-1}^2 \leq r^2 - z^2, \; x_n = z\} \qquad (3\text{-}14)$$

(Exercise 4); in other words, it is an $(n - 1)$-dimensional ball centered at the point $(0, 0, \ldots, 0, z)$ and of radius $\sqrt{r^2 - z^2}$. For $x_n = z$, $f(\mathbf{x})$ is a continuous function of its first $n - 1$ variables; thus, by the induction hypothesis, the integral of f (as a function of $x_1, \ldots, x_{n-1}$) over the $(n - 1)$-dimensional ball (3-14) is well defined, and is denoted by

$$F_{n-1}(\sqrt{r^2 - z^2}, z).$$

The integral of f over the whole n-dimensional ball is obtaining by integration over z:

$$F_n(r) = \int_{-r}^{r} F_{n-1}(\sqrt{r^2 - z^2}, z)\, dz; \tag{3-15}$$

this corresponds to adding the integrals over all the "slices" (3-14). By means of the extension of Theorem 3-3, it can be shown that this integral is the same for all orders of integration over the various coordinates.

Example 3-3. Let us find the explicit formula for the integral of a continuous function of four variables over the ball of radius r in R^4. By Formula (3-13), the integral of $f(x_1, x_2, x_3, z)$ over the three-dimensional ball of radius $\sqrt{r^2 - z^2}$ is

$$\int_{-\sqrt{r^2-z^2}}^{\sqrt{r^2-z^2}} \int_0^{2\pi} \int_0^{\sqrt{r^2-z^2-u^2}} f(\rho \cos\theta, \rho \sin\theta, u, z)\rho\, d\rho\, d\theta\, du$$

$$= F_3(\sqrt{r^2 - z^2}, z).$$

The integral over the four-dimensional ball is then obtained by integration over z from $-r$ to r. ■

Example 3-4. When $f(x)$ is taken to be the function identically equal to 1, the integral over the ball is called the *volume* of the ball. From (3-13) we obtain the familiar formula $(4/3)\pi r^3$ for the volume of the three-dimensional ball. From this and from the result of Example 3-3 we get the volume of the four-dimensional ball:

$$(4\pi/3)\int_{-r}^{r} (r^2 - z^2)^{3/2}\, dz = \pi^2 r^4/2. \quad ■$$

Later we shall derive the general formula for the volume of the n-dimensional ball.

There is an important connection, to be repeatedly used, between rotation and integration: it is that the integral of a continuous function over a sphere is invariant under the rotation of the variable of integration. This is a consequence of the fact that *volume* is invariant under rotation.

THEOREM 3-4

Let $f(\mathbf{x})$ be a continuous function, B a ball in R^n centered at $\mathbf{0}$, and $\mathbf{x}'$ the image of $\mathbf{x}$ under a given rotation; then the integrals of the functions $f(\mathbf{x})$ and $f(\mathbf{x}')$ over B are equal:

$$\int \cdots \int_B f(\mathbf{x})\,d\mathbf{x} = \int \cdots \int_B f(\mathbf{x}')\,d\mathbf{x}.$$

Proof: Consider first the case $n = 2$, and a rotation of angle ω. The integral of $f(\mathbf{x}) = f(x_1, x_2)$ over a disk of radius r is, in polar coordinates,

$$\int_0^{2\pi} \int_0^r f(\rho\cos\theta, \rho\sin\theta)\rho\,d\rho\,d\theta.$$

If $x_1 = \rho\cos\theta$, $x_2 = \rho\sin\theta$, then

$$x_1' = \rho\cos(\theta + \omega), \qquad x_2' = \rho\sin(\theta + \omega);$$

therefore, the integral of $f(x_1', x_2')$ is, in polar coordinates,

$$\int_0^{2\pi} \int_0^r f(\rho\cos(\theta + \omega), \rho\sin(\theta + \omega))\rho\,d\rho\,d\theta.$$

This is equal to the former double integral because the sine and cosine are periodic: $\sin(\theta + 2\pi) = \sin\theta$, $\cos(\theta + 2\pi) = \cos\theta$ (Exercise 5).

We turn to the case of a rotation in R^n. If the conclusion of the theorem is shown to be true for each of two rotations, then it is also true for their product: if $\mathbf{x}'$ is the image of $\mathbf{x}$ under the first rotation, and $\mathbf{x}''$ the image of $\mathbf{x}'$ under the second, then

$$\int \cdots \int_B f(\mathbf{x}'')\,d\mathbf{x} = \int \cdots \int_B f(\mathbf{x}')\,d\mathbf{x} = \int \cdots \int_B f(\mathbf{x})\,d\mathbf{x}.$$

For this reason it is sufficient to prove the theorem in the case of a *plane* rotation because every rotation is a product of these. For simplicity we consider a plane rotation in the first and second coordinates. Write $f(\mathbf{x})$ as $f(x_1, x_2, \ldots, x_n)$, and change the first two variables to polar coordinates:

$$f(\mathbf{x}) = f(\rho\cos\theta, \rho\sin\theta, x_3, \ldots, x_n)$$

$$f(\mathbf{x}') = f(\rho\cos(\theta + \omega), \rho\sin(\theta + \omega), x_3, \ldots, x_n).$$

By virtue of the construction of the integral of f over the ball as an iterated integral (3-15), and by the proved invariance in the special case $n = 2$, the integrals of $f(\mathbf{x})$ and $f(\mathbf{x}')$ are equal. The invariance under a plane rotation in any other pair of coordinates follows because, as remarked above, the integral over the ball is the same for all orders of integration.

It has been tacitly supposed in this proof that $f(\mathbf{x}')$ is a continuous function of the original variable $\mathbf{x}$; the truth of this assumption can be verified (Exercise 6). ■

EXERCISES

1. *Prove:* If $f(x_1, x_2)$ is continuous, then, for fixed x_2, it is continuous as a function of the single variable x_1.

2. For a fixed value of θ, consider the double integral of the integrand in (3-13) with respect to (ρ, z) over the region $0 \le \rho \le \sqrt{r^2 - z^2}$, $-r \le z \le r$. Transform to polar coordinates by means of the substitution

$$\rho = t \cos \phi, \quad z = t \sin \phi, \quad 0 \le t \le r, \quad -\pi/2 \le \phi \le \pi/2.$$

Write the resulting triple integral.

3. Let y be a point in $H(z)$ not equal to $(0, \ldots, 0, z)$. Prove that the segment connecting these two points is perpendicular to the segment connecting the origin and $(0, \ldots, 0, z)$.

4. Prove that the intersection of $H(z)$ and the ball of radius r is the set (3-14).

5. Prove that

$$\int_0^{2\pi} f(\rho \cos (\theta + \omega), \rho \sin (\theta + \omega))\, d\theta$$

is the same for all ω. (*Hint:* Change the variable of integration from θ to $\theta - \omega$ and the limits of integration; then apply the periodicity of the sine and cosine.)

6. *Prove:* If $f(\mathbf{x})$ is continuous, then so is $f(\mathbf{x}')$ as a function of $\mathbf{x}$. (*Hint:* Consider first $n = 2$; then, plane rotations in R^n; finally, show that the truth of the conclusion for two rotations implies its truth for their product.)

3-4. REGULAR SETS

In this section we are going to describe a large and useful class of sets in R^n for which the integral (3-10) has a meaning. This is the class of *regular* sets. It will be shown that many of the familiar sets—balls, rectangles, half-spaces, etc.—belong to this class. The important feature of this class is that it is closed under the elementary set operations.

Let A be any set in R^n for which the integral (3-10) has already been defined (rectangle, ball). For any two continuous functions f_1 and f_2, the following assertion is true:

If $f_1(\mathbf{x}) \leq f_2(\mathbf{x})$ *for all* $\mathbf{x}$, *then*

$$\int_A \cdots \int f_1(\mathbf{x})\, d\mathbf{x} \leq \int_A \cdots \int f_2(\mathbf{x})\, d\mathbf{x}; \tag{3-16}$$

this follows by iterated integration from the known property of the integral in one variable.

The volume of the ball in R^n, defined in Example 3-4 as

$$\operatorname{vol}(B) = \int_B \cdots \int 1\, d\mathbf{x},$$

is finite; this is a consequence of the recursive relation (3-15). (Later we shall compute this volume.) From (3-16) we deduce: If $f(\mathbf{x})$ is a continuous function on R^n such that

$$0 \leq f(\mathbf{x}) \leq 1, \quad \text{for} \quad \mathbf{x} \in R^n, \tag{3-17}$$

and if B is a ball, then

$$0 \leq \int_B \cdots \int f(\mathbf{x})\, d\mathbf{x} \leq \operatorname{vol}(B). \tag{3-18}$$

The class of regular sets is defined in the following manner. A criterion is applied to an arbitrary set: if the set satisfies the condition prescribed by the criterion, then it is called *regular*; if not, it is *not regular*. In this way the regular sets are defined as the class of sets satisfying given quantitative conditions. The concepts of least upper bound and greatest lower bound are essential to the definition, and should be reviewed by the reader who is unsure of these.

Let M be a given set in R^n. Let $\mathfrak{U}(M)$ be the class of continuous functions on R^n such that

$$0 \leq f(\mathbf{x}) \leq 1, \quad \mathbf{x} \in R^n; \quad \text{and} \quad f(\mathbf{x}) \equiv 1, \quad \mathbf{x} \in M; \tag{3-19}$$

in other words, f assumes values between 0 and 1 inclusive everywhere, but is strictly equal to 1 on M. Such a function certainly exists; for example, there is the function identically equal to 1. Let $\mathfrak{L}(M)$ be the class of continuous functions such that

$$0 \leq f(\mathbf{x}) \leq 1, \quad \mathbf{x} \in R^n; \quad \text{and} \quad f(\mathbf{x}) \equiv 0, \quad \mathbf{x} \notin M; \tag{3-20}$$

f is between 0 and 1, inclusive, everywhere but strictly equal to 0 outside M. Such a function is the one identically equal to 0. It follows from (3-19) and (3-20) that

$$f_1(\mathbf{x}) \leq 1 = f_2(\mathbf{x}), \quad \text{for} \quad \mathbf{x} \in M$$

$$f_1(\mathbf{x}) = 0 \leq f_2(\mathbf{x}), \quad \text{for} \quad \mathbf{x} \notin M, \quad f_1 \in \mathfrak{L}(M), \quad f_2 \in \mathfrak{U}(M);$$

therefore,

$$f_1(\mathbf{x}) \leq f_2(\mathbf{x}), \quad \text{for} \quad \mathbf{x} \in R^n, \quad f_1 \in \mathfrak{L}(M), \quad f_2 \in \mathfrak{U}(M). \tag{3-21}$$

$\mathfrak{L}(M)$ and $\mathfrak{U}(M)$ are called the *lower* and *upper* class functions, respectively. It follows that for every ball B we have

$$\int_B \cdots \int f_1(\mathbf{x})\, d\mathbf{x} \leq \int_B \cdots \int f_2(\mathbf{x})\, d\mathbf{x},$$

$$\text{for} \quad f_1 \in \mathfrak{L}(M), \quad f_2 \in \mathfrak{U}(M). \tag{3-22}$$

The set of real numbers consisting of the values of the definite integrals

$$\int_B \cdots \int f_1(\mathbf{x})\, d\mathbf{x}, \quad f_1 \in \mathfrak{L}(M)$$

is nonempty. By (3-18) it is bounded above by the volume of B. Since it is a nonempty set of real numbers bounded above, it has, by the completeness axiom of the real-number system, a least upper bound L. The set of real numbers consisting of the values of the definite integrals

$$\int_B \cdots \int f_2(\mathbf{x})\, d\mathbf{x}, \quad f_2 \in \mathfrak{U}(M)$$

is nonempty; and, by (3-18) it is bounded below by 0; thus, by the completeness axiom, it has a greatest lower bound U. By (3-22) every number in the first set is less than or equal to every one in the second set; therefore,

$$L \leq U. \tag{3-23}$$

(Exercise 1).

DEFINITION 3-1

A set M is called regular if for every ball B centered at 0 there is strict equality in (3-23). The volume of M is the least upper bound (finite or infinite) of the common L, U value over all balls.

The intuitive interpretation of the regularity of a set is that its boundary is not too "ragged": among the continuous functions vanishing outside M and the ones equal to 1 on M, we can find two (one from each) which are "nearly" equal. Before giving examples of regular sets, we present an example of one which is not regular.

Example 3-4. The set of all rational numbers on the real line is not regular. The only function in $\mathfrak{U}(M)$ is the one identically equal to 1, and the only function in $\mathfrak{L}(M)$ is the one identically equal to 0; indeed, a continuous function equal to 1 on all rationals is identically equal to 1, and a continuous function equal to 0 on the set of irrational numbers is identically 0 (Exercises 2, 3). It follows that there is strict inequality in (3-23). This example can be extended to R^n: the set of points with rational coordinates is not regular. ■

Many familiar sets in one and two dimensions—intervals, rectangles, half-spaces and disks—are regular.

Example 3-5. The disk in R^2 centered at the origin and of radius $r > 0$ is regular, and its "volume" is, as expected, equal to πr^2. The proof in this and other examples involves the use of two similar continuous functions on the line:

$$\begin{aligned} g_1(x) &= 1, \quad x \leq r - \epsilon \\ &= 0, \quad x \geq r, \end{aligned}$$

and varies linearly from 1 to 0 on $[r - \epsilon, r]$;

$$g_2(x) = 1, \quad x \leq r$$
$$= 0, \quad x \geq r + \epsilon,$$

and varies linearly from 1 to 0 on $[r, r + \epsilon]$, where $\epsilon > 0$ is a small fixed positive number (Fig. 3-6).

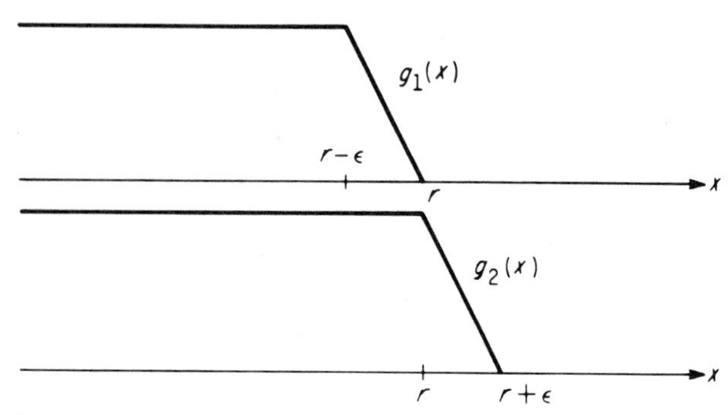

Fig. 3-6. The continuous functions g_1 and g_2.

Define two continuous functions on the plane in terms of g_1 and g_2:

$$f_1(x_1, x_2) = g_1(\sqrt{x_1^2 + x_2^2})$$
$$f_2(x_1, x_2) = g_2(\sqrt{x_1^2 + x_2^2});$$

then f_1 and f_2 assume values between 0 and 1, inclusive; the former vanishes outside the disk of radius r, and so belongs to $\mathfrak{L}$, the lower class for the disk; and the latter, f_2, is equal to 1 on the disk, and so belongs to $\mathfrak{U}$, the upper class. The functions f_1 and f_2 differ only in the narrow annular band between the circles of radii $r \pm \epsilon$; therefore, the difference between the integrals of f_2 and f_1 is at most equal to the area between the circles, $\pi[(r + \epsilon)^2 - (r - \epsilon)^2] = 4\pi r\epsilon$; therefore, the same is, of course, true for the difference between the integrals of f_2 and f_1 over any ball (disk). Since $\epsilon > 0$ is arbitrary, and $|L - U| \leq \epsilon$ for every ball, it follows that there is strict equality in (3-23); therefore, the disk is regular. It is also evident that the common value of L and U with respect to a large ball is equal to πr^2 (Exercise 4). ■

Example 3-6. The half-space in the plane, $\{x_1, x_2) : x_1 \leq r\}$ is regular. For the proof, consider its intersection with a ball, and functions

$$f_i(x_1,x_2) = g_i(x_1), \qquad i = 1,2,$$

where g_1 and g_2 are defined in Example 3-5 (Exercise 5). The volume is infinite. The same is true of the open half-sapce $\{(x_1, x_2): x_1 < r\}$. ■

In the proof of the next result (as outlined in the exercises) as well as in several later proofs we shall associate with each rotation and each function f a "rotated" function f^*. It is derived from f as follows. Let $\mathbf{x}'$ be the image of $\mathbf{x}$ under the rotation. We define f^* at the point $\mathbf{x}'$ to be equal to f at the point $\mathbf{x}$:

$$f^*(\mathbf{x}') = f(\mathbf{x}).$$

This defines f^* everywhere because each point in the space is the image of some point under the rotation. If f is continuous, then so is f^*. By Theorem 3-4, the integrals of f and f^* over each ball are equal. Now we prove that the class of regular sets is closed under the operation of rotation:

THEOREM 3-5
The image of a regular set under a rotation is also regular, and the volume of the set and its image are equal.

Proof: By virtue of the invariance under rotation of the integral over a ball (Theorem 3-4), the integrals of the functions of the lower and upper classes appearing in the criterion of regularity (Definition 3-1) are the same for a set and its image, respectively (Exercise 6). ■

Example 3-7. In R^2, the half-plane

$$\{(x_1, x_2): x_1 + x_2 \leq c\}$$

is regular; indeed, this set is obtained from the half-space

$$\{(x_1,x_2): x_1 \leq c/\sqrt{2}\}$$

by a rotation of 45°, and the latter set is regular (Example 3-6). ■

We prove the other closure properties of the class of regular sets.

THEOREM 3-6

The complement of a regular set is regular.

Proof: For any continuous function f, let $1 - f$ be the function whose value at $\mathbf{x}$ is $1 - f(\mathbf{x})$. Let M be a regular set and N its complement. A function f belongs to $\mathfrak{L}(M)$ if and only if $1 - f$ belongs to $\mathfrak{U}(N)$; similarly, f belongs to $\mathfrak{U}(M)$ if and only if $1 - f$ belongs to $\mathfrak{L}(N)$ (Exercise 7). We have to prove: For any ball B,

$$\operatorname*{lub}_{f\in\mathfrak{L}(N)} \int_B\cdots\int f(\mathbf{x})\,d\mathbf{x} = \operatorname*{glb}_{f\in\mathfrak{U}(M)} \int_B\cdots\int f(\mathbf{x})\,d\mathbf{x},$$

where the "lub" and "glb" stand for least upper and greatest lower bounds, respectively. This follows from these relations:

$$\begin{aligned}
&\operatorname*{lub}_{f\in\mathfrak{L}(N)} \int_B\cdots\int f(\mathbf{x})\,d\mathbf{x} \\
&= \operatorname*{lub}_{f\in\mathfrak{U}(M)} \int_B\cdots\int (1 - f(\mathbf{x}))\,d\mathbf{x} \\
&= \int_B\cdots\int d\mathbf{x} - \operatorname*{glb}_{f\in\mathfrak{U}(M)} \int_B\cdots\int f(\mathbf{x})\,d\mathbf{x} \\
&= \int_B\cdots\int d\mathbf{x} - \operatorname*{lub}_{f\in\mathfrak{L}(M)} \int_B\cdots\int f(\mathbf{x})\,d\mathbf{x} \quad \text{(by regularity of } M\text{)} \\
&= \operatorname*{glb}_{f\in\mathfrak{L}(M)} \int_B\cdots\int (1 - f(\mathbf{x}))\,d\mathbf{x} \\
&= \operatorname*{glb}_{f\in\mathfrak{U}(N)} \int_B\cdots\int f(\mathbf{x})\,d\mathbf{x}. \quad \blacksquare
\end{aligned}$$

THEOREM 3-7

The space R^n is a regular set.

Proof: The only function in the upper class $\mathfrak{U}(R^n)$ is the one $f(\mathbf{x}) \equiv 1$. This function also belongs to the lower class $\mathfrak{L}(R^n)$: the condition that $f(\mathbf{x}) = 0$ for $\mathbf{x} \notin R^n$ is vacuously satisfied; thus $L = U$ for every ball. ■

THEOREM 3-8
The intersection of finitely many regular sets is regular.

Proof: We give the proof for the intersection of two regular sets M and N; the proof for an arbitrary number n follows by induction (Exercise 8). Let $M \cap N$ be the intersection. If f and g belong to $\mathfrak{L}(M)$ and $\mathfrak{L}(N)$, respectively, then fg belongs to $\mathfrak{L}(M \cap N)$ (Exercise 9); similarly, if f' and g' belong to $\mathfrak{U}(M)$ and $\mathfrak{U}(N)$, respectively, then $f'g'$ belongs to $\mathfrak{U}(M \cap N)$; therefore, it is sufficient to prove:

$$\operatorname*{lub}_{f \in \mathfrak{L}(M),\, g \in \mathfrak{L}(N)} \left[\int_B \cdots \int fg \, d\mathbf{x}\right] = \operatorname*{glb}_{f' \in \mathfrak{U}(M),\, g' \in \mathfrak{U}(N)} \left[\int_B \cdots \int f'g' \, d\mathbf{x}\right] \tag{3-24}$$

for any ball B.

Since these functions assume values between 0 and 1, inclusive, and, by (3-21),

$$f(\mathbf{x}) \leq f'(\mathbf{x}), \quad g(\mathbf{x}) \leq g'(\mathbf{x}),$$

it follows that

$$0 \leq f'g' - fg = f'(g' - g) + g(f' - f) \leq (g' - g) + (f' - f).$$

Integrate over B:

$$\begin{aligned} \int_B \cdots \int f'g' \, d\mathbf{x} - \int_B \cdots \int fg \, d\mathbf{x} & \\ \leq \left[\int_B \cdots \int g' \, d\mathbf{x} - \int_B \cdots \int g \, d\mathbf{x}\right] & \\ + \left[\int_B \cdots \int f' \, d\mathbf{x} - \int_B \cdots \int f \, d\mathbf{x}\right]. & \end{aligned} \tag{3-25}$$

By the assumed regularity of M and N, the differences on the right-hand side of (3-25) can be made arbitrarily small by a suitable choice of f and f' in $\mathfrak{L}(M)$ and $\mathfrak{U}(M)$, respectively, and of g and g' in $\mathfrak{L}(N)$ and $\mathfrak{U}(N)$, respectively; therefore, the same must be true of the difference on the left-hand side; thus the greatest lower bound of the latter difference is 0. This implies (3-24). ∎

As an application of these theorems we show that another important type of set is regular.

Example 3-8. Any rectangle in the plane is regular. If the sides are not parallel to the axes, then a suitable rotation makes them parallel; thus, by the invariance of regularity under rotation (Theorem 3-5), it suffices to prove regularity when the sides are parallel to the axes. Let the rectangle be $\{(x_1, x_2): a_1 \leq x_1 \leq b_1, a_2 \leq x_2 \leq b_2\}$. The set $\{(x_1, x_2): a_1 \leq x_1 \leq b_1\}$ is regular because it is the intersection of the regular set $\{(x_1, x_2): x_1 \leq b_1\}$ and the complement of the regular set $\{(x_1, x_2): x_1 < a_1\}$. By analogy, the set $\{(x_1, x_2): a_2 \leq x_2 \leq b_2\}$ is regular. The rectangle is the intersection of the two regular sets $\{(x_1, x_2): a_i \leq x_i \leq b_i\}$, $i = 1,2$, and so is regular. It also remains regular when any of the bounding segments are excluded from the set (Exercise 10). ∎

THEOREM 3-9
The union of a finite number of regular sets is regular.

Proof: A class of sets closed under intersection and complementation is also closed under union (Exercises 13, 14). ∎

EXERCISES

1. Let A and B be nonempty sets of real numbers such that every member of A is less than or equal to every member of B. *Prove:* The least upper bound of A is less than or equal to the greatest lower bound of B. [*Hint:* Every member of $A(B)$ is a lower (upper) bound for $B(A)$.]

2. Let $f(x)$ be a continuous function equal to 1 on all rationals. Prove that $f(x) = 1$ for all real x. (*Hint:* If x is irrational, let $x_1, x_2, \ldots$ be its sequence of decimal approximations.)

3. Let $f(x)$ be a continuous function equal to 1 on all irrational x. Prove that $f(x) = 1$ for all rational x. (*Hint:* Every rational number x is the limit of a sequence of irrational numbers $x_n = x + \sqrt{2}/n, n \geq 1$.)

4. Prove that the volume of a disk in the plane of radius r (Definition 3-1) is equal to πr^2.

5. Prove that the half-space $\{(x_1, x_2): x_1 \leq r\}$ (Example 3-6) is regular.

6. Let M' and $\mathbf{x}'$ be the images of the set M and the point $\mathbf{x}$ under a given rotation. *Prove:* The continuous function $f(\mathbf{x})$ belongs to $\mathfrak{L}(M)$ or $\mathfrak{U}(M)$ if and only if the function f^* defined at $\mathbf{x}'$ as

$f^*(\mathbf{x}') = f(\mathbf{x})$ belongs to $\mathcal{L}(M')$ or $\mathcal{U}(M')$, respectively. (*Hint:* $\mathbf{x}$ belongs to M if and only if $\mathbf{x}'$ belongs to M'.)

7. *Prove:* f belongs to $\mathcal{L}(M)$ if and only if $1 - f$ belongs to $\mathcal{U}(N)$, where N is the complement of M; and belongs to $\mathcal{U}(M)$ if and only if $1 - f$ belongs to $\mathcal{L}(N)$.

8. Suppose that for some $n \geq 2$ every union of n regular sets is regular; show that the same is true for the union of $n + 1$ regular sets.

9. *Prove:* If f and g belong to $\mathcal{L}(M)$ and $\mathcal{L}(N)$, respectively, then fg belongs to $\mathcal{L}(M \cap N)$.

10. Show that the open rectangle $\{(x_1, x_2): a_1 < x_1 < b_1,\ a_2 < x_2 < b_2\}$ is regular; also the semiopen rectangle $\{(x_1, x_2): a_1 < x_1 \leq b_1,\ a_2 \leq x_2 \leq b_2\}$.

11. Consider the line segment from $(0, 0)$ to $(1, 0)$ in the plane. Show that it is regular and that its volume is 0.

12. Show that the lines bounding a rectangle in the plane form a regular set.

13. Prove the De Morgan law of sets: The complement of the intersection of sets is the union of their complements.

14. Use the result of Exercise 13 to prove: If the complement of a regular set is regular (Theorem 3-6), and the intersection of a finite number of such sets is regular (Theorem 3-8), then the same is true of the union of a finite number of such sets.

3-5. INTEGRATION OVER REGULAR SETS

We shall define the integral (3-10) for every nonnegative continuous function and every regular set A; later we shall extend the definition to a class of continuous functions assuming positive and negative values. The definition is based on the following result:

LEMMA 3-1

Let $g(\mathbf{x})$ be a nonnegative continuous function, and B a ball centered at the origin; then, for any regular set M, we have

$$\underset{f_1 \in \mathcal{L}(M)}{lub} \int \cdots \int_B f_1(\mathbf{x}) g(\mathbf{x})\, d\mathbf{x} = \underset{f_2 \in \mathcal{U}(M)}{glb} \int \cdots \int_B f_2(\mathbf{x}) g(\mathbf{x})\, d\mathbf{x}. \tag{3-26}$$

Proof: The difference between the integrals in (3-26) is

$$\int_B \cdots \int (f_2(\mathbf{x}) - f_1(\mathbf{x}))g(\mathbf{x})\,d\mathbf{x}.$$

The integrand is nonnegative by (3-21). The function $g(x)$ is bounded above by some positive constant G because, by a well-known theorem of the calculus, a continuous function is bounded on every ball; therefore, as a consequence of the preservation of inequality under integration (3-16), the integral above is at most equal to

$$G\left[\int_B \cdots \int f_2(\mathbf{x})\,d\mathbf{x} - \int_B \cdots \int f_1(\mathbf{x})\,d\mathbf{x}\right].$$

By the assumed regularity of M, the difference between the two integrals above can be made arbitrarily small by the appropriate choice of f_1 and f_2; therefore, the same is true of the difference between the integrals in (3-26). ■

For each ball B, the numbers appearing on either side of (3-26) are nonnegative, and increase, or at least do not decrease, with the radius of the ball. As the radius tends to infinity, these numbers either converge to a positive limit or else tend to infinity (Exercise 1).

DEFINITION 3-2
The integral

$$\int_M \cdots \int g(\mathbf{x})\,d\mathbf{x} \tag{3-27}$$

of the nonnegative continuous function g over the regular set M is defined as the common limit of both members of (3-26) as the radius of B tends to infinity; the limit is finite or infinite.

The positivity and linearity of the integral on the real line is extended to the integral (3-27): If $g_1 \le g_2$, then

$$\int_M \cdots \int g_1(\mathbf{x})\,d\mathbf{x} \le \int_M \cdots \int g_2(\mathbf{x})\,d\mathbf{x},$$

and, for any constants α and β,

$$\int_M \cdots \int (\alpha g_1(\mathbf{x}) + \beta g_2(\mathbf{x}))\, d\mathbf{x}$$

$$= \alpha \int_M \cdots \int g_1(\mathbf{x})\, d\mathbf{x} + \beta \int_M \cdots \int g_2(\mathbf{x})\, d\mathbf{x}.$$

As a "function" on the class of regular sets, the integral (3-27) has this property:

THEOREM 3-10
If M_1 and M_2 are regular, and M_1 is contained in M_2, then

$$\int_{M_1} \cdots \int g(\mathbf{x})\, d\mathbf{x} \le \int_{M_2} \cdots \int g(\mathbf{x})\, d\mathbf{x}.$$

Proof: (Exercise 2.) ■

A second property is that of "finite additivity":

THEOREM 3-11
If $M_1, \ldots, M_k$ are disjoint regular sets, and M is their union, which, by Theorem 3-9, is regular, then

$$\int_M \cdots \int g(\mathbf{x})\, d\mathbf{x} = \sum_{i=1}^{k} \int_{M_i} \cdots \int g(\mathbf{x})\, d\mathbf{x}.$$

Proof: We give the proof just for $k = 2$; the general case follows by induction (Exercise 3). Let M_1 and M_2 be disjoint regular sets, and f_1, f_1', f_2 and f_2' functions in $\mathfrak{L}(M_1)$, $\mathfrak{U}(M_1)$, $\mathfrak{L}(M_2)$, and $\mathfrak{U}(M_2)$, respectively. Since M_1 and M_2 are disjoint, $f_1(x)$ and $f_2(x)$ are never positive at the same point, and both vanish outside the union of M_1 and M_2; therefore,

$$f_1 + f_2 \in \mathfrak{L}(M \cup N).$$

It follows from this and from the definition of the integral (over $M \cup N$) that

$$\int_B \cdots \int f_1(\mathbf{x})\, g(\mathbf{x})\, d\mathbf{x} + \int_B \cdots \int f_2(\mathbf{x})\, g(\mathbf{x})\, d\mathbf{x}$$

$$= \int_B \cdots \int (f_1(\mathbf{x}) + f_2(\mathbf{x}))\, g(\mathbf{x})\, d\mathbf{x} \le \int_{M \cup N} \cdots \int g(\mathbf{x})\, d\mathbf{x}.$$

Since f_1 and f_2 are arbitrary members of the lower classes, the inequality involving the extreme members of this inequality is preserved for the least upper bounds:

$$\operatorname*{lub}_{f_1 \in \mathfrak{L}(M_1)} \int_B \cdots \int f_1(\mathbf{x})\, g(\mathbf{x})\, d\mathbf{x} + \operatorname*{lub}_{f_2 \in \ (M_2)} \int_B \cdots \int f_2(\mathbf{x})\, g(\mathbf{x})\, d\mathbf{x}$$

$$\le \int_{M_1 \cup M_2} \cdots \int g(\mathbf{x})\, d\mathbf{x}.$$

Let the radius of B tend to infinity; by Definition 3-1, we have

$$\int_{M_1} \cdots \int g(\mathbf{x})\, d\mathbf{x} + \int_{M_2} \cdots \int g(\mathbf{x})\, d\mathbf{x} \le \int_{M_1 \cup M_2} \cdots \int g(\mathbf{x})\, d\mathbf{x}. \quad (3\text{-}28)$$

Now we get the reverse inequality. For any functions f_1' and f_2' in the upper classes, the function $f_1'(\mathbf{x}) + f_2'(\mathbf{x})$ is at least equal to 1 on $M_1 \cup M_2$. The function

$$\min\,(1, f_1'(\mathbf{x}) + f_2'(x))$$

is continuous (Exercise 4), and is equal to 1 on $M_1 \cup M_2$; therefore, it belongs to $\mathfrak{U}(M_1 \cup M_2)$. It follows that for any ball B, we have

$$\int_B \cdots \int f_1'(\mathbf{x})\, g(\mathbf{x})\, d\mathbf{x} + \int_B \cdots \int f_2'(\mathbf{x})\, g(\mathbf{x})\, d\mathbf{x}$$

$$\ge \int_B \cdots \int \min\,(1, f_1'(\mathbf{x}) + f_2'(\mathbf{x}))\, g(\mathbf{x})\, d\mathbf{x}$$

$$\ge \operatorname*{glb}_{f \in \mathfrak{U}(M_1 \cup M_2)} \int_B \cdots \int f(\mathbf{x})\, g(\mathbf{x})\, d\mathbf{x}.$$

Pass to the glb in the first member of this inequality, let the radius of B tend to infinity, and apply the definition of the integral:

$$\int_{M_1} \cdots \int g(\mathbf{x})\, d\mathbf{x} + \int_{M_2} \cdots \int g(\mathbf{x})\, d\mathbf{x}$$

$$\ge \operatorname*{glb}_{f \in \mathfrak{U}(M_1 \cup M_2)} \int_B \cdots \int f(\mathbf{x})\, g(\mathbf{x})\, d\mathbf{x}.$$

Let the radius of B tend to infinity on the right-hand side, and apply the definition of the integral:

$$\int_{M_1}\cdots\int g(\mathbf{x})\,d\mathbf{x} + \int_{M_2}\cdots\int g(\mathbf{x})\,d\mathbf{x} \geq \int_{M_1\cup M_2}\cdots\int g(\mathbf{x})\,d\mathbf{x}. \quad \blacksquare \qquad (3\text{-}29)$$

EXERCISES

1. *Prove:* If $\{a_n, n \geq 1\}$ is a nondecreasing sequence of real numbers, then it either converges to a finite limit or tends to $+\infty$. (*Hint:* Consider the cases where the sequence is bounded or not.)

2. *Prove:* If M_1 is contained in M_2, then the class $\mathfrak{L}(M_1)$ is contained in $\mathfrak{L}(M_2)$. Use this to prove Theorem 3-10.

3. Suppose that the statement of Theorem 3-11 is true for some $k \geq 2$; prove that it is true for $k + 1$.

4. Show that if f is a continuous function then so is min $(1, f(x))$. [*Hint:*

$$\min(a, b) = \tfrac{1}{2}(a + b - |a - b|).]$$

5. The rectangle $\{(x_1, x_2): a_1 \leq x_1 \leq b_1, a_2 \leq x_2 \leq b_2\}$ in the plane is regular (Example 3-8). Show that Definition 3-2 of the integral of a non-negative continuous function over this set coincides with the earlier definition as an iterated integral.

6. Repeat Exercise 5 for the disk.

3-6. THE NORMAL SAMPLE SPACE

We extend the notions of Sample point and Sample space from the case of the finite population (Sec. 1-3) to that of the normal population, defined in Chapter 2. In defining a system of probabilities, we replace the likelihood function in (1-8) by the product of normal density functions:

$$L(\mathbf{x}; \mu, \sigma) = \prod_{i=1}^{n} \frac{1}{\sigma}\,\phi\left(\frac{x_i - \mu}{\sigma}\right)$$

$$= (2\pi\sigma^2)^{-n/2} \exp\left[-\frac{1}{2\sigma^2}\sum_{i=1}^{n}(x_i - \mu)^2\right]; \qquad (3\text{-}30)$$

this is also called the *likelihood function*.

As in the one-dimensional case (Sec. 2-6), we avoid rigorously defining the Sample point $(X_1, \ldots, X_n)$ in the normal Sample space.

It may intuitively be thought to be a "random point" in R^n selected in accordance with a given system of probabilities, defined below. In accordance with the notation of Sec. 3-1, the Sample point is denoted by $\mathbf{X}$, in boldface type. It is customary to call it a "Sample of n observations from the normal population with mean μ and standard deviation σ." For the purpose of rigor we have only to define the *probabilities* that $\mathbf{X}$ falls in various sets. The *sum* (1-10) defining the probability that a Sample from a finite population falls in a given set is, in the normal case, replaced by the *integral* of the likelihood function L over the set. While the former sum is defined for *every* set in R^n, the latter integral is not, as it has been defined only for *regular* sets.

DEFINITION 3-3

Let $\mathbf{X}$ be a Sample of n observations from a normal population with mean μ and standard deviation σ. For any regular set M in R^n, the probability that $\mathbf{X}$ falls in M is defined as

$$P(M) = \int_M \cdots \int L(\mathbf{x}; \mu, \sigma)\, d\mathbf{x},$$

where the integral is given by Definition 3-2. Endowed with this system of probabilities, R^n is called a normal Sample space.

As a function on the class of regular sets $P(\cdot)$ satisfies:

$$0 \leq P(M) \leq P(R^n) = 1 \qquad \text{(Exercise 1);} \tag{3-31}$$

If M_1 is contained in M_2, then

$$P(M_1) \leq P(M_2) \quad \text{(by Theorem 3-10);} \tag{3-32}$$

If M is the union of the disjoint regular sets $M_1, \ldots, M_k$, then

$$P(M) = \sum_{i=1}^{k} P(M_i) \quad \text{(by Theorem 3-11).} \tag{3-33}$$

We shall make use of the following representations of $P(M)$:

$$P(M) = \operatorname*{lub}_{f \in \mathfrak{L}(M)} \int_{R^n} \cdots \int f(\mathbf{x})\, L(\mathbf{x}; \mu, \sigma)\, d\mathbf{x} \tag{3-34}$$

$$P(M) = \operatorname*{glb}_{f \in \mathfrak{U}(M)} \int_{R^n} \cdots \int f(\mathbf{x})\, L(\mathbf{x}; \mu, \sigma)\, d\mathbf{x}. \tag{3-35}$$

(Exercises 2, 3).

The following sets are among the regular ones: the ball, the rectangle $\{\mathbf{x}: a_i \leq x_i \leq b_i,\ i = 1, \ldots, n\}$ and the half-spaces $\{\mathbf{x}: x_k \leq b\}$ and $\{\mathbf{x}: x_1 + \cdots + x_n \leq c\}$. The proofs were given in the case $n = 2$ in Sec. 3-4; they are similar in the general case. The integrals over these sets, defined in Definition 3-1, are equivalent to the iterated integrals defined earlier (Exercises 5 and 6, Sec. 3-5 for the case $n = 2$).

There is a suggestive special case of $P(M)$ which illustrates the meaning of the likelihood function L. Let M be the infinitesimally small rectangle $\{\mathbf{x}: y_i \leq x_i \leq y_i + dy_i, i = 1, \ldots, n\}$; then

$$P(M) = \prod_{i=1}^{n} \int_{y_i}^{y_i + dy_i} \frac{1}{\sigma} \phi\left(\frac{x_i - \mu}{\sigma}\right) dx_i.$$

By the mean value theorem for integrals, the ith integral in the above product is approximately equal to

$$\frac{1}{\sigma} \phi\left(\frac{y_i - \mu}{\sigma}\right) dy_i;$$

thus the probability that $\mathbf{X}$ falls in this rectangle is approximately equal to the value of L at $\mathbf{y}$ *times* the volume $\prod_{i=1}^{n} dy_i$ of the rectangle.

EXERCISES

1. Verify the relations (3-31).

2. (a) Prove that $P(M)$ is at least equal to the right-hand side of (3-34). (*Hint:*

$$P(M) \geq \operatorname*{lub}_{f \in \mathfrak{L}(M)} \int \cdots \int_B f(\mathbf{x})\, L(\mathbf{x}; \mu, \sigma)\, d\mathbf{x}$$

for any ball B.)

(b) Prove that $P(M)$ is at most equal to the right hand side. (*Hint:*

$$\operatorname*{lub}_{f \in \mathfrak{L}(M)} \int \cdots \int_B f(\mathbf{x})\, L(\mathbf{x}; \mu, \sigma)\, d\mathbf{x} \leq \operatorname*{lub}_{f \in \mathfrak{L}(M)} \int \cdots \int_{R^n} f(\mathbf{x})\, L(\mathbf{x}; \mu, \sigma)\, d\mathbf{x}$$

for any ball B.)

3. Verify (3-35) by the method of Exercise 2.

3-7. THE SAMPLE SPACE FOR A GENERAL POPULATION

The particular form of the likelihood function is not important in the definition of the Sample space. It is only because ϕ is the normal density that we call the space with the particular function L a *normal* Sample space. Let $h(x)$ be an arbitrary density function, and define L as

$$L(\mathbf{x}) = \prod_{i=1}^{n} h(x_i).$$

The integral $P(M)$ in Definition 3-3 has, for the general density h, the same properties as in the special case $h = \phi$ (Exercise 1).

The family of normal densities is a two-parameter class of densities indexed by the pair (μ, σ). A general parametric family of densities is defined by a given function $h(x; \theta)$, where θ is a parameter of one or more dimensions, and, for each θ, $h(x; \theta)$ is a density function with respect to x. [For the normal family θ is two-dimensional: $\theta = (\mu, \sigma)$.] The likelihood function is denoted by

$$L(\mathbf{x}; \theta) = \prod_{i=1}^{n} h(x_i; \theta). \tag{3-36}$$

Example 3-9. Let $h(x; \theta)$ be the uniform density on $[0, \theta]$:

$$h(x, \theta) = \frac{1}{\theta} \quad \text{if} \quad 0 < x < \theta$$
$$= 0, \quad \text{elsewhere;}$$

then

$$L(\mathbf{x}, \theta) = \theta^{-n}, \quad \text{if} \quad 0 < \min x_i, \max x_i < \theta$$
$$= 0, \quad \text{elsewhere.} \quad \blacksquare$$

Example 3-10. Let $h(x; \theta)$ be the exponential density with mean θ:

$$h(x; \theta) = 0, \qquad x \leq 0$$
$$= \frac{1}{\theta} e^{-(x/\theta)}, \quad x > 0;$$

then

$$L(\mathbf{x};\theta) = 0 \quad \text{if} \quad \min x_i \leq 0$$

$$= \theta^{-n} \exp\left[-\sum_{i=1}^{n} x_i/\theta\right], \quad \text{if} \quad \min x_i > 0. \quad ■$$

All of the preceding material on regularity and integration applies to the general likelihood function L; however, the theory of rotations is specifically meant for the normal Sample space.

EXERCISE

1. Show in detail that the integral in Definition 3-3 has exactly the same properties as when ϕ is replaced by h.

BIBLIOGRAPHY

Integration over n-dimensional space is discussed in M. E. Munroe, *Modern Multidimensional Calculus* (Reading, Mass.: Addison-Wesley, 1963).

A brief but excellent introduction to advanced integration theory is contained in the first part of Harald Cramér, *Mathematical Methods of Statistics* (Princeton, N.J.: Princeton U. P., 1946).

CHAPTER 4

Statistics and Their Distributions

4-1. THE DEFINITION AND ROLE OF THE STATISTIC

In a typical statistical problem involving a normal population, an investigator has to make a decision about the population in the absence of complete knowledge of the values of μ and σ. He gains "information" about μ and σ by taking a Sample $\mathbf{X}$ of n observations from the population. This information is "summarized" by taking a fixed function $f(\mathbf{x})$, $\mathbf{x} \in R^n$, and evaluating it at the Sample point $\mathbf{x} = \mathbf{X}$; the computed number $f(\mathbf{X})$ is the summary. As an illustration, we mention the problem of estimating an unknown mean μ; a more systematic description of estimation and testing is given in the later chapters. The investigator takes the Sample point $\mathbf{X}$, and computes the average of the coordinates; then, he uses this computed value as the estimate of μ. In this case, the function f is $f(\mathbf{x}) = (1/n)(x_1 + \cdots + x_n)$; the information about μ is summarized in the average of the observations, $(1/n)(X_1 + \cdots + X_n)$.

For a fixed function f, its value at $\mathbf{X}$ is not predictable with certainty because the Sample point itself does not have a predictable value; the exception is, of course, the trivial case when f is a constant function. The probability function $P(M)$ on the Sample space influences—actually determines—the probability that $f(\mathbf{X})$ will assume certain values. As an example, consider the two-dimensional normal Sample space, and the function $f(\mathbf{x}) = \sqrt{x_1^2 + x_2^2}$. Its value at $\mathbf{X}$ is the distance of the Sample point from the origin. The number $f(\mathbf{X})$ takes on a value less than 2 if and only if $\mathbf{X}$ falls within the disk of radius 2; thus, we *define* the probability that $f(\mathbf{X})$ is less than 2 as the probability that $\mathbf{X}$ falls within this disk, namely,

$$\iint_{\{(x_1,x_2):x_1^2+x_2^2<4\}} \prod_{i=1}^{2} \frac{1}{\sigma} \phi\left(\frac{x_i - \mu}{\sigma}\right) \prod_{i=1}^{2} dx_i$$

Here the probability that $f(\mathbf{X})$ assumes a value less than 2 is defined as the probability that $\mathbf{X}$ itself falls in

$$\{\mathbf{x}: f(\mathbf{x}) < 2\},$$

the preimage of the semi-infinite line $(-\infty, 2)$ under f.

The above example is generalized to an arbitrary function $f(\mathbf{x})$ on R^n. For any real number y, the random number $f(\mathbf{X})$ assumes some value less than or equal to y if and only if $\mathbf{X}$ falls in the set

$$\{\mathbf{x}: f(\mathbf{x}) \leq y\}. \tag{4-1}$$

The probability that $\mathbf{X}$ falls in this set is well defined if the set is regular; this motivates the identification of a class of functions which we call *statistics*:

DEFINITION 4-1

A statistic is a continuous function $f(\mathbf{x})$ on R^n such that for every y the set (4-1) is regular.

Examples of statistics are: $x_1 + \cdots + x_n$, $x_1^2 + \cdots + x_n^2$, x_1, $(1/n)(x_1 + \cdots + x_n)$, $(x_1^2 + \cdots + x_n^2)^{1/2}$; this follows from the regularity of the sets indicated in Sec. 3-6.

The regularity of the set (4-1) for every y implies that for every y_1 and y_2 each of the following sets are also regular:

$$\{\mathbf{x}: y_1 < f(\mathbf{x}) \leq y_2\}, \{\mathbf{x}: f(\mathbf{x}) < y_2\} \tag{4-2}$$

$$\{\mathbf{x}: y_1 \leq f(\mathbf{x}) \leq y_2\}, \{\mathbf{x}: f(\mathbf{x}) \geq y_2\}$$

(and others).

Here is the proof for the first set: By hypothesis, the sets $\{\mathbf{x}: f(\mathbf{x}) \leq y_1\}$ and $\{\mathbf{x}: f(\mathbf{x}) \leq y_2\}$ are regular. The complement of the latter, $\{\mathbf{x}: f(\mathbf{x}) > y_1\}$, is regular (Theorem 3-6); therefore, its intersection with $\{\mathbf{x}: f(\mathbf{x}) \leq y_2\}$ is regular and is equal to $\{\mathbf{x}: y_1 < f(\mathbf{x}) \leq y_2\}$. The proofs of the other relations also depend on the closure properties of the regular sets (Exercise 1).

EXERCISE

1. Prove that the last three sets in (4-2) are regular.

4-2. DISTRIBUTION AND DENSITY FUNCTION OF A STATISTIC

We define the system of probabilities governing the "random value" $f(\mathbf{X})$:

DEFINITION 4-2
The distribution function $F(y)$ of the statistic f is defined at the point y as

$$F(y) = \int \cdots \int_{\{\mathbf{x}: f(\mathbf{x}) \leq y\}} L(\mathbf{x}; \mu, \sigma)\, d\mathbf{x}.$$

If it is differentiable, its derivative $F'(y)$ is called the density function of f.

This is a formal consequence of the earlier observation that $f(\mathbf{X})$ assumes a value less than or equal to y if and only if $\mathbf{X}$ falls in (4-1).

The distribution function of a statistic has these properties (cf. Sec. 2-8):

1. *It is nondecreasing.* This is a consequence of the property (3-32): if $y_1 < y_2$, then $\{\mathbf{x}: f(\mathbf{x}) < y_1\}$ is contained in $\{\mathbf{x}: f(\mathbf{x}) < y_2\}$, and so the probability of the former set is smaller.

2. $\lim_{y \to \infty} F(y) = 1$. The idea in this is that f is finite so that the set $\{\mathbf{x}: f(\mathbf{x}) < \infty\}$ holds all the probability; the details of the formal proof are left as an exercise (Exercise 9).

3. $\lim_{y \to -\infty} F(y) = 0$. The proof is similar to that of the previous property (Exercise 10).

Here are examples of statistics and their distributions:

Example 4-1. For $a > 0$, and any b, consider the statistic $f(x) = ax + b$ on R^1. The distribution function $F(y)$ is

$$\int_{\{x: ax+b \leq y\}} \frac{1}{\sigma}\, \phi\left(\frac{x-\mu}{\sigma}\right) dx = \int_{-\infty}^{(y-b)/a} \frac{1}{\sigma}\, \phi\left(\frac{x-\mu}{\sigma}\right) dx$$

$$= \Phi\left(\frac{y - a\mu - b}{a\sigma}\right)$$

which is the normal distribution with mean $a\mu + b$ and standard deviation $a\sigma$. ■

Example 4-2. Put $f(x) = x^2/\sigma^2$, and assume that $\mu = 0$. The distribution of f is

$$F(y) = 0, \qquad y \le 0$$
$$= \int_{\{x: x^2 \le y\sigma^2\}} \phi(x/\sigma)\,dx/\sigma = \int_{-\sqrt{y}}^{\sqrt{y}} \phi(x)\,dx, \quad y > 0.$$

By the chain rule, the density is

$$F'(y) = 0, \qquad y \le 0$$
$$= (1/\sqrt{y})\,\phi(\sqrt{y}), \quad y > 0. \quad \blacksquare$$

Example 4-3. Suppose $n = 2$, $f(\mathbf{x}) = x_1 + x_2$, and $L = (1/\sigma)^2\phi((x_1 - \mu)/\sigma)\,\phi((x_2 - \mu)/\sigma)$. We shall prove that f has a distribution which is normal with mean 2μ and standard deviation $\sqrt{2}\sigma$. In accordance with Definition 4-2, the distribution of f is

$$F(y) = \iint_{\{(x_1, x_2): x_1 + x_2 \le y\}} (2\pi\sigma^2)^{-1} \exp\left\{-\frac{1}{2\sigma^2}[(x_1 - \mu)^2 + (x_2 - \mu)^2]\right\} dx_1 dx_2$$

Substitute variables: $u_i = (x_i - \mu)/\sigma$, $i = 1, 2$:

$$F(y) = \iint_{\{(u_1, u_2): u_1 + u_2 \le (y - 2\mu)/\sigma\}} (2\pi)^{-1} \exp\left[-\tfrac{1}{2}(u_1^2 + u_2^2)\right] du_1\, du_2.$$

We shall identify this integral by means of a geometric argument that will be used later. The domain of integration is the half-plane bounded by the line $u_1 + u_2 = (y - 2\mu)/\sigma$ (see Fig. 4-1).

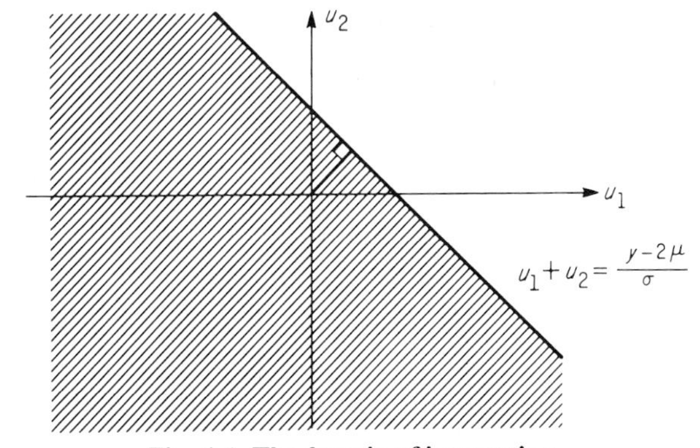

Fig. 4-1. The domain of integration.

The length of the perpendicular from the origin to the line is $(y - 2\mu)/\sqrt{2}\sigma$. The integrand depends on the variables (u_1, u_2) as a function of the square of the distance from the origin: it is constant over circles about the origin. It follows that the integral is unchanged if the region below the line $u_1 + u_2 = (y - 2\mu)/\sigma$ is rotated through an angle of $-45°$, as in Fig. 4-2.

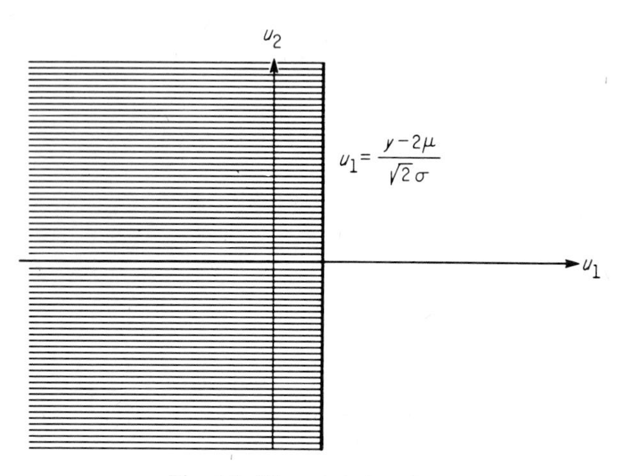

Fig. 4-2. The rotated region.

The rotated region is the half-plane $\{(u_1, u_2): u_1 \leq (y - 2\mu)/\sqrt{2}\sigma$; the integral over it is

$$F(y) = \int_{-\infty}^{(y-2\mu)/\sqrt{2}\sigma} \int_{-\infty}^{\infty} \phi(u_1)\,\phi(u_2)\,du_2\,du_1$$

$$= \Phi\left(\frac{y - 2\mu}{\sqrt{2}\sigma}\right).$$

This is the normal distribution with mean 2μ and standard deviation $\sqrt{2}\sigma$. ■

Example 4-4. Suppose $n = 2$, $f(x_1, x_2) = x_1^2 + x_2^2$, and L is $\phi(x_1)\,\phi(x_2)$. The distribution F of f is 0 for $y \leq 0$, and

$$F(y) = \iint_{\{(x_1,x_2):\,x_1^2+x_2^2 \leq y\}} (2\pi)^{-1} \exp\left[-\tfrac{1}{2}(x_1^2 + x_2^2)\right] dx_1\,dx_2, \quad y > 0.$$

Transforming the integral to polar coordinates, we obtain:

$$F(y) = 1 - e^{-y/2}, \quad y > 0.$$

(Exercise 7.) This is the exponential distribution with mean $\theta = 2$. ■

Now we turn to an important theoretical application of the distribution function. Let $f(x)$ be a continuous function on an interval $[a, b]$, with a continuous derivative. For any continuous g, the integral of the composite function,

$$\int_{-\infty}^{\infty} g(f(x))\, dx,$$

can, by a change of variable $y = f(x)$, be transformed into an integral of $g(y)$ over the range of f after multiplication by $(d/dy)f^{-1}(y)$. (We assume that f is nondecreasing or nonincreasing.) This is now generalized to the class of statistics f over R^n, and multiple integrals over R^n. In the place of the simple integral of a composite function, we consider multiple integrals

$$\int \cdots \int_{R^n} g(f(\mathbf{x}))\, L(\mathbf{x}; \mu, \sigma)\, d\mathbf{x}, \tag{4-3}$$

where g is a function on the line. Under certain conditions on g and f, it is possible to transform this integral by means of the "change of variable" $y = f(\mathbf{x})$ into a single integral of the form

$$\int_{-\infty}^{\infty} g(y)\, F'(y)\, dy,$$

where F' is the density function of f.

Before stating this result, we have to take up a technical point in the definition of the multiple integral (4-3). The integral of a continuous function over a regular set was defined only for nonnegative functions; now we extend it to continuous functions which may assume positive and negative values. If $f(\mathbf{x})$ is a continuous function it can be uniquely represented as the difference of two nonnegative continuous functions:

$$\begin{aligned} f(\mathbf{x}) &= \tfrac{1}{2}(f(\mathbf{x}) + |f(\mathbf{x})|) - \tfrac{1}{2}(|f(\mathbf{x})| - f(\mathbf{x})) \\ &= f^+(\mathbf{x}) - f^-(\mathbf{x}). \end{aligned}$$

(It is easy to verify that the function

$$\tfrac{1}{2}[\,|f(\mathbf{x})| \pm f(\mathbf{x})] = f^{\pm}(\mathbf{x})$$

is nonnegative and continuous.) The integrals of these nonnegative functions over a regular set are well defined (Definition 3-2).

DEFINITION 4-3

Let $f(\mathbf{x})$ be a continuous function, and M a regular set. If the integrals of $f^{+}(\mathbf{x})$ and $f^{-}(\mathbf{x})$ over M are finite, then $f(\mathbf{x})$ is said to be integrable over M, and its integral over M is defined as

$$\int_M \cdots \int f(\mathbf{x})\, d\mathbf{x} = \int_M \cdots \int f^{+}(\mathbf{x})\, d\mathbf{x} - \int_M \cdots \int f^{-}(\mathbf{x})\, d\mathbf{x}.$$

There is a simple criterion for the integrability of f over M. It follows from the definition of $f^{\pm}$ that the absolute value of f may be expressed as

$$|f(\mathbf{x})| = f^{+}(\mathbf{x}) + f^{-}(\mathbf{x});$$

thus the integrals of f^{+} and f^{-} over M are finite if and only if the same is true of the integral of $|f(\mathbf{x})|$:

$f(\mathbf{x})$ is integrable over M if and only if

$$\int_M \cdots \int |f(\mathbf{x})|\, d\mathbf{x} < \infty.$$

THEOREM 4-1

Let $f(\mathbf{x})$ be a statistic on R^n with a continuous density $F'(y)$. Let $g(y)$ be a continuous function such that

$$\int_{R^n} \cdots \int |g(f(\mathbf{x}))|\; L(\mathbf{x}; \mu, \sigma)\, d\mathbf{x} < \infty;$$

then

$$\int_{R^n} \cdots \int g(f(\mathbf{x})) L(\mathbf{x}; \mu, \sigma)\, d\mathbf{x} = \int_{-\infty}^{\infty} g(y) F'(y)\, dy. \quad (4\text{-}4)$$

Proof: Let m be an arbitrary positive integer. The set $\{\mathbf{x}: -m < f(\mathbf{x}) \leq m\}$ is regular because it is the intersection of the regular sets $\{\mathbf{x}: f(\mathbf{x}) \leq m\}$ and $\{\mathbf{x}: f(\mathbf{x}) > -m\}$. It is representable, for any $h > 0$, as the union of a finite number of disjoint regular sets of the form $\{\mathbf{x}: k\,h < f(\mathbf{x})\} \le (k+1)h$, where k assumes integer values, and

$$-\frac{m}{h} \le k \le \frac{m}{h} - 1.$$

By the finite additivity property (Theorem 3-11) it follows that the integral

$$\underset{\{\mathbf{x}:\, -m < f(\mathbf{x}) \le m\}}{\int \cdots \int} g(f(\mathbf{x}))\, L(\mathbf{x}; \mu, \sigma)\, d\mathbf{x}$$

is the sum of a finite number of integrals of the form

$$\underset{\{\mathbf{x}:\, y < f(\mathbf{x}) \le y+h\}}{\int \cdots \int} g(f(\mathbf{x}))\, L(x; \mu, \sigma)\, d\mathbf{x}. \tag{4-5}$$

By the order preserving property of integration, this integral is not diminished when $g(f(x))$ is replaced by the maximum value of g on $[y, y+h]$; similarly, it is not augmented when $g(f(\mathbf{x}))$ is replaced by its minimum value. It follows from the intermediate value theorem for continuous functions ("If f is continuous it assumes every value between its maximum and minimum.") that there exists a number y', $y \le y' \le y + h$, such that the above integral is equal to

$$g(y') \cdot \underset{\{\mathbf{x}:\, y < f(\mathbf{x}) \le y+h\}}{\int \cdots \int} L(x; \mu, \sigma)\, dx$$

$$= g(y') \left[\underset{\{\mathbf{x}:\, f(\mathbf{x}) \le y+h\}}{\int \cdots \int} L\, d\mathbf{x} - \underset{\{\mathbf{x}:\, f(\mathbf{x}) \le y\}}{\int \cdots \int} L\, d\mathbf{x} \right]$$

$$= g(y')[F(y+h) - F(y)].$$

By the law of the mean, there exists a number y'' between y and $y + h$ such that the last expression is equal to

$$g(y')\, F'(y'')\, h. \tag{4-6}$$

Now sum over all regions $\{\mathbf{x}: y < f(\mathbf{x}) \le y + h\}$, and then let $h \to 0$: the sum of the terms (4-6) is an approximating sum for the integral

$$\int_{-m}^{m} g(y)\, F'(y)\, dy,$$

and converges to it as $h \to 0$; therefore:

$$\int \cdots \int_{\{\mathbf{x}:\, -m < f(\mathbf{x}) \leq m\}} g(f(\mathbf{x}))\, L(\mathbf{x}; \mu, \sigma)\, d\mathbf{x} = \int_{-m}^{m} g(y)\, F'(y)\, dy.$$

Let $m \to \infty$ on each side of this equation. The right hand member converges to the corresponding member of (4-4). By the argument preceding Example 4-1 (property 2 of $F(y)$), the left-hand side of the equation above also converges to the corresponding member of (4-4). ■

There is a useful interpretation of the density function which assists in its calculation. As in the proof of Theorem 4-1, the difference $F(y + h) - F(y)$ is representable as

$$\int \cdots \int_{\{\mathbf{x}:\, y < f(\mathbf{x}) \leq y+h\}} L(\mathbf{x}; \mu, \sigma)\, d\mathbf{x};$$

thus, if $F'(y)$ exists, then it is equal to the limit, as $h \to 0$ of

$$(1/h) \int \cdots \int_{\{\mathbf{x}:\, y < f(\mathbf{x}) \leq y+h\}} L(\mathbf{x}; \mu, \sigma)\, d\mathbf{x}. \tag{4-7}$$

Consider the ratio

$$\text{Volume of } \{\mathbf{x}:\, y < f(\mathbf{x}) \leq y + h\}/h$$

(Definition 3-1). If it converges to a limit as $h \to 0$, the limit is called the "area of the surface $\{\mathbf{x}: f(\mathbf{x}) = y\}$." If the ratio

$$\frac{\displaystyle\int \cdots \int_{\{\mathbf{x}:\, y < f(\mathbf{x}) \leq y+h\}} L(x; \mu, \sigma)\, d\mathbf{x}}{\text{Volume of } \{\mathbf{x}:\, y < f(\mathbf{x}) \leq y + h\}}$$

converges to a limit as $h \to 0$, the limit is called the "average of L over the surface $\{\mathbf{x}: f(\mathbf{x}) = y\}$." If the average of L over the surface and the area of the surface exist and are finite and the latter positive, then $F'(y)$ is the product of these.

Example 4-5. Let us find the density of the statistic $f(\mathbf{x}) = x_1^2 + x_2^2 + x_3^2$ in sampling from a standard normal population. The area of the surface $\{\mathbf{x}: x_1^2 + x_2^2 + x_3^2 = y\}$ is

$$\lim_{h\to 0} (1/h) \cdot \text{volume between spheres of radii } \sqrt{y+h} \text{ and } \sqrt{y}$$

$$= \lim_{h\to 0} (1/h)(4\pi/3)[(y+h)^{3/2} - y^{3/2}] = 2\pi y^{1/2}.$$

The average of L over the sphere of radius y is $(2\pi)^{-3/2}e^{-y/2}$. The reasoning is as follows. L depends on $\mathbf{x}$ only through the sum of the squares $x_1^2 + x_2^2 + x_3^2$; thus it is equal to $(2\pi)^{-3/2}e^{-y/2}$ on the sphere of radius $\sqrt{y}$. The average of L is the limit of the ratio

$$\frac{\text{Integral of } L \text{ over the region } \{\mathbf{x}: y < x_1^2 + x_2^2 + x_3^2 \le y + h\}}{\text{Volume of the region}}.$$

The maximum of L on this region is attained on the inner sphere (of radius $\sqrt{y}$), and the minimum value on the outer sphere (of radius $\sqrt{y+h}$); thus the integral of L over the region is at least equal to

$$(2\pi)^{-3/2}e^{-(y+h)/2} \cdot \text{Volume of region}$$

and at most equal to

$$(2\pi)^{-3/2}e^{-y/2} \cdot \text{Volume of region};$$

therefore the ratio above is at least $(2\pi)^{-3/2}e^{-(y+h)/2}$ and at most $(2\pi)^{-3/2}e^{-y/2}$. For $h \to 0$ the limit of this ratio is $(2\pi)^{-3/2}e^{-y/2}$; therefore, the density of this statistic is

$$\begin{aligned} &(2\pi)^{-1/2}y^{1/2}e^{-y/2}, && y > 0 \\ &0, && y \le 0. \quad \blacksquare \end{aligned}$$

EXERCISES

1. *Prove:* If $f(\mathbf{x})$ is a statistic with the distribution $F(y)$, then the function $cf(\mathbf{x})$, $c > 0$, is also a statistic and has the distribution $F(y/c)$. What if $c < 0$?

2. *Prove:* If $f(\mathbf{x})$ is a positive statistic with the distribution $F(y)$, then $(f(\mathbf{x}))^2$ is a statistic with the distribution $F(\sqrt{y})$.

3. Repeat Exercise 2 for a positive statistic $f(\mathbf{x})$ and the transformed statistic $(f(\mathbf{x}))^{1/2}$.

4. Show that the results of Examples 4-3 and 4-4 are unchanged if the Sample space is n-dimensional, $n > 2$, but the statistics f depend on $\mathbf{x}$ only through the first two coordinates: $f(\mathbf{x}) = x_1 + x_2$ and $f(\mathbf{x}) = x_1^2 + x_2^2$, respectively.

5. Derive the density in Example 4-4 by the method of Example 4-5.

6. Find the distribution and density of the statistic $\sqrt{x_1^2 + x_2^2}$ for a Sample of two observations from a standard normal population.

7. Carry out the details of the double integration in Example 4-4.

8. Find the distribution of $f(\mathbf{x}) = ax_1 + bx_2$, ($a, b$ constants) in a Sample of two observations from a standard normal population.

9. Let $f(\mathbf{x})$ be a statistic. As a continuous function, it is bounded on a ball B of radius r. Put y equal to the maximum of f on this ball; then

$$F(y) \geq \int \cdots \int_B L(\mathbf{x}; \mu, \sigma)\, d\mathbf{x}.$$

Use this to prove that $F(y) \to 1$ for $y \to \infty$.

10. By a similar argument, prove that $F(y) \to 0$ for $y \to -\infty$.

4-3. THE CHI-SQUARE DISTRIBUTION

One of the important statistics is the sum of the squares of the coordinates, $x_1^2 + \cdots + x_n^2$, in sampling from a standard normal population, or, the *sum of the squares of the observations*. Its distribution is called the "chi-square distribution with n degrees of freedom," and its density is similarly identified. The term *n degrees of freedom* refers to the dimensionality of the Sample space, and has no other connotation here. The chi-square density function, denoted ψ_n, is

$$\begin{aligned} \psi_n(y) &= 0, & y \leq 0 \\ &= \frac{1}{2^{n/2}\Gamma(n/2)}\, y^{(n/2)-1} e^{-y/2}, & y > 0, \end{aligned} \tag{4-8}$$

where $\Gamma(\cdot)$ is the gamma function, defined in (4-9) below. The distribution function is the indefinite integral of ψ_n. In Table II, Appendix, there are given values of x for which the integral

$$\int_x^\infty \psi_n(y)\, dy$$

is specified. We note that Examples 4-2, 4-4, and 4-5 contain derivations of the chi-square density in the particular cases $n = 1$, 2, and 3, respectively. (See Fig. 4-3.)

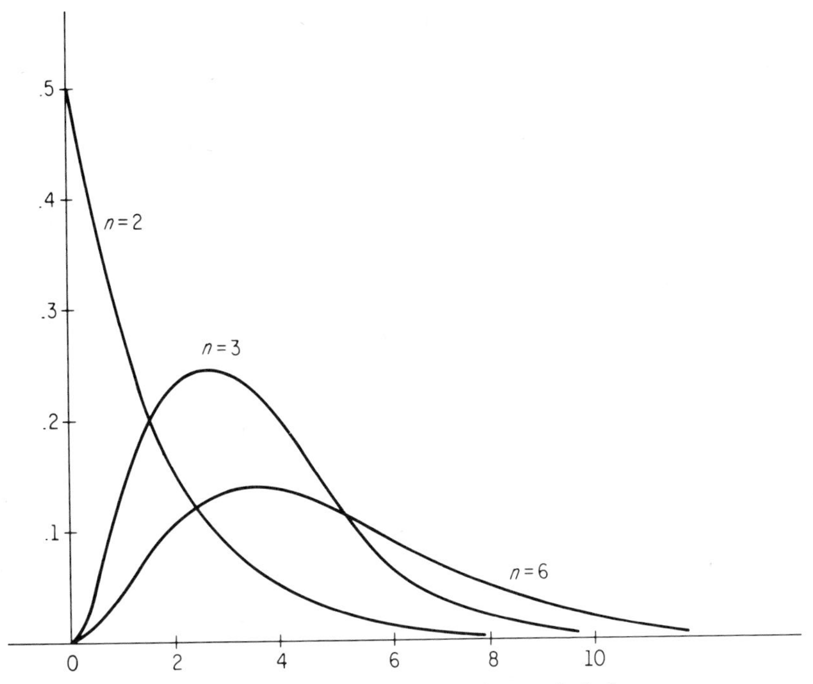

Fig. 4-3. Chi-square density for $n = 2, 3, 6$.

The gamma function $\Gamma(u)$ is defined for $u > 0$ as the improper integral

$$\Gamma(u) = \int_0^\infty x^{u-1} e^{-x} \, dx. \tag{4-9}$$

It satisfies the functional equation

$$\Gamma(u + 1) = u\Gamma(u), \quad u > 0, \tag{4-10}$$

and $\Gamma(1) = 1$; Eq. (4-10) is verified by integration by parts (Exercise 1). If n is a positive integer, then, by iteration of (4-10), we find that $\Gamma(n + 1) = n!$ The gamma function can also be explicitly evaluated for half-integer values of u. For $u = 1/2$, we have

$$\Gamma(1/2) = \sqrt{\pi} \quad \text{(Exercise 2)}$$

The values of $\Gamma(n/2)$, $n > 1$, are obtained from this and (4-10) by iteration; for example,

$$\Gamma(5/2) = (3/2)\,\Gamma(3/2) = (3/4)\sqrt{\pi}$$

Here is a formal statement and proof of the main result of this section:

THEOREM 4-2
In sampling from a standard normal population, the statistic $f(\mathbf{x}) = \|\mathbf{x}\|^2$ *has the chi-square distribution with n degrees of freedom.*

Proof: We use the method of "averages," outlined after Theorem 4-1. The surface $\{\mathbf{x}: f(\mathbf{x}) = y\}$, $y > 0$, is the sphere centered at the origin and of radius $\sqrt{y}$. The likelihood function

$$L(\mathbf{x}; 0, 1) = (2\pi)^{-n/2} e^{-(1/2)\|\mathbf{x}\|^2}$$

is constant and equal to $(2\pi)^{-n/2}e^{-y/2}$ on the sphere; therefore, by the same argument as for the case $n = 3$ in Example 4-5, the average of L over the sphere has the same value $(2\pi)^{-n/2}e^{-y/2}$. Now we compute the "area" of the sphere. Let C_n be the volume of the unit ball in R^n; then the volume of the ball of radius r is $C_n r^n$ (Exercises 3, 4). The volume between the spheres of radii $\sqrt{y+h}$ and $\sqrt{y}$, respectively, is

$$C_n[(y + h)^{n/2} - y^{n/2}].$$

Divide this by h, and let $h \to 0$: the area of the sphere is

$$C_n(n/2)\, y^{(n/2)-1}.$$

The density function is the product of the average of L and the surface area:

$$(2\pi)^{-n/2} e^{-y/2} C_n(n/2)\, y^{(n/2)-1}.$$

Except for the undetermined constant, this coincides with (4-8) for $y > 0$. Since the integral of a density over the whole line must be equal to 1, it follows that C_n is uniquely determined and is equal to

$$\frac{\pi^{n/2}}{\Gamma((n/2) + 1)} \tag{4-11}$$

(Exercise 5). ■

For later reference, we record the extension of Theorem 4-2 to the case of a Sample from a normal population with arbitrary mean and standard deviation:

EXTENSION OF THEOREM 4-2

In Sampling from a normal population with mean μ and standard deviation σ, the statistic

$$\sum_{i=1}^{n} (x_i - \mu)^2/\sigma^2$$

has the chi-square distribution with n degrees of freedom.

Proof: (Exercise 6). ■

EXERCISES

1. Verify (4-10) by integration by parts: in the integral (4-9) replace u by $u + 1$, differentiate x^u and integrate e^{-x}.

2. Evaluate $\Gamma(1/2)$ by the substitution $x = y^2$ in (4-9), and by the use of fact that $\int_0^\infty \phi(x)\,dx = 1/2$.

3. Let $C_n(r)$ be the volume of the ball of radius r in R^n (the integral of the constant function 1 over the ball). Prove that the recursion formula

$$C_n(r) = \int_{-r}^{r} C_{n-1}(\sqrt{r^2 - z^2})\,dz, \quad n \geq 3,$$

holds. (*Hint:* Eq. 3-15.)

4. For $n = 2$, $C_2(r)$ factors into the product $C_2(1) \cdot r^2 = C_2 \cdot r^2$, where C_2 is the volume of the unit ball (area of disk). Prove by induction on n that $C_n(r) = C_n \cdot r^n$, where C_n = volume of unit ball in R^n. [*Hint:* Suppose that $C_{n-1}(r) = C_{n-1} \cdot r^{n-1}$, substitute on the right-hand side of the recursion formula in Exercise 3, and change the variables of integration.)

5. Show that $\int_0^\infty \psi_n(y)\,dy = 1$. From this conclude that (4-11) is the volume of the unit ball.

6. Prove the extension of Theorem 4-2 by expressing the distribution of the statistic in terms of the defining multiple integral, changing the variables of integration from x_i to $(x_i - \mu)/\sigma$, and applying the original version of Theorem 4-2.

7. Find the distribution of the statistic $(x_1^2 + \cdots + x_n^2)^{1/2}$ in sampling from a standard normal population.

8. Verify the recursive relation

$$\psi_{n+1}(y) = \int_0^y \psi_n(y-u)\psi_1(u)\,du, \quad n \geq 1,\ y > 0.$$

4-4. THE FUNDAMENTAL SUM-OF-SQUARES DECOMPOSITION AND ITS GEOMETRIC MEANING

We state and prove an elementary identity about real numbers which has significant applications in the theory of statistics for normal populations.

Let $x_1, \ldots, x_n$ be real numbers; put

$$\bar{x} = \frac{1}{n}(x_1 + \cdots + x_n);$$

then

$$\sum_{i=1}^{n} x_i^2 = \sum_{i=1}^{n} (x_i - \bar{x})^2 + n\bar{x}^2. \tag{4-12}$$

In words: The sum of the squares of n real numbers is equal to the sum of the squares about the average *plus* n times the square of the average. Equation 4-12 is verified by expanding the squares in the sum on the right-hand side,

$$\sum_{i=1}^{n} (x_i - \bar{x})^2 = \sum_{i=1}^{n} x_i^2 - 2\bar{x}\sum_{i=1}^{n} x_i + n\bar{x}^2;$$

using the identity $\sum_{i=1}^{n} x_i = n\bar{x}$ in the middle term above; cancelling terms $\pm n\bar{x}^2$; then comparing the resulting terms on the right-hand side of (4-12) with that on the left.

Another form of (4-12) is: For an arbitrary real number d,

$$\sum_{i=1}^{n} (x_i - d)^2 = \sum_{i=1}^{n} (x_i - \bar{x})^2 + n(\bar{x} - d)^2. \tag{4-13}$$

This is obtained from (4-12) by substituting $x_i - d$ for x_i (Exercise 1).

The identity (4-12) has a geometric interpretation. Consider first the case $n = 2$ (Fig. 4-4).

Let (x_1, x_2) be a point in the plane. The identity (4-12) states

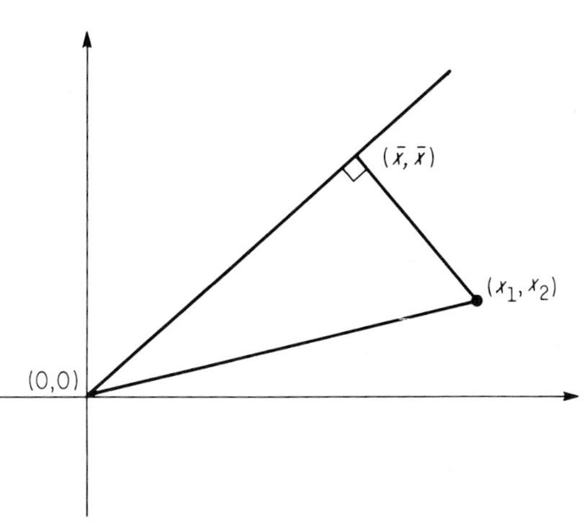

Fig. 4-4. Geometric interpretation of Eq. 4-12 in R^2.

that the square of the distance from the origin to (x_1, x_2), $x_1^2 + x_2^2$, is equal to the square of the distance from (x_1, x_2) to $(\bar{x}, \bar{x})$, namely $(x_1 - \bar{x})^2 + (x_2 - \bar{x})^2$, *plus* the square of the distance from $(\bar{x}, \bar{x})$ to the origin, $2\bar{x}^2$. This implies that the points $(0, 0)$, $(\bar{x}, \bar{x})$ and (x_1, x_2) determine a right triangle with right angle at $(\bar{x}, \bar{x})$. In R^n, $n > 2$, the interpretation is similar. Let $\mathbf{1}$ stand for the point with all coordinates equal to 1, so that $\bar{x}\mathbf{1}$ is the point with all coordinates equal to $\bar{x}$. In terms of this notation we find

$$\sum_{i=1}^{n} (x_i - \bar{x})^2 = \| \mathbf{x} - \bar{x}\mathbf{1} \|^2, \quad \| \bar{x}\mathbf{1} \|^2 = n\bar{x}^2$$

(Exercise 2); thus (4-12) may be written as

$$\| \mathbf{x} \|^2 = \| \mathbf{x} - \bar{x}\mathbf{1} \|^2 + \| \bar{x}\mathbf{1} \|^2;$$

this means that the line segment from $\mathbf{0}$ to $\bar{x}\mathbf{1}$ is perpendicular to the one from $\bar{x}\mathbf{1}$ to $\mathbf{x}$ (Sec. 3-1).

For any constant c the set of points $\mathbf{x}$ in R^n

$$\{\mathbf{x}: x_1 + \cdots + x_n = c\} \tag{4-14}$$

is called a *hyperplane* with equal intercepts c; for example, in R^2 it is a line of slope -1 and x_2-intercept c; and in R^3 it is a plane cutting each of the three coordinate axes at points of distance c from the origin (Fig. 4-5). Here is a characterization of the hyperplane in terms of perpendicularity:

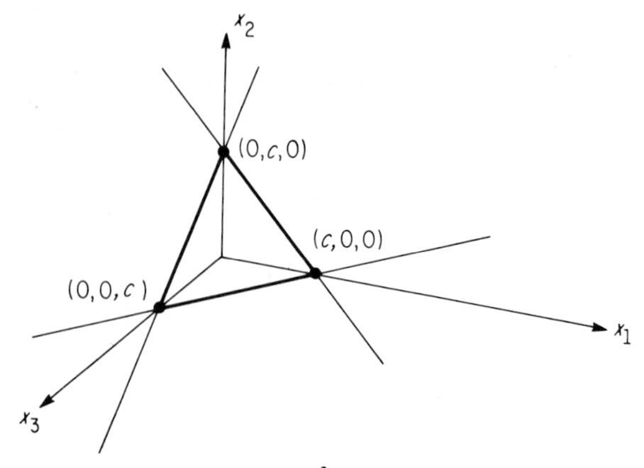

Fig. 4-5. A plane in R^3 of equal intercepts c.

LEMMA 4-1

If $c = 0$, $\mathbf{x}$ belongs to the hyperplane (4-14) if and only if

(a) $\mathbf{x} = \mathbf{0}$, or

(b) $\mathbf{x} \neq \mathbf{0}$ and the line segment from $\mathbf{0}$ to $\mathbf{x}$ is perpendicular to the one from $\mathbf{0}$ to $\mathbf{1}$.

Proof: The case $\mathbf{x} = \mathbf{0}$ is trivial.

Suppose $\mathbf{x} \neq \mathbf{0}$. If $\mathbf{x}$ belongs to (4-14), then $\bar{x} = 0$. Apply (4-13) with $d = 1$, $\bar{x} = 0$:

$$\| \mathbf{x} - \mathbf{1} \|^2 = \| \mathbf{x} \|^2 + \| \mathbf{1} \|^2;$$

therefore, the segments in (b) are perpendicular. Conversely, if the latter equation holds, then, by the expansion of the sum of squares on the left-hand side, we get

$$-2 \sum_{i=1}^{n} x_i = 0;$$

therefore, $\mathbf{x}$ belongs to (4-14). ■

LEMMA 4-2

If $c \neq 0$, then $\mathbf{x}$ belongs to the hyperplane (4-14) if and only if the line segment from $\mathbf{0}$ to $(c/n)\mathbf{1}$ is perpendicular to the one from $(c/n)\mathbf{1}$ to $\mathbf{x}$.

Proof: If $\mathbf{x}$ belongs to (4-14), then $\bar{x} = c/n$, and from (4-12) we obtain

$$\|\mathbf{x}\|^2 = \|\mathbf{x} - (c/n)\mathbf{1}\|^2 + \|(c/n)\mathbf{1}\|^2.$$

Conversely, if the latter equation holds, then expand the first sum of squares on the right-hand side, and solve the equation for $\sum_{i=1}^{n} x_i$; we get

$$\sum_{i=1}^{n} x_i = c,$$

and so $\mathbf{x}$ belongs to the hyperplane. ■

Now we apply a rotation to the space. By Theorem 3-1 there is a rotation which carries the point $\mathbf{1}$ onto the point $(\sqrt{n}, 0, \ldots, 0) = \sqrt{n}\mathbf{e}_1$ because these are on the sphere of radius $\sqrt{n}$. We shall call this a *diagonal rotation* because it transforms the diagonal line (the one having points with equal coordinates) onto the first axis; indeed, a rotation transforms lines into lines. This is the fundamental rotation used in the analysis of the normal Sample space; its effect on the hyperplane (4-14) is described in the following theorem:

THEOREM 4-3

A point belongs to the hyperplane (4-14) if and only if the first coordinate of its image under a diagonal rotation is $c/\sqrt{n}$; in other words, the set

$$\{\mathbf{x}: x_1 = c/\sqrt{n}\} \tag{4-15}$$

is the image of (4-14).

Proof: Let $\mathbf{x}'$ be the image of $\mathbf{x}$ under the diagonal rotation; then $\mathbf{1}'$, the image of $\mathbf{1}$, is $\sqrt{n}\mathbf{e}_1$.

Suppose $c = 0$. By Lemma 4-1, $\mathbf{x}$ belongs to the hyperplane if and only if

$$\|\mathbf{x} - \mathbf{1}\|^2 = \|\mathbf{x}\|^2 + \|\mathbf{1}\|^2.$$

The equation continues to hold when $\mathbf{x}$ and $\mathbf{1}$ are replaced by their images because rotation preserves distance:

$$\|\mathbf{x}' - \mathbf{1}'\|^2 = \|\mathbf{x}'\|^2 + \|\mathbf{1}'\|^2.$$

(The latter equation also implies the former.) Express this equation in terms of the coordinates of the points, and apply the fact that all the coordinates of $\mathbf{1}'$, except the first, vanish:

$$(x_1' - \sqrt{n})^2 = x_1'^2 + n.$$

The only solution of this equation is $x_1' = 0$.

Suppose $c \neq 0$. By Lemma 4-2, $\mathbf{x}$ belongs to the hyperplane if and only if

$$\| \mathbf{x} \|^2 = \| \mathbf{x} - (c/n)\mathbf{1} \|^2 + c^2/n;$$

therefore the same is true for $\mathbf{x}'$ and $\mathbf{1}'$:

$$\| \mathbf{x}' \|^2 = \| \mathbf{x}' - (c/n)\mathbf{1}' \|^2 + c^2/n.$$

(The latter equation also implies the former.) Express this in terms of the coordinates:

$$x_1'^2 = (x_1' - (c/\sqrt{n}))^2 + c^2/n.$$

For $c \neq 0$ the only solution of this equation is $x_1' = c/\sqrt{n}$. ■

EXERCISES

1. Show that (4-13) is a consequence of (4-12).

2. Prove that $\| \bar{x}\mathbf{1} \|^2 = n\bar{x}^2$.

3. Prove that the square of the average is less than or equal to the average of the squares:

$$(\bar{x})^2 \leq (1/n) \sum_{i=1}^{n} x_i^2.$$

(*Hint:* Apply 4-12.)

4. Show that for given $x_1, \ldots, x_n$ the sum of squares $\sum_{i=1}^{n} (x_i - d)^2$ is minimized with respect to d at the point $d = \bar{x}$. Give a geometric interpretation of this result in R^2.

4-5. THE REPRODUCTIVE PROPERTY OF THE NORMAL DISTRIBUTION: THE DISTRIBUTION OF THE SUM OF THE COORDINATES OF THE SAMPLE POINT

In this section we have our first application of the connection between rotation and integration. The diagonal rotation will be applied in the derivation of the distribution of the statistic $\sum_{i=1}^{n} x_i$ in sampling from a normal population.

Let M be a regular set in R^n, and M' its image under some rotation; then M' is also regular. The following theorem states that in a standard normal Sample space the probabilities attached to M and M', respectively, are equal.

THEOREM 4-4

Let $\mathbf{X}$ be a Sample from a standard normal population, and $P(M)$ and $P(M')$ the probabilities that $\mathbf{X}$ falls in M and M', respectively; then

$$P(M) = P(M'). \tag{4-16}$$

Proof: (This result was implicitly used in the case $n = 2$ in Example 4-3.) By Eq. 3-34, $P(M)$ may be expressed as

$$\operatorname*{lub}_{f \in \mathfrak{L}(M)} (2\pi)^{-n/2} \int_{R^n} \cdots \int f(\mathbf{x}) e^{-(1/2)\|\mathbf{x}\|^2} d\mathbf{x}.$$

Let f^* be the function defined as $f^*(\mathbf{x}') = f(\mathbf{x})$. By the invariance of integration under rotation (Theorem 3-4), the above integral is unchanged when f^* is substituted for f in the integrand. (While Theorem 3-4 was stated for a ball of finite radius, it extends to the whole space R^n by letting the radius become infinite.) The expression for $P(M)$ becomes

$$\operatorname*{lub}_{f \in \mathfrak{L}(M)} (2\pi)^{-n/2} \int_{R^n} \cdots \int f^*(\mathbf{x}) e^{-(1/2)\|\mathbf{x}\|^2} d\mathbf{x}.$$

(Note: $\|\mathbf{x}\|^2 = \|\mathbf{x}'\|^2$.) Now f belongs to $\mathfrak{L}(M)$ if and only if the function f^* belongs to $\mathfrak{L}(M')$ (Exercise 6, Sec. 3-4); therefore, the least upper bound of the integrals of $f(\mathbf{x})$ over all f in $\mathfrak{L}(M)$ is the same as the least upper bound of the integrals of $f^*(\mathbf{x})$ over all f^* in $\mathfrak{L}(M')$:

$$\operatorname*{lub}_{f^* \in \mathfrak{L}(M')} (2\pi)^{-n/2} \int_{R^n} \cdots \int f^*(\mathbf{x}) e^{-(1/2)\|x\|^2} d\mathbf{x}.$$

By Eq. 3-34, this is $P(M')$. ■

The following theorem asserts that the sum of the coordinates of the Sample point in sampling from a normal population has the same (normal) distribution except that the mean μ and standard deviation σ are replaced by $n\mu$ and $\sqrt{n}\sigma$, respectively. This was proved for $n = 2$ in Example 4-3.

THEOREM 4-5

In a Sample of n observations from a normal population with mean μ and standard deviation σ, the statistic $f(\mathbf{x}) = \sum_{i=1}^{n} x_i$ has a normal distribution with mean $n\mu$ and standard deviation $\sqrt{n}\sigma$.

Proof: We shall give the proof first for the standard normal population. The distribution of f at the point y is the probability that $\mathbf{X}$ falls in the set

$$M = \{\mathbf{x}: x_1 + \cdots + x_n \leq y\},$$

which is regular (Sec. 3-6). By Theorem 4-3 the image of this set under the diagonal rotation is

$$M' = \{\mathbf{x}: x_1 \leq y/\sqrt{n}\}.$$

It follows from the equality of the probabilites of M and M' (Theorem 4-4) that the distribution of $x_1 + \cdots + x_n$ at y is equal to that of x_1 at the point $y/\sqrt{n}$. The latter distribution is normal with mean 0 and standard deviation $\sqrt{n}\sigma$:

$$\int \cdots \int_{\{\mathbf{x}: x_1 \leq y/\sqrt{n}\}} \prod_{i=1}^{n} \phi(x_i)\, d\mathbf{x}$$

$$= \int_{-\infty}^{y/\sqrt{n}} \phi(x_1) \left[\int_{-\infty}^{\infty} \cdots \int_{-\infty}^{\infty} \prod_{i=2}^{n} \phi(x_i) \cdot \prod_{i=2}^{n} dx_i \right] dx_1$$

$$= \int_{-\infty}^{y/\sqrt{n}} \phi(x_1)\, dx_1 = \Phi(y/\sqrt{n}).$$

This completes the proof in the standard case.

In the general case where

$$L(\mathbf{x}; \mu, \sigma) = \prod_{i=1}^{n} \frac{1}{\sigma} \phi\left(\frac{x_i - \mu}{\sigma}\right),$$

transform each of the variables of integration from x_i to $(x_i - \mu)/\sigma$ in the integral

$$\int_M \cdots \int L(\mathbf{x}; \mu, \sigma)\, d\mathbf{x}.$$

The integrand is transformed into $\prod_{i=1}^{n} \phi(x_i)$ and the domain of integration M into

$$\{\mathbf{x}\colon x_1 + \cdots + x_n \leq (y - n\mu)/\sigma\}.$$

The distribution of $\sum_{i=1}^{n} x_i$ at the point y is seen to be that in the standard case at the point $(y - n\mu)/\sigma$. ■

There are two useful variations of this theorem which are stated as corollaries.

COROLLARY 1
The statistic $\overline{x} = (1/n)(x_1 + \cdots + x_n)$—the Sample mean—has a normal distribution with mean μ and standard deviation $\sigma/\sqrt{n}$.

This follows from Theorem 4-2 and Example 4-1; $f(\mathbf{x}) = x_1 + \cdots + x_n$ has a normal distribution with mean $n\mu$ and standard deviation $\sqrt{n}\sigma$, so that $\overline{x} = f(\mathbf{x})/n$ has the normal distribution with mean μ and standard deviation $\sigma/\sqrt{n}$. ■

COROLLARY 2
The "standardized" Sample mean,

$$\sqrt{n}\,(\overline{x} - \mu)/\sigma,$$

has a standard normal distribution.

The proof is similar to that of Corollary 1. ■

EXERCISES

1. Show that Theorem 4-4 is valid even when σ is not equal to 1 but that $\mu = 0$ is an essential condition.

2. Find the distribution of the statistic $f(\mathbf{x}) = x_1 \cos\theta + x_2 \sin\theta$ in a Sample of two observations from a standard normal population, where θ is a fixed number between 0 and 2π. (*Hint:* Use an appropriate rotation to evaluate the double integral.)

4-6. STATISTICS AND THEIR DISTRIBUTIONS FOR MORE GENERAL SAMPLE SPACES (OPTIONAL)

A statistic on a Sample space generated by a population with the density $h(x)$ is defined in exactly the same way as for the normal space (Definition 4-1). Its distribution $F(y)$ is also given by Definition 4-2 with the modification that the normal likelihood function is replaced by the general likelihood function. Theorem 4-1 is still valid. In computing distributions the technique of rotation is useful only in the normal case; however, other methods of integration are sufficient in many nonnormal cases.

Example 4-6. Consider the Sample space of two observations from a population with the exponential distribution with mean $\theta = 1$:

$$H(x) = 1 - e^{-x}, \quad x > 0; \; = 0, \; x \leq 0.$$

Let us find the distribution of the statistic $f(x_1, x_2) = x_1 + x_2$. The likelihood function is

$$\begin{aligned} h(x_1)h(x_2) &= 0 \qquad \text{if either } x_1 < 0 \text{ or } x_2 < 0 \\ &= e^{-x_1 - x_2} \quad \text{if both } x_1 > 0,\ x_2 > 0. \end{aligned}$$

The distribution function of f at the point y is the integral of $h(x_1)h(x_2)$ over the half-space below and left of the line in the x_1x_2-plane: $x_1 + x_2 = y$. This integral is equal to 0 if $y < 0$. For $y > 0$, it is equal to the double integral of $e^{-x_1-x_2}$ over the triangle formed by the coordinate axes and the line $x_1 + x_2 = y$ (Fig. 4-6).

By iterated integration, the double integral is equal to

$$\int_0^y \left(\int_0^{y-x_2} e^{-x_1}\, dx_1 \right) e^{-x_2}\, dx_2 = 1 - e^{-y} - ye^{-y};$$

thus the density function is

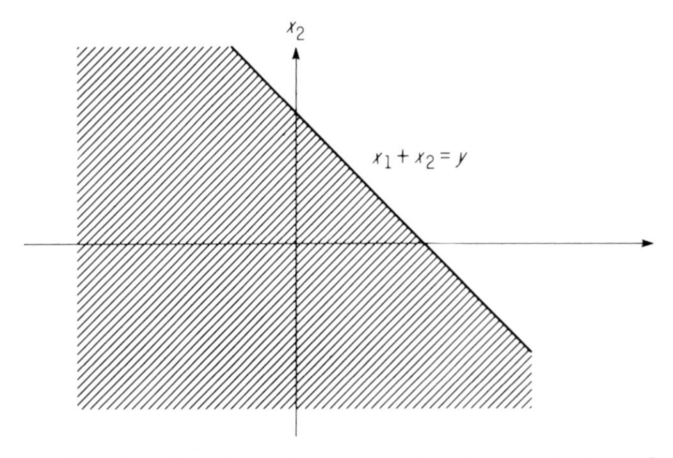

Fig. 4-6. Domain of integration for the distribution of $x_1 + x_2$.

$$
\begin{aligned}
F'(y) &= 0, & y < 0 \\
&= ye^{-y}, & y > 0,
\end{aligned}
$$

and is arbitrarily defined at $y = 0$. ■

Example 4-7. This is a generalization of the previous example. Consider the distribution function $F(y)$ of the statistic $f = x_1 + x_2$ on the Sample space of two observations from the population with the density $h(x)$:

$$F(y) = \iint_{\{(x_1, x_2): x_1 + x_2 \le y\}} h(x_1)h(x_2)\,dx_1\,dx_2$$

The integral is equivalent to the iterated integral

$$\int_{\infty}^{\infty} \left(\int_{\infty}^{y - x_2} h(x_1)\,dx_1 \right) h(x_2)\,dx_2.$$

Supposing that the inner integral may be differentiated with respect to y under the outer integral sign, we find the density

$$F'(y) = \int_{-\infty}^{\infty} h(y - x)h(x)\,dx. \tag{4-17}$$

The integral on the right-hand side is called the *convolution* of the density h with itself. When $h(x)$ vanishes for $x \le 0$, the convolution is of the special form

$$F'(y) = 0, \qquad y \le 0$$
$$= \int_0^y h(y - x)h(x)\,dx, \quad y > 0. \tag{4-18}$$

The result of Example 4-6 is a special case of this. ■

Example 4-8. Consider the statistic $f(\mathbf{x}) = \min(x_1, \ldots, x_n)$ on the Sample space of n observations from the population with the density function $h(x)$. The set $\{\mathbf{x}: \min(x_1, \ldots, x_n) \le y\}$ is the complement of the intersection of the (regular) sets $\{\mathbf{x}: x_i > y\}$, $i = 1, \ldots, n$. The integral of the likelihood function over the intersection

$$\int \cdots \int_{\{\mathbf{x}:\, x_i > y,\, i = 1, \ldots, n\}} h(x_1) \cdots h(x_n)\,dx_1 \cdots dx_n,$$

is equivalent to the iterated integral

$$\int_y^\infty \cdots \int_y^\infty h(x_1) \cdots h(x_n)\,dx_1 \cdots dx_n = (1 - H(y))^n.$$

It follows (by subtraction from 1) that the distribution function of f is

$$F(y) = 1 - (1 - H(y))^n, \tag{4-19}$$

and so the density function is

$$F'(y) = n(1 - H(y))^{n-1}h(y). \tag{4-20}$$

If h is the exponential density with mean θ, then it follows from (4-20) that f has the exponential density with mean θ/n (Exercise 1). ■

Now we state without proof one of the most remarkable theorems of probability theory—the central limit theorem—on the distribution of the statistic

$$f(\mathbf{x}) = \sqrt{n}(\bar{x} - \mu)/\sigma \tag{4-21}$$

on a general Sample space, where μ and σ are the mean and standard deviation, respectively, of the density of the population. (It is assumed that μ and σ are well defined and finite (cf. Sec. 2-8).)

Corollary 2 of Theorem 4-5 states that if the underlying population is normal, then the statistic (4-21) has a standard normal distribution. The central limit theorem asserts that if n is very large then the distribution of the statistic (4-21)—*for any underlying population*—is approximately a standard normal one. The precise form of the statement of the theorem is that for every y the distribution $F(y)$ of the statistic (4-21) converges to $\Phi(y)$ as $n \to \infty$. Results on the universal rate of convergence are also known. This theorem explains the prevalence of the normal distribution in numerical data: many measurements are sums or averages of "independent observations" and so follow a normal distribution.

The central limit theorem also implies the "robustness" of certain statistical procedures. Most of the classical statistical procedures of estimation and testing were devised for normal populations. During the past forty years theoretical statisticians have found methods of treating populations which are either not normal or else of unknown form. Methods requiring few or no assumptions about the population are known as *nonparametric*. It has been found that some of the classical procedures constructed specifically for normal populations are almost as good for many nonnormal populations as the nonparametric methods. The reason for the robustness of these procedures can often be found in the central limit theorem: they are often based on a statistic like (4-21) which, for a large sample, has a distribution approximately equal to that in the normal case.

EXERCISES

1. Derive the distribution and density of min $(x_1, \ldots, x_n)$ in sampling from an exponential population with mean θ.

2. Show that $H^n(y)$ is the distribution of $f(\mathbf{x}) = \max(x_1, \ldots, x_n)$ in sampling from a population with the density $h(x)$.

3. Find the distribution and density of $f(x_1, x_2) = x_1 + x_2$ in a sample of two observations from populations with densities of the following kind. (Wherever possible perform the integration.)

(a) uniform distribution on the unit interval (Example 2-6)

(b) normal distribution with mean μ and standard deviation σ (use the method of this section and compare with Example 4-3)

(c) chi-square distribution with one degree of freedom (Example 4-2)
(d) Cauchy distribution (Example 2-10).

4. Extend the convolution formula to the sum of three coordinates: Prove that the density of $x_1 + x_2 + x_3$ is equal to

$$F'(y) = \int_{-\infty}^{\infty} \int_{-\infty}^{\infty} h(y - x - z)h(x)h(z)\,dz\,dx.$$

5. By induction extend the result of Exercise 4 to the density of $x_1 + \cdots + x_n$:

$$F'(y) = \int_{-\infty}^{\infty} \cdots \int_{-\infty}^{\infty} h(y - x_1 - \cdots x_{n-1})h(x_1)\cdots h(x_{n-1})\,dx_1\cdots dx_{n-1}.$$

6. Consider the statistic $f(x_1, x_2) = x_1 x_2$ in a Sample of two observations from a population with the density $h(x)$. Write the integral formula for the density function of f in terms of h.

4-7. NUMERICAL ILLUSTRATIONS OF THE DISTRIBUTIONS OF STATISTICS (OPTIONAL)

Example 4-9. We illustrate the distribution of the statistic $x_1^2 + x_2^2 + x_3^2$ in a Sample of three observations from a standard normal population. A Sample of 75 normal deviates was taken from Table VI, Appendix. Each number was squared, and then successive triples were summed, so that 25 sums were obtained, and rounded to two decimal places:

2.18	.58	.70	1.66	2.30
2.47	16.70	2.72	6.56	.82
4.75	1.48	4.29	.84	.82
1.63	3.72	1.89	5.08	.72
1.25	4.65	4.01	2.94	5.14

These numbers are plotted as points on the horizontal axis in Fig. 4-7, and the theoretical density, the chi-square with three degrees of freedom, is plotted in the xy-plane. (Note that the observation 16.70 is too far out to be recorded.) ■

Example 4-10. The central limit theorem is true not only for observations from "continuous" populations, that is, those with

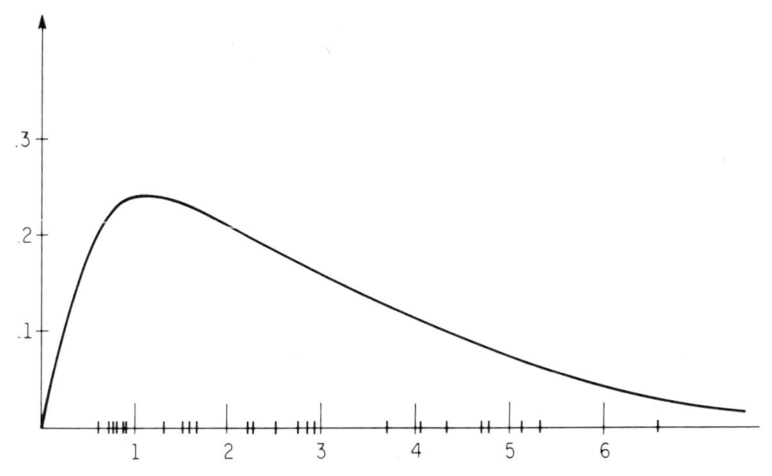

Fig. 4-7. Twenty-five values of the statistics $x_1^2 + x_2^2 + x_3^2$ drawn on the horizontal axis of the chi-square density with three degrees of freedom.

density functions, but also for populations with a fixed finite set of labels $x_1, \ldots, x_k$ in proportions $p_1, \ldots, p_k$. The mean $m = \Sigma x_i p_i$ and the variance $v = \Sigma(x_i - m)^2 p_i$, defined in Sec. 2-2, are used in place of m and σ^2, respectively. We shall illustrate the central limit theorem in sampling from the table of random digits; this corresponds to a population with labels $0, 1, \ldots, 9$ in equal proportions 1/10. The mean of the population is

$$m = (1/10)(1 + \cdots + 9) = 4.5$$

and the variance is

$$v = (1/10)(1^2 + \cdots + 9^2) - (4.5)^2 = 8.25,$$

and so the standard deviation is

$$\sqrt{v} = 2.87.$$

Fifty samples of sums of 50 random digits were taken from Table V, Appendix. Each row on the second page of the table was summed; the fifty sums are given in Table 4-1. In order to avoid an excessive rounding error we compute the statistic $\sqrt{n}(\bar{x} - m)/\sqrt{v}$ in the form

$$\left(\sum_{i=1}^{50} x_i - 50m\right) / \sqrt{50v}, \tag{4-22}$$

TABLE 4-1
Fifty Sums of Fifty Random Digits.

226	236	203	247	254
223	206	245	253	209
224	223	195	242	212
220	187	195	242	217
207	224	185	282	249
249	219	225	254	224
241	237	211	249	222
195	200	200	230	228
190	231	219	209	223
208	294	222	234	204

where $50m = 225$, and $\sqrt{50v} = 20.3$. The values of (4-22) corresponding to the sums in Table 4-1 are given in Table 4-2.

TABLE 4-2
The Values of $\sqrt{n}(\bar{x} - m)/\sqrt{v}$ Corresponding to the Sums in Table 4-1.

.049	.542	−1.084	1.084	1.429
−.098	−.934	.985	1.379	−.788
−.049	−.099	−1.478	.837	−.640
−.246	−1.87	−1.478	.837	−.394
−.887	−.049	−1.970	2.808	1.182
1.182	−.296	.000	1.429	−.049
.788	.591	−.690	1.182	−.148
−1.478	−1.232	−1.232	.246	.148
−1.724	.296	−.296	−.788	−.098
−.837	−1.527	−.148	.542	−1.034

In Table 4-3 the numbers in Table 4-2 are grouped and their proportions are compared to the corresponding probabilities for the standard normal distribution, as in Table 2-1. ■

TABLE 4-3
Distribution of the Values of $\sqrt{n}(\bar{x} - m)/\sqrt{v}$ Compared to the Standard Normal.

Interval	Frequency	Relative Frequency	Standard Normal Probability
[−3, −2)	—	.00	.0215
[2, −1)	11	.22	.1359
[−1, 0)	19	.38	.3413
[0, 1)	12	.24	.3413
[1, 2)	7	.14	.1359
[2, 3)	1	.02	.0215

EXERCISES (Optional)

1. Using 75 normal deviates from Table VI, Appendix, construct an empirical distribution of the statistic $x_1 + x_2 + x_3$ for 25 triples (x_1, x_2, x_3). Compare the observed frequencies to the probabilities for the normal distribution with mean 0 and standard deviation $\sigma = \sqrt{3}$ (cf. Table 4-3). Use the six equal intervals from -3σ to 3σ.

2. Repeat Example 4-10 for 100 Samples of 50 random digits.

BIBLIOGRAPHY

Distributions of statistics are derived by various methods in M. G. Kendall, *The Advanced Theory of Statistics* (London: Griffin, 1943, 1946).

A complete proof of the central limit theorem is given in Harald Cramér, *Mathematical Methods of Statistics* (Princeton, N.J.: Princeton U. P., 1946).

Derivations of distributions of statistics by the method of moment generating functions are found in Alexander M. Mood and Franklin A. Graybill, *Introduction to the Theory of Statistics*. 2nd ed. (New York: McGraw-Hill, 1963).

CHAPTER 5

Joint Distributions of Several Statistics: Independence

5-1. JOINT DISTRIBUTION OF TWO STATISTICS: INDEPENDENCE

One statistic does not provide enough information about the population in some statistical problems; thus the investigator must choose two or more statistics to summarize the data in the Sample. (When both the mean and standard deviation of the population are unknown, the Sample mean $\bar{x}$ and the Sample standard deviation s (defined in Sec. 5-2) are used to give information about these unknown parameters). In this section we discuss the joint distribution of two statistics. This is generalized to more than two in Sec. 5-3.

Let $f_1(\mathbf{x})$ and $f_2(\mathbf{x})$ be two statistics on R^n—that is, continuous functions such that for every pair of real numbers y_1 and y_2, the sets

$$\{\mathbf{x}: f_1(\mathbf{x}) \leq y_1\} \qquad \text{and} \qquad \{\mathbf{x}: f_2(\mathbf{x}) \leq y_2\}$$

are regular. The intersection of the two sets,

$$\{\mathbf{x}: f_1(\mathbf{x}) \leq y_1, \quad f_2(\mathbf{x}) \leq y_2\}$$

is regular (Theorem 3-8). The *joint distribution* function of the pair f_1 and f_2 at the point (y_1, y_2) is defined as the integral of the likelihood function over this intersection:

$$F(y_1, y_2) = \int \cdots \int_{\{\mathbf{x}: f_1(\mathbf{x}) \leq y_1, f_2(\mathbf{x}) \leq y_2\}} L(\mathbf{x}; \mu, \sigma)\, d\mathbf{x}. \tag{5-1}$$

The statistical interpretation of the joint distribution is a direct extension of that for the distribution of a single statistic: If $\mathbf{X}$ is a Sample from a normal population, then $F(y_1, y_2)$ represents the probability that $\mathbf{X}$ falls in the intersection of the sets where $f_1(\mathbf{x}) \leq y_1$ and $f_2(\mathbf{x}) \leq y_2$, respectively.

The joint distribution of two statistics is also called a *bivariate distribution*. If in (5-1) we let $y_2 \to \infty$, then the integral approaches that of L over $\{\mathbf{x}: f_1(\mathbf{x}) \leq y_1\}$ (Exercise 6). The latter represents the distribution of the single statistic $f_1(\mathbf{x})$ at the point y_1; this distribution is called the *marginal distribution* of f_1. By analogy, the limit of $F(y_1, y_2)$ for $y_1 \to \infty$ is the marginal distribution of f_2. We denote these marginal distributions by $F(y_1, \infty)$ and $F(\infty, y_2)$, respectively. $F(y_1, y_2)$ satisfies these relations:

$$\lim_{y_1 \to -\infty} F(y_1, y_2) = \lim_{y_2 \to -\infty} F(y_1, y_2) = 0$$

$$\lim_{y_1 \to \infty, y_2 \to \infty} F(y_1, y_2) = 1 \tag{5-2}$$

$$F(y_1 + h, y_2 + k) - F(y_1 + h, y_2) - F(y_1, y_2 + k) + F(y_1, y_2) \geq 0, \qquad h, k > 0.$$

The first two of these follow by the reasoning used for corresponding relations for the distribution of one statistic (Sec. 4-2). The third relation is a consequence of the fact that the second order difference on the left hand side is equal to the integral of L over $\{\mathbf{x}: y_1 < f_1(\mathbf{x}) \leq y_1 + h, y_2 < f_2(\mathbf{x}) \leq y_2 + k\}$.

The second-order mixed derivative,

$$\frac{\partial^2 F}{\partial y_1 \partial y_2},$$

when it exists, is called the *joint density function* of the pair f_1 and f_2. By the last relation in (5-2) it is nonnegative; and, by the first two relations, if it is continuous, then

$$\int_{-\infty}^{\infty} \int_{-\infty}^{\infty} \frac{\partial^2 F}{\partial y_1 \partial y_2} \, dy_1 \, dy_2 = 1.$$

The joint distribution is used in the reduction of certain multiple integrals. Here is a generalization of Theorem 4-1:

THEOREM 5-1

Let f_1 and f_2 be statistics on R^n with a continuous density $\partial^2 F/\partial y_1 \partial y_2$. Let $g(y_1, y_2)$ be a continuous function of two *variables such that*

$$\int_{R^n}\cdots\int |g(f_1(\mathbf{x}), f_2(\mathbf{x}))| \, L(\mathbf{x}; \mu, \sigma) \, d\mathbf{x} \leq \infty;$$

then

$$\int_{R^n}\cdots\int g(f_1(\mathbf{x}), f_2(\mathbf{x})) \, L(\mathbf{x}; \mu, \sigma) \, d\mathbf{x}$$

$$= \int_{-\infty}^{\infty} \int_{-\infty}^{\infty} g(y_1, y_2)(\partial^2 F/\partial y_1 \partial y_2) \, dy_1 \, dy_2.$$

Proof: The proof is similar to that of Theorem 4-1: the space is decomposed into sets

$$\{\mathbf{x}: y_1 < f_1(\mathbf{x}) \leq y_1 + h, \; y_2 < f_2(\mathbf{x}) \leq y_2 + h\}$$

for small $h > 0$, the integral over each is estimated, and these estimates are summed (Exercise 1). ■

The joint density has an interpretation as a surface average similar to that for the univariate (single variable) density. For $h, k > 0$, the second order difference quotient of F at (y_1, y_2) is

$$(1/hk)[F(y_1 + h, y_2 + k) - F(y_1 + h, y_2) - F(y_1, y_2 + k)$$

$$+ F(y_1, y_2)] = (1/hk) \int_{\{\mathbf{x}: y_1 < f_1(\mathbf{x}) \leq y_1 + h, \; y_2 < f_2(\mathbf{x}) \leq y_2 + k\}}\cdots\int L(\mathbf{x}; \mu, \sigma) \, d\mathbf{x}. \tag{5-3}$$

The limit as $h, k \to 0$, if it exists, is the joint density. If the quotient

$$(1/hk) \int_{\{\mathbf{x}: y_1 < f_1(\mathbf{x}) \leq y_1 + h, \; y_2 < f_2(\mathbf{x}) \leq y_2 + k\}}\cdots\int d\mathbf{x} \tag{5-4}$$

(the ratio of the volume of the set to hk) has a limit for $h, k \to 0$, the latter is called the "area of the intersection of the surfaces $\{\mathbf{x}: f_1(\mathbf{x}) = y_1\}$ and $\{\mathbf{x}: f_2(\mathbf{x}) = y_2\}$." The limit of the ratio of (5-3) to (5-4) is called the *average* of L over the intersection of the surfaces; thus, the joint density is representable as the average of L over the intersection of the surfaces *times* the area of the intersection.

Most of the joint distributions arising in this book are products of their marginal distributions:

$$F(y_1, y_2) = F(y_1, \infty)\, F(\infty, y_2). \tag{5-5}$$

This is the defining relation of the concept of *independence*.

DEFINITION 5-1

Two statistics are called independent if their joint distribution factors into a product of their marginal distributions.

If the joint distribution $F(y_1, y_2)$ factors into a product of functions of y_1 and y_2,

$$F(y_1, y_2) = G_1(y_1)\, G_2(y_2),$$

then it necessarily factors as in (5-5); indeed, passing to limits in the above equation, we obtain

$$F(y_1, \infty) = G_1(y_1)\, G_2(\infty), \quad F(\infty, y_2) = G_1(\infty)\, G_2(y_2)$$
$$1 = F(\infty, \infty) = G_1(\infty)\, G_2(\infty),$$

so that $G_1(y_1)\, G_2(y_2) = F(y_1, \infty)\, F(\infty, y_2)$.

When the joint density exists and is continuous, the condition (5-5) is equivalent to the factorization of the joint density:

$$\frac{\partial^2 F}{\partial y_1 \partial y_2} = \frac{\partial}{\partial y_1} F(y_1, \infty) \cdot \frac{\partial}{\partial y_2} F(\infty, y_2) \qquad \text{(Exercise 2).} \tag{5-6}$$

The significance of the independence of two statistics is that the value assumed by one does not "affect" that of the other. Suppose f_1 and f_2 are independent. For any y_1 and y_2, $F(y_1, y_2)$ represents the proportion of the underlying population for which $f_1(\mathbf{x}) \leq y_1$ and $f_2(\mathbf{x}) \leq y_2$. The *relative* proportion for which $f_1(\mathbf{x}) \leq y_1$—*relative to the subpopulation for which* $f_2(\mathbf{x}) \leq y_2$—is, by the same logic, equal to

$$\frac{F(y_1, y_2)}{F(\infty, y_2)}.$$

(Among all those satisfying $f_2(\mathbf{x}) \leq y_2$, we identify those also satisfying $f_1(\mathbf{x}) \leq y_1$.) By the defining relation of independence, Eq. 5-5, the numerator in the above ratio factors into $F(y_1, \infty)\cdot$

$F(\infty, y_2)$, and the fraction reduces to $F(y_1, \infty)$, which is precisely the *proportion of the entire population for which* $f_1(\mathbf{x}) \le y_1$. From this we infer that the relative proportion is equal to the proportion for the entire population *for every* y_2. In this sense, the values of f_1 are not affected by those of f_2.

Example 5-1. Put $f_1(\mathbf{x}) = x_1$ and $f_2(\mathbf{x}) = x_2$; then the joint distribution $F(y_1, y_2)$ is

$$\int_{-\infty}^{\infty} \cdots \int_{-\infty}^{\infty} \int_{-\infty}^{y_2} \int_{-\infty}^{y_1} L(\mathbf{x}; \mu, \sigma)\, dx_1\, dx_2\, dx_3 \cdots dx_n.$$

Since L is the product of n normal densities, it follows that

$$F(y_1, y_2) = \int_{-\infty}^{y_2} \int_{-\infty}^{y_1} \frac{1}{\sigma} \phi\left(\frac{x_1 - \mu}{\sigma}\right) \cdot \frac{1}{\sigma} \phi\left(\frac{x_2 - \mu}{\sigma}\right) dx_1\, dx_2$$

$$= \Phi\left(\frac{y_1 - \mu}{\sigma}\right) \Phi\left(\frac{y_2 - \mu}{\sigma}\right);$$

thus f_1 and f_2 are independent. ■

Example 5-2. Assume $\mu = 0$, $\sigma = 1$ and put $f_1(\mathbf{x}) = (x_1 - x_2)/\sqrt{2}$, $f_2(\mathbf{x}) = (x_1 + x_2)/\sqrt{2}$. As in the previous example, the integral of L over the variables $x_3, \ldots, x_n$ is 1, so that the joint distribution of f_1 and f_2 may be computed as in a two-dimensional Sample space:

$$F(y_1, y_2) = \frac{1}{2\pi} \iint_{\{(x_1, x_2): x_1 - x_2 \le \sqrt{2} y_1,\, x_1 + x_2 \le \sqrt{2} y_2\}} e^{-(1/2)(x_1^2 + x_2^2)}\, dx_1\, dx_2 \qquad (5\text{-}7)$$

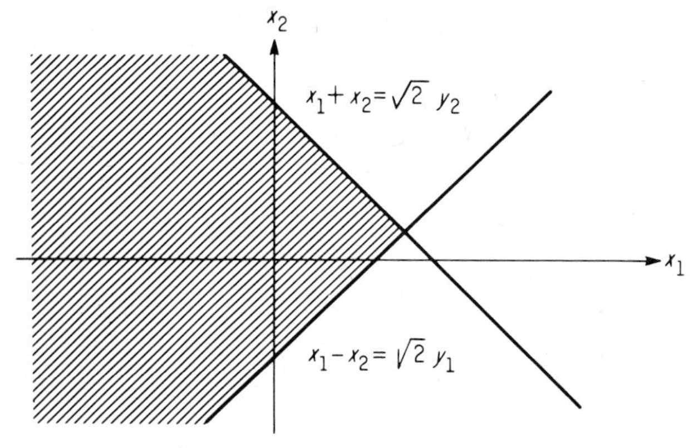

Fig 5-1. The domain of integration.

The domain of integration is the quarter-plane bounded by the lines $x_1 - x_2 = \sqrt{2}y_1$ and $x_1 + x_2 = \sqrt{2}y_2$ (Fig. 5-1).

Rotate the region through an angle of 45° so that the line $x_1 + x_2 = \sqrt{2}y_2$ becomes horizontal and the line $x_1 - x_2 = \sqrt{2}y_1$ vertical. The region becomes the quarter-plane bounded above by the line $x_2 = y_2$ and on the right by the line $x_1 = y_1$ (Fig. 5-2).

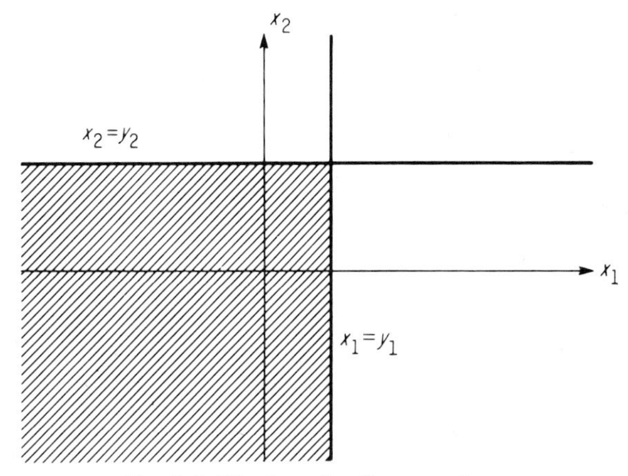

Fig. 5-2. The domain after rotation.

The integrand in (5-7) is invariant under rotations because it is constant on circles; thus, the integrals over the original domain and its image are equal. The latter is

$$\frac{1}{2\pi}\int_{-\infty}^{y_2}\int_{-\infty}^{y_1} \phi(x_1)\,\phi(x_2)\,dx_1\,dx_2 = \Phi(y_1)\,\Phi(y_2);$$

thus f_1 and f_2 are independent.

Examples 5-1 and 5-2 are special cases of the following two theorems, respectively:

THEOREM 5-2

Let f_1 and f_2 be statistics on R^n depending on distinct nonoverlapping sets of coordinates; for example, $f_1(\mathbf{x})$ depends only on $x_1, \ldots, x_k$, and $f_2(\mathbf{x})$ on $x_{k+1}, \ldots, x_n$, $1 \leq k < n$. Then f_1 and f_2 are independent.

Proof: Since the function L is a product of n functions, each depending on a single coordinate, L can be factored into a product of two functions L_1 and L_2, the first a product of functions of $x_1, \ldots, x_k$, and the second a product of functions of $x_{k+1}, \ldots, x_n$.

The integral (5-1) factors into a product of two multiple integrals: the first is the integral of L_1 over $\{\mathbf{x}: f_1(\mathbf{x}) \leq y_1\}$, and the second the integral of L_2 over $\{\mathbf{x}: f_2(\mathbf{x}) \leq y_2\}$; indeed, by hypothesis, these sets are determined by the first k and last $n - k$ coordinates, respectively. These integrals depend on y_1 and y_2, respectively, so that the joint distribution factors into a product of functions of y_1 and y_2; therefore, by the remarks following Definition 5-1, f_1 and f_2 are independent. ∎

THEOREM 5-3

Let $f_1(\mathbf{x})$ and $f_2(\mathbf{x})$ be statistics on a standard normal Sample space. For a given rotation let $\mathbf{x}'$ be the image of $\mathbf{x}$, and let f_i^* and f_2^* be statistics defined at the point $\mathbf{x}'$ as

$$f_1^*(\mathbf{x}') = f_1(\mathbf{x}), \qquad f_2^*(\mathbf{x}') = f_2(\mathbf{x});$$

in other words, the value of f_i^* at a point is that of f_i at the preimage under rotation $i = 1, 2$. Then f_1^* and f_2^* have the same joint distribution as f_1 and f_2.

Proof: First of all we have to show that f_i^* is actually a statistic, $i = 1, 2$. It is continuous (Exercise 6, Sec. 3-3). The image of $\{\mathbf{x}: f_i(\mathbf{x}) \leq y_i\}$ is $\{\mathbf{x}: f_i^*(\mathbf{x}) \leq y_i\}$ (Exercise 3); thus the latter is the image of the former under the rotation and so is regular (Theorem 3-5).

Now we show that the joint distributions are the same. Let M be the set in R^n:

$$M = \{\mathbf{x}: f_1(\mathbf{x}) \leq y_1, f_2(\mathbf{x}) \leq y_2\}.$$

By the reasoning above, its image is

$$M' = \{\mathbf{x}: f_1^*(\mathbf{x}) \leq y_1, f_2^*(\mathbf{x}) \leq y_2\}.$$

The joint distribution of f_1^* and f_2^* at (y_1, y_2) is the integral of L over M. By the rotation-invariance of the integral (Theorem 4-4), it is also equal to the integral of L over M'; the latter integral is the joint distribution of f_1 and f_2 at (y_1, y_2). ∎

We remark that this result is valid even when the population has a standard deviation not equal to 1 (cf. Exercise 1, Sec. 4-5).

In Example 5-2, the functions f_i^* are

$$f_1^*(\mathbf{x}') = x_1' = \frac{x_1 - x_2}{\sqrt{2}}$$

$$f_2^*(\mathbf{x}') = x_2' = \frac{x_1 + x_2}{\sqrt{2}},$$

and are shown to be independent.

EXERCISES

1. Give the details of the outlined proof of Theorem 5-1.
2. Show that the relation (5-6) is equivalent to that of independence.
3. Show that the image of $\{\mathbf{x}: f_i(\mathbf{x}) \le y_i\}$ is $\{\mathbf{x}: f_i^*(\mathbf{x}) \le y_i\}$.
4. *Prove:* In a Sample of two observations from a standard normal population, the statistics $x_1^2 + x_2^2$ and x_1/x_2 are independent. Use the method of averages. (It can be shown by means of a simple argument that the second function has a well-defined distribution even though it is not continuous at points where $x_2 = 0$.)
5. Use the result of Example 5-2 to prove that in a Sample of two from a standard normal population, the statistics

$$\bar{x} = \tfrac{1}{2}(x_1 + x_2) \qquad \text{and} \qquad (x_1 - \bar{x})^2 + (x_2 - \bar{x})^2$$

are independent. What are the marginal distributions?

6. Show that the distribution of f_1 is obtained when $y_2 \to \infty$.

5-2. JOINT DISTRIBUTION OF THE SAMPLE MEAN AND VARIANCE

In sampling from a normal population the two fundamental statistics used in procedures of testing and estimation are the Sample mean,

$$\bar{x} = \frac{1}{n}(x_1 + \cdots + x_n),$$

and the Sample variance,

$$s^2 = \frac{1}{n-1}\sum_{i=1}^{n}(x_i - \bar{x})^2.$$

The latter is $n/(n-1)$ times the average sum of squares about $\bar{x}$. Its square root s is the Sample standard deviation. The importance of these statistics will be demonstrated in subsequent chapters. In this section we prove one of the basic theorems of mathematical statistics, concerned with the joint distribution of $\bar{x}$ and s^2. (The particular case $n = 2$ and for the standard normal population is contained in Exercise 5, Sec. 5-1.)

THEOREM 5-4

In sampling from a normal population with mean μ and standard deviation σ, the statistics $\sqrt{n}(\bar{x} - \mu)/\sigma$ and $(n-1)s^2/\sigma^2$ are independent. The former has a standard normal distribution (Theorem 4-5 and its corollaries), and the latter has a chi-square distribution with $n - 1$ degrees of freedom.

Proof: The proof of the general case is reducible to that for the *standard* normal population (Exercise 1); thus, we shall assume the latter in the proof.

The idea of the proof is that the diagonal rotation (the one carrying **1** onto $\sqrt{n}\mathbf{e}_1$) associates with $\bar{x}$ and s^2 particular functions f_1^* and f_2^* which satisfy the hypothesis of Theorem 5-2. We shall prove: If $\mathbf{x}'$ is the image of $\mathbf{x}$ under this rotation, and x_i' and x_i are the ith coordinates of these points, respectively, then the coordinates are related by the equations

$$\sqrt{n}\bar{x} = x_1', \qquad \sum_{i=1}^{n} (x_i - \bar{x})^2 = \sum_{i=2}^{n} x_i'^2. \tag{5-8}$$

The second of these incidentally proves that s^2 is actually a statistic.

To prove the first of these, suppose that $\bar{x}$ has the value c; then $\mathbf{x}$ belongs to the hyperplane $\{\mathbf{x}: x_1 + \cdots + x_n = nc\}$; therefore, by Theorem 4-3 (with nc in place of c), the first coordinate of $\mathbf{x}'$ is equal to $nc/\sqrt{n} = \sqrt{n}c$; consequently, $x_1' = \sqrt{n}c = \sqrt{n}\bar{x}$. The reasoning is reversible: if $x_1' = \sqrt{n}c$, then $\bar{x} = c$.

To prove the second equation in (5-8), note that, by the preservation of distance under rotation, we have

$$\sum_{i=1}^{n} x_i^2 = \sum_{i=1}^{n} x_i'^2.$$

Apply the sum of squares decomposition (4-12) to the left-hand side:

$$\sum_{i=1}^{n} (x_i - \bar{x})^2 + n\bar{x}^2 = \sum_{i=1}^{n} x_i'^2.$$

Substitute from the first equation in (5-8):

$$\sum_{i=1}^{n} (x_i - \bar{x})^2 + x_1'^2 = \sum_{i=1}^{n} x_i'^2.$$

Cancel $x_1'^2$ on both sides. This completes the proof of (5-8).

Let f_1^* be the function defined at $\mathbf{x}'$ as the value of $f_1(\mathbf{x}) = \sqrt{n}\bar{x}$ at $\mathbf{x}$, and f_2^* the function defined at $\mathbf{x}'$ as the value of $f_2(\mathbf{x}) = (n - 1)s^2$ at x. Equations 5-8 imply:

$$f_1^*(\mathbf{x}') = x_1', \qquad f_2^*(\mathbf{x}') = \sum_{i=2}^{n} x_i'^2.$$

Since this is true for all $\mathbf{x}'$, and since every point in R^n is the image of some other point under the rotation, it follows that

$$f_1^*(\mathbf{x}) = x_1, \qquad f_2^*(\mathbf{x}) = \sum_{i=2}^{n} x_i^2, \qquad \mathbf{x} \in R^n.$$

By Theorem 5-3, f_1^* and f_2^* have the same joint distribution as $\sqrt{n}\bar{x}$ and $(n - 1)s^2$. By Theorem 5-2, f_1^* and f_2^* are independent because the former depends only on the first coordinate, and the latter on the last $n - 1$ coordinates. f_1^* has a standard normal distribution because it is equal to the single coordinate x_1. f_2^* has the chi-square distribution with $n - 1$ (in place of n) degrees of freedom because it is the sum of the squares of $n - 1$ coordinates (Theorem 4-2). ■

It is interesting to compare the distribution of $(n - 1)s^2/\sigma^2$ to that of $\Sigma(x_i - \mu)^2/\sigma^2$. Both have chi-square distributions: by Theorem 5-4, the first has $n - 1$ degrees of freedom, and by the extension of Theorem 4-2, the second has n degrees of freedom. In the terminology of the statistician we say that one degree of freedom is "lost" when the population mean μ is replaced by its "estimated value" $\bar{x}$.

While Theorem 5-4 states that $\sqrt{n}(\bar{x} - \mu)/\sigma$ and $(n - 1)s^2/\sigma^2$

are independent, it follows immediately that $\bar{x}$ and s^2, and also $\bar{x}$ and s are independent (Exercise 2).

EXERCISES

1. By changing the variables of integration in the multiple integral defining the joint distribution of $\sqrt{n}(\bar{x} - \mu)/\sigma$ and $(n - 1)s^2/\sigma^2$, show that the joint distribution is the same for all μ and σ—in particular, for $\mu = 0$ and $\sigma = 1$.

2. Show that Theorem 5-4 implies the independence of $\bar{x}$ and s^2, and of $\bar{x}$ and s.

3. Find the marginal distributions of s^2 and s, respectively. For the latter, use Exercise 3, Sec. 4-2.)

5-3. DISTRIBUTION OF THE QUOTIENT OF TWO INDEPENDENT STATISTICS

Several classical statistical procedures depend on ratios of independent statistics; for example, we shall see that confidence intervals and tests of significance may be based on the ratio of the Sample mean to the Sample standard deviation. For this reason it is convenient to have a general formula for the distribution of the ratio of two independent statistics. For simplicity we shall consider only ratios with nonnegative denominators; this will cover all our applications.

Before formally deriving the expression for the distribution, we have to consider the technical questions of regularity and integration. Suppose that f_1 and f_2 are independent statistics, $f_2 \geq 0$, and we want the distribution of the ratio f_1/f_2. Even though f_1 and f_2 are statistics (by Definition 4-1), it does not follow that their ratio is also a statistic: it is not necessarily continuous because the denominator may vanish at some points where the numerator does not; furthermore, it is not evident that the regularity of the sets $\{\mathbf{x}: f(\mathbf{x})_i \leq y_i\}$, $i = 1, 2$, implies that of the set

$$\{\mathbf{x}: f_1(\mathbf{x}) \leq y f_2(\mathbf{x})\} \tag{5-9}$$

for all y.

We are interested not in the general problem of quotients of statistics but only in several particular examples. In each of these we can independently verify that the set (5-9) is regular; therefore, the assumption of the regularity of this set will be included in the hypothesis of the general theorem to be stated. Now the definition of the *distribution* of a statistic (Definition 4-2) depends only on the *regularity* condition but not on the continuity of the statistic; the latter hypothesis is made only to simplify statements and proofs of theorems. For this reason the distribution of f_1/f_2 is defined if (5-9) is regular—even if f_2 vanishes somewhere. We begin with a simple example to illustrate the ideas of the derivation of the general case.

Example 5-3. For $n = 2$, and a standard normal population, put $f_1(x_1, x_2) = x_1$, and $f_2(x_1, x_2) = |x_2|$; these are independent (Theorem 5-2). We derive the distribution of the quotient $x_1/|x_2|$. For any real y, the set in the plane,

$$\{(x_1, x_2) \colon x_1 \leq |x_2| y\}$$

is regular (Fig. 5-3). Note that the inequality $x_1/|x_2| \leq y$ is the same as $x_1 \leq |x_2| y$ because $|x_2| \geq 0$.

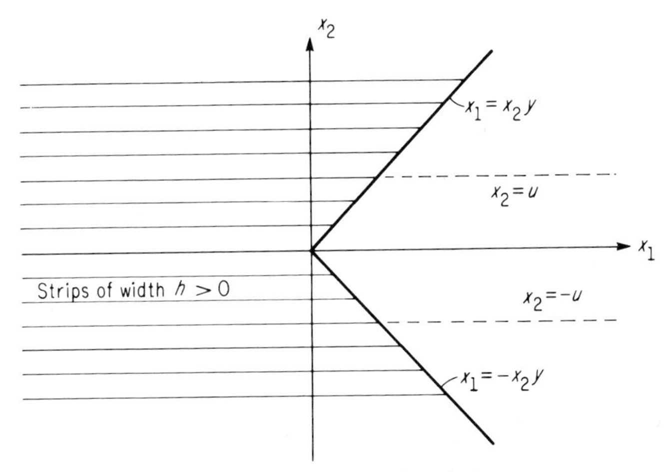

Fig. 5-3. The set $\{(x_1, x_2)\colon x_1 \leq |x_2| y\}$ for $y > 0$.

The distribution of $x_1/|x_2|$ at y is the double integral

$$F(y) = \frac{1}{2\pi} \iint_{\{(x_1, x_2)\colon x_1 \leq |x_2| y\}} e^{-(1/2)(x_1^2 + x_2^2)}\, dx_1\, dx_2. \tag{5-10}$$

The domain may be split up into a union of sets of the form

$$\{(x_1, x_2): x_1 \leq |x_2| y, \qquad u < |x_2| \leq u + h\}$$

where $h > 0$ and u is a positive multiple of h. This set includes

$$\{(x_1, x_2): x_1 \leq uy, \qquad u < |x_2| \leq u + h\}$$

and is included in

$$\{(x_1, x_2): x_1 \leq (u + h) y, \qquad u < |x_2| \leq u + h\}.$$

The integrals of $(2\pi)^{-1}e^{-(1/2)(x_1^2+x_2^2)}$ over these sets are equal to

$$\Phi(uy) \cdot 2[\Phi(u + h) - \Phi(u)] \tag{5-11}$$

and

$$\Phi((u + h) y) \cdot 2[\Phi(u + h) - \Phi(u)], \tag{5-12}$$

respectively (Exercise 2). It follows that the integral (5-10) is bounded above by the sum of terms (5-12) over u, and below by the sum of terms (5-11). Letting $h \to 0$, we find that the upper and lower sums have the common limit

$$2 \int_0^\infty \Phi(uy)\, \phi(u)\, du \tag{5-13}$$

because $\Phi(u + h) - \Phi(u) \sim \phi(u) h$; thus $F(y)$ is given by (5-13). Let us formally differentiate under the sign of integration to get the density:

$$2 \int_0^\infty u\phi(uy)\, \phi(u)\, du.$$

This is equal to

$$\frac{1}{\pi(1 + y^2)},$$

and is known as the *Cauchy density* (Exercise 3). ∎

THEOREM 5-5

Let f_1 and f_2 be independent statistics, $f_2 \geq 0$, having continuous densities F_1' and F_2', respectively. If, for every y, the set

$$\{\mathbf{x}: f_1(\mathbf{x}) \leq f_2(\mathbf{x}) y\} \tag{5-14}$$

is regular, then the integral of L over it is equal to

$$F(y) = \int_0^\infty F_1(uy)\,F_2'(u)\,du. \tag{5-15}$$

This is the distribution of f_1/f_2.

Proof: The set (5-14) is the union of sets of the form

$$\{\mathbf{x}: f_1(\mathbf{x}) \le y f_2(\mathbf{x}),\ nh < f_2(\mathbf{x}) \le (n+1)h\},$$
$$h > 0, \qquad \text{and} \qquad n \ge 0, \tag{5-16}$$

and the set

$$\{\mathbf{x}: f_1(\mathbf{x}) \le 0,\ \ f_2(\mathbf{x}) = 0\}. \tag{5-17}$$

The integral of L over (5-17) is 0 because f_2 has a density (Exercise 4). Put $u = nh$. The set (5-16) is included in

$$\{\mathbf{x}: f_1(\mathbf{x}) \le (u+h)y,\ \ u < f_2(\mathbf{x}) \le u+h\}$$

and includes

$$\{\mathbf{x}: f_1(\mathbf{x}) \le uy,\ \ u < f_2(\mathbf{x}) \le u+h\}.$$

Since f_1 and f_2 are independent the integral of L over the former set is the product of the integrals of L over $\{\mathbf{x}: f_1(\mathbf{x}) \le (u+h)y\}$ and $\{\mathbf{x}: u < f_2(\mathbf{x}) \le u+h\}$, respectively, namely

$$F_1((u+h)y)[F_2(u+h) - F_2(u)] \tag{5-18}$$

(Exercise 5). By the same reasoning the integral over the second set displayed above is

$$F_1(uy)[F_2(u+h) - F_2(u)]. \tag{5-19}$$

Sum (5-18) and (5-19) over all n, and let $h \to 0$: these have the common limit (5-15) because $F_2(u+h) - F_2(u) \sim hF_2'(u)$.

We are justified in calling (5-15) the *distribution of the ratio* because the integral over the portion (5-17) where the ratio is undefined is equal to 0. ■

Example 5-4. We shall find the distribution of the quotient

$$\frac{x_1}{[(x_2^2 + \cdots + x_n^2)/(n-1)]^{1/2}} \tag{5-20}$$

in sampling from a standard normal population. (Example 5-3 covers the case $n = 2$.) Although we shall not furnish the details, the regularity of the region

$$\{\mathbf{x}: x_1 \leq y[(x_2^2 + \cdots + x_n^2)/(n-1)]^{1/2}\}$$

can be established by the method in the proof of Theorem 5-5: the set is cut into thin "disks":

$$\{\mathbf{x}: x_1 \leq y[(x_2^2 + \cdots + x_n^2)/(n-1)]^{1/2}, \qquad u < x_2^2 + \cdots + x_n^2 \leq u + h\}$$

and the integrals of continuous functions are approximated by those over the regular sets

$$\{\mathbf{x}: x_1 \leq yu/\sqrt{n-1}, \quad u < x_2^2 + \cdots + x_n^2 \leq u + h\}.$$

Now we obtain the density of (5-20) by formal substitution in (5-15). The numerator x_1 has the distribution $\Phi(x)$. By Theorem 4-2, $x_2^2 + \cdots + x_n^2$ has the density $\psi_{n-1}(y)$ (chi-square with $n-1$ degrees of freedom); therefore

$$(x_2^2 + \cdots + x_n^2)/(n-1) \quad \text{has density} \quad (n-1)\psi_{n-1}((n-1)y)$$

(Exercise 1, Sec. 4-2); and so

$$[(x_2^2 + \cdots + x_n^2)/(n-1)]^{1/2} \quad \text{has the density}$$
$$2(n-1)y\psi_{n-1}((n-1)y^2) \qquad \text{(Exercise 3, Sec. 4-2).}$$

Since the numerator and denominator in (5-20) depend on different coordinates, they are independent (Theorem 5-2). The distribution of the quotient is

$$\int_0^\infty \Phi(uy)2(n-1)u\psi_{n-1}((n-1)u^2)\,du.$$

Differentiate under the sign of integration to get the density; this formal operation can be justified (Exercise 3). We obtain

$$2\int_0^\infty u^2(n-1)\phi(uy)\psi_{n-1}((n-1)u^2)\,du.$$

By substituting the particular forms of the functions in the integrand, and then applying the definition of the gamma function, we find the integral above to be

$$s_{n-1}(y) = \frac{1}{\sqrt{\pi(n-1)}} \frac{\Gamma(n/2)}{\Gamma((n-1)/2)} \left(1 + \frac{y^2}{n-1}\right)^{-n/2}$$

(Exercise 6). (5-21)

This is the density of "Student's distribution," or the "t-distribution," with $n - 1$ degrees of freedom. (See Fig. 5-4.) It was dis-

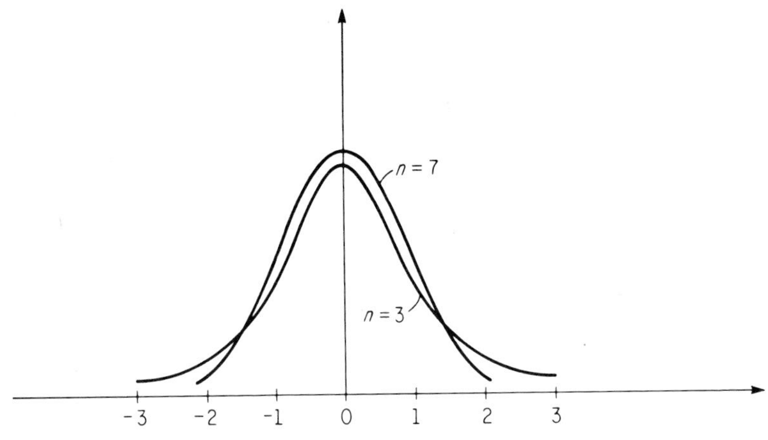

Fig. 5-4. t-density with $n = 3, 7$.

covered by William S. Gossett (1908) who wrote under the name of "Student." It is used in construction of tests and confidence intervals for the mean of a normal population with unknown variance (Chapters 8, 9), and in prediction (Chapter 12). Note that the Cauchy distribution, derived in Example 5-3, is the t-distribution with one degree of freedom. ■

Example 5-5. Let us determine the distribution of the quotient

$$\frac{(x_1^2 + \cdots + x_k^2)/k}{(x_{k+1}^2 + \cdots + x_n^2)/(n-k)} \tag{5-22}$$

in sampling from a standard normal population. By the same reasoning as in Example 5-4, the functions $x_1^2 + \cdots + x_k^2$ and $x_{k+1}^2 + \cdots + x_n^2$ are independent with densities ψ_k and ψ_{n-k}, respectively. By the same formal steps as before, we find the distribution of their quotient to be

$$\int_0^\infty \left(\int_0^{uy} \psi_k(z)\, dz \right) \psi_{n-k}(u)\, du.$$

The distribution of (5-22) is obtained from this expression by replacing y by $yk/(n - k)$ (Exercise 1, Sec. 4-2); the density is then obtained by formal differentiation with respect to y:

$$\int_0^\infty \frac{uk}{(n - k)} \psi_k(uy)\, \psi_{n-k}(u)\, du.$$

Substitute the particular forms of ψ, and integrate; we obtain

$$C \cdot y^{(k/2)-1} \left(1 + \frac{ky}{n - k}\right)^{n/2}, \tag{5-23}$$

where

$$C = \frac{\Gamma(n/2)}{\Gamma(k/2)} \frac{[k/(n - k)]^{k/2}}{\Gamma((n - k)/2)}$$

(Exercise 7). This is the density of the "F-distribution with k and $n - k$ degrees of freedom." (See Fig. 5-5.) The degrees of freedom

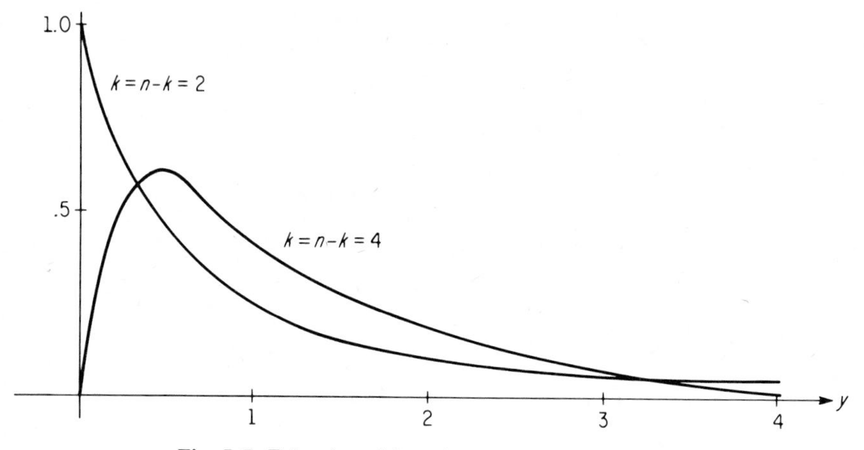

Fig. 5-5. *F*-density with various degrees of freedom.

refer to the numbers of squares in the numerator and denominator, respectively, in (5-22). It was named the F-distribution in honor of the famous English statistician, R. A. Fisher. It is used in the analysis of variance (Chapter 13). ■

EXERCISES

1. Show that the domain of integration in Example 5-3 is a regular set. (*Hint:* Theorem 3-9.)

2. Verify (5-11) and (5-12).

3. By actual computation of the limiting difference quotient, justify the formal operation of differentiation under the integration sign in (5-13). [*Hint:* If $y > 0$, $h > 0$, then

$$uh\phi(u(y + h)) < \Phi(u(y + h)) - \Phi(uy) < uh\phi(uy).$$

There is a similar inequality for $y < 0$.]

4. Show that the integral of L over (5-17) is equal to 0.

5. Use the defining equation of independence, Eq. 5-5, to verify (5-18) and (5-19).

6. Verify the expression (5-21).

7. Verify the expression (5-23) and the given form of C.

8. By further investigation of the gamma function it can be shown that

$$\frac{\Gamma(x + h)}{x^h \Gamma(x)} \to 1, \qquad x \to \infty$$

for fixed $h > 0$. Use this to show that the t-density with $n - 1$ degrees of freedom converges to the standard normal as $n \to \infty$.

9. Let f_1 and f_2 be independent statistics with densities F_1' and F_2', respectively. By modifying the proof of Theorem 5-5, show that the sum $f_1(\mathbf{x}) + f_2(\mathbf{x})$ has the distribution

$$F(y) = \int_{-\infty}^{\infty} F_1(y - u)\, F_2'(u)\, du.$$

5-4. JOINT DISTRIBUTION OF TWO OR MORE STATISTICS

Now we extend the concept of joint distribution of two statistics to that of several statistics. Let $f_i(\mathbf{x})$, $i = 1, \ldots, k$ be statistics. For any k-tuple of real numbers, we define the joint distribution of $f_1, \ldots, f_k$ at $(y_1, \ldots, y_k)$ as the integral of the likelihood function over the intersection of the sets

$$\{\mathbf{x}: f_i(\mathbf{x}) \leq y_1\}, \quad i = 1, \ldots, k.$$

This represents the probability that $\mathbf{X}$ falls in this intersection.

The joint distribution of $f_1, \ldots, f_{k-1}$ (f_k excluded) is obtained from the joint distribution of $f_1, \ldots, f_k$ by putting $y_k = +\infty$; this follows as in the particular case $k = 2$. More generally, the joint distribution of any subset $\{f_\alpha\}$ of the set $\{f_1, \ldots, f_k\}$ is obtained by putting $y_\beta = +\infty$ for all indices β not equal to one of the α's. This is the "consistency" property of the joint distribution.

The kth order partial derivative,

$$\frac{\partial^k F}{\partial y_1 \cdots \partial y_k},$$

when it exists, is called the *joint density* of $f_1, \ldots, f_k$. By similarity to the particular case $k = 2$, the joint density is the limit of the quotient of the integral of L over the intersection of the sets

$$\{\mathbf{x}: y_i < f_i(\mathbf{x}) \leq y_i + h\}$$

divided by h^k, $h \to 0$. The statistics $f_1, \ldots, f_k$ are called *mutually independent*, or, more briefly, *independent*, if their joint distribution is the product of their marginal distributions. Theorems 5-1, 5-2, and 5-3 have direct extensions to the case of several statistics. We shall not record the formal statement because the most general version needed here is for the case $k = 3$.

EXERCISES

1. What are the forms of the relations (5-2) for three statistics?

2. State Theorem 5-1 for three statistics.

5-5. JOINT DISTRIBUTIONS OF STATISTICS FOR A GENERAL POPULATION (OPTIONAL)

Suppose that $f_1(\mathbf{x})$ and $f_2(\mathbf{x})$ are statistics on a Sample space generated by a population with the density function $h(x)$. The definitions and many of the properties of statistics on the normal space extend to the general space: Theorems 5-1 and 5-2 are valid under the appropriate definition of the likelihood function. The only properties that do not carry over to the general case are those related to rotations (Theorems 5-3 and 5-4). These are based on the invariance of the normal likelihood function under rotation (Theorem 4-4). (It can also be shown that the normal Sample space is the *only* one for which Theorem 4-4 is true (Exercises 1, 2).) The technique of rotation is responsible for the unified and extensive development of normal sampling theory; by contrast, there are few general results on sampling from other populations.

EXERCISES

1. Let $h(x)$ be a density function such that the corresponding likelihood function $L(\mathbf{x}) = \prod_{i=1}^{n} h(x_i)$ depends on $\mathbf{x}$ through the single variable $\|\mathbf{x}\|^2$ for every n—that is, there exists a function of one variable, $g(t)$, such that

$$\prod_{i=1}^{n} h(x_i) = g\left(\sum_{i=1}^{n} x_i^2\right), \qquad n \geq 1.$$

Prove: $h(x) = g(x^2)$, and the equation

$$g(x + y) = g(x)g(y)$$

holds for all x and y.

2. It is known that the only positive functions $g(x)$, $x \geq 0$, which are bounded and which satisfy the last equation in Exercise 1 are those of the form $g(x) = be^{-cx}$, where $b \geq 0$ and $c \geq 0$. From this and from the result of Exercise 1 deduce that the normal density is the only bounded and continuous one whose likelihood function is invariant under rotations for all $n \geq 1$.

BIBLIOGRAPHY

A geometric derivation of the joint distribution of the Sample mean and variance is to be found in M. G. Kendall, *The Advanced Theory of Statistic* (London: Griffin, 1943, 1946).

Most other books which prove this theorem use the algebraic method or the moment-generating function.

CHAPTER 6

Spherical Averages, Sufficiency, and Square-Integrable Functions

6-1. THE SPHERICAL AVERAGE OF A CONTINUOUS FUNCTION

Much of the mathematical theory of statistics is about the analysis of multiple integrals over the Sample space. In Theorems 4-1 and 5-1 it is shown that certain multiple integrals are reducible to single or double integrals; these multiple integrals must be expressible as integrals of composite functions of one or two statistics. One of the most important tools of the theory of statistics is the concept of *sufficiency*. In the present setting it means that every multiple integral

$$\int_{R^n}\cdots\int f(\mathbf{x})\, L(\mathbf{x}; \mu, \sigma)\, d\mathbf{x}$$

may be expressed as the integral of a related function $\bar{f}$ of two variables,

$$\int_{R^n}\cdots\int \bar{f}(\sqrt{n}\bar{x}, (n-1)s^2)\, L(\mathbf{x}; \mu, \sigma)\, d\mathbf{x};$$

the latter, by Theorems 5-1 and 5-4, is reducible to a double integral of $\bar{f}(u, v)$ with respect to the normal and chi-square densities. This greatly simplifies the analysis.

The concept of sufficiency is based on the notion of the spherical average of a continuous function on R^n. It is similar to the "average of L" over surfaces defined by functions, described in Sec. 4-2 in connection with densities. In the latter case the function averaged was L, and the surface arbitrary; however, in the present case, the averaged function f is arbitrary but the surface is always a sphere about the origin. Although the concept of spherical average is well known in analysis, particularly potential theory, we shall now present a self-contained description of the elements we need.

Let $f(x_1, x_2)$ be a continuous function on the plane. For $r > 0$ and $h > 0$, consider the annular region between the circles of radii r and $r + h$, respectively, with centers at the origin; its area is $\pi((r + h)^2 - r^2)$. The integral of f over the region, in polar coordinates, is

$$\int_0^{2\pi} \int_r^{r+h} f(\rho \cos\theta, \rho \sin\theta)\, \rho\, d\rho\, d\theta.$$

Divide this by the area, and let $h \to 0$: the quotient converges to

$$(2\pi)^{-1} \int_0^{2\pi} f(r\cos\theta, r\sin\theta)\, d\theta = \bar{f}(r^2). \tag{6-1}$$

(Although the left-hand side depends on r, we write the function on the right-hand side as a function of r^2; this is permissible because $r \geq 0$.) $\bar{f}$ is a continuous function of r^2 (and r); indeed, $f(r \cos\theta, r\sin\theta)$ is continuous in r and θ, and $\bar{f}$ is obtained from f by integrating out one of the variables (Theorem 3-3). We call $\bar{f}(r^2)$ the spherical average of f over the two-dimensional sphere of radius r; it is actually a "circular average."

Our interest in $\bar{f}$ is based entirely on the following result. The integral of $\bar{f}$ over a disk about the origin is equal to the integral of the spherical average:

$$\iint_{\{x_1^2 + x_2^2 \leq r^2\}} f(x_1, x_2)\, dx_1\, dx_2 = \iint_{\{x_1^2 + x_2^2 \leq r^2\}} \bar{f}(x_1^2 + x_2^2)\, dx_1\, dx_2. \tag{6-2}$$

To prove this, change to polar coordinates and apply the definition (6-1). The right-hand side is

$$\int_0^{2\pi} \int_0^r \frac{1}{2\pi} \int_0^{2\pi} \left(f(\rho\cos\theta', \rho\sin\theta')\, d\theta' \right) \rho\, d\rho\, d\theta;$$

the integrand does not depend on θ; thus, integrating over θ, we obtain

$$\int_0^r \int_0^{2\pi} f(\rho\cos\theta', \rho\sin\theta')\, d\theta'\, \rho\, d\rho,$$

which is equal to the left-hand side of (6-2) in (ρ, θ') coordinates. If f is a continuous function on the plane such that

$$\int_{-\infty}^{\infty} \int_{-\infty}^{\infty} |f(x_1, x_2)| \, dx_1 \, dx_2 < \infty,$$

then, by letting $r \to \infty$ in (6-2), we obtain

$$\int_{-\infty}^{\infty} \int_{-\infty}^{\infty} f(x_1, x_2) \, dx_1 \, dx_2 = \int_{-\infty}^{\infty} \int_{-\infty}^{\infty} \bar{f}(x_1^2 + x_2^2) \, dx_1 \, dx_2.$$

Now we extend this to spherical averages over R^n. Let $f(\mathbf{x})$ be continuous, and let $F_n(r)$ be its integral over the ball of radius r (Sec. 3-3). The difference $F_n(r + h) - F_n(r)$, $h > 0$, represents the integral of f over the region between the spheres of radii $r + h$ and h, respectively. The volume of this region is the difference of the volumes of the corresponding balls; in accordance with the notation of Sec. 4-3, it is equal to $C_n[(r + h)^n - r^n]$, where C_n is the volume of the unit ball. Form the quotient

$$\frac{F_n(r + h) - F_n(r)}{C_n[(r + h)^n - r^n]};$$

divide the numerator and denominator by h; and then let $h \to 0$. The quotient converges to

$$\frac{F_n'(r)}{nC_n r^{n-1}} = \bar{f}(r^2), \tag{6-3}$$

which we call the *spherical average of f over the sphere of radius r.* One point has to be justified: the differentiability of $F_n(r)$ with respect to r. In the case $n = 2$, $F_2'(r)$ is equal to $2\pi\bar{f}(r^2)$, which, as we have shown, is continuous. The continuous differentiability of $F_n(r)$ for $n > 2$ follows by induction from the recursion formula (3-15) (Exercise 4) because F_n is expressed as an integral of F_{n-1}; in particular, $\bar{f}$ is continuous.

The version of Eq. 6-2 for R^n is

$$\underset{\{\mathbf{x}: \|\mathbf{x}\| \le r\}}{\int \cdots \int} f(\mathbf{x}) \, d\mathbf{x} = \underset{\{\mathbf{x}: \|\mathbf{x}\| \le r\}}{\int \cdots \int} \bar{f}(\|\mathbf{x}\|^2) \, d\mathbf{x}. \tag{6-4}$$

Since the integral over the ball is representable as the sum of integrals over the thin concentric shells $\{\mathbf{x}: \rho < \|\mathbf{x}\| \le \rho + h\}$ composing the ball, it suffices to compare the integrals of f and

$\bar{f}$ over such a shell in order to prove (6-4). By definition, we have

$$\int \cdots \int_{\{\mathbf{x}:\rho<\|\mathbf{x}\|\leq\rho+h\}} f(\mathbf{x})\,d\mathbf{x} = F_n(\rho + h) - F_n(\rho),$$

which, for small h, is approximately $hF_n'(\rho)$. The integral

$$\int \cdots \int_{\{\mathbf{x}:\rho<\|\mathbf{x}\|\leq\rho+h\}} \bar{f}(\|\mathbf{x}\|^2)\,d\mathbf{x}$$

is at most equal to

$$\max_{\rho\leq y\leq\rho+h} \bar{f}(y^2)\cdot C_n[(\rho + h)^n - \rho^n],$$

and at least equal to

$$\min_{\rho\leq y\leq\rho+h} \bar{f}(y^2)\cdot C_n[(\rho+h)^n - \rho^n].$$

Since $\bar{f}$ is continuous, each of these expressions is approximately

$$\bar{f}(\rho^2)\cdot C_n[(\rho + h)^n - \rho^n],$$

which, by (6-3), is approximately $hF_n'(\rho)$; thus (6-4) is verified.

If $f(\mathbf{x})$ is a continuous function such that

$$\int \cdots \int_{R^n} |f(\mathbf{x})|\,d\mathbf{x} < \infty,$$

then, by letting $r \to \infty$ in (6-4), we obtain

$$\int \cdots \int_{R^n} f(\mathbf{x})\,d\mathbf{x} = \int \cdots \int_{R^n} \bar{f}(\|\mathbf{x}\|^2)\,d\mathbf{x}. \tag{6-5}$$

Our first application is to multiple integrals of continuous functions over the standard normal Sample space,

$$\int \cdots \int_{R^n} f(\mathbf{x})(2\pi)^{-n/2}e^{-(1/2)\|\mathbf{x}\|^2}\,d\mathbf{x}.$$

Since the exponential depends only on the distance of $\mathbf{x}$ from the origin, the spherical average of the integrand above over a sphere of radius r is equal to

$$2\pi^{-n/2}\,\bar{f}(r^2)\,e^{-(1/2)r^2};$$

this follows from the defining relation (6-3) (Exercise 5). Apply

(6-5) to $f(\mathbf{x})\,e^{-(1/2)\|\mathbf{x}\|^2}$ in place of $f(\mathbf{x})$:

$$\int_{R^n}\cdots\int f(\mathbf{x})(2\pi)^{-n/2}e^{-(1/2)\|\mathbf{x}\|^2}\,d\mathbf{x}$$

$$= \int_{R^n}\cdots\int \bar{f}(\|\mathbf{x}\|^2)(2\pi)^{-n/2}e^{-(1/2)\|\mathbf{x}\|^2}\,d\mathbf{x}. \tag{6-6}$$

Apply Theorem 4-1 with $\bar{f}$ and $\|\mathbf{x}\|^2$ in place of g and f, respectively:

$$\int\cdots\int \bar{f}(\|\mathbf{x}\|^2)(2\pi)^{-n/2}e^{-(1/2)\|\mathbf{x}\|^2}\,d\mathbf{x}$$

$$= \int_0^\infty \bar{f}(u)\,\psi_n(u)\,du; \tag{6-7}$$

indeed, by Theorem 4-2, the statistic $\|\mathbf{x}\|^2$ has the chi-square distribution with n degrees of freedom in sampling from a standard normal population. We see that every multiple integral of the type appearing on the left-hand side of (6-6) is reducible to the single integral on the right-hand side of (6-7) involving the density of $\|\mathbf{x}\|^2$. This result is hardly changed when the standard deviation is σ (not necessarily 1) but the mean still 0:

$$\int_{R^n}\cdots\int f(\mathbf{x})(2\pi\sigma^2)^{-n/2}e^{-\|\mathbf{x}\|^2/2\sigma^2}\,d\mathbf{x}$$

$$= \int_{R^n}\cdots\int \bar{f}(\|\mathbf{x}\|^2)(2\pi\sigma^2)^{-n/2}e^{-\|\mathbf{x}\|^2/2\sigma^2}\,d\mathbf{x}$$

$$= \int_0^\infty \bar{f}(u)\,\psi_n\left(\frac{u}{\sigma^2}\right)\frac{du}{\sigma^2}. \tag{6-8}$$

The last equation is obtained by writing

$$\bar{f}(\|\mathbf{x}\|^2) = \bar{f}\left(\frac{\sigma^2\|\mathbf{x}\|^2}{\sigma^2}\right),$$

applying the extended version of Theorem 4-2 to $\|\mathbf{x}\|^2/\sigma^2$, and then changing the variable from u to $u\sigma^2$ in the third integral in (6-8). This equation is the precise form of the statistical principle that $\|\mathbf{x}\|^2$ is a "sufficient statistic" in sampling from a normal population with mean 0 and arbitrary standard deviation σ.

EXERCISES

1. Find the explicit double integral defining the spherical average in R^3.

2. Repeat Exercise 1 for R^4.

3. Show by the argument supporting (6-4) that

$$\underset{R^n}{\int \cdots \int} \bar{f}(\|\mathbf{x}\|^2)\, d\mathbf{x} = nC_n \int_0^\infty r^{n-1}\bar{f}(r^2)\, dr.$$

4. By an examination of the limit of the difference quotient show that Eq. 3-15 may be differentiated on both sides with respect to r, and becomes

$$F_n'(r) = \int_{-r}^{r} F_{n-1}'(\sqrt{r^2 - z^2}, z) \frac{r}{\sqrt{r^2 - z^2}}\, dz$$

where $F_{n-1}'(u, z) = (\partial/\partial u) F_{n-1}(u, z)$; start with $n = 3$, and proceed by induction.

5. Let $f(\mathbf{x})$ and $g(\mathbf{x})$ be continuous functions such that g depends on $\mathbf{x}$ only through $\|\mathbf{x}\|^2$: $g = g(\|\mathbf{x}\|^2)$. Show that the spherical average of $f(\mathbf{x}) \cdot g(\|\mathbf{x}\|^2)$ is equal to $\bar{f}(r^2) \cdot g(r^2)$.

6. Show that spherical averaging is a *linear operation*: if f_1 and f_2 are functions, and α and β numbers, then the spherical average of the linear combination $\alpha f_1(\mathbf{x}) + \beta f_2(\mathbf{x})$ is equal to $\alpha \bar{f_1} + \beta \bar{f_2}$.

6-2. CONDITIONAL EXPECTATION AND SUFFICIENCY

Now we generalize the previous results to integrals of functions over the general normal Sample space—with mean μ and standard deviation σ. We want to reduce the integral

$$\underset{R^n}{\int \cdots \int} f(\mathbf{x})\, L(\mathbf{x}; \mu, \sigma)\, d\mathbf{x} \tag{6-9}$$

to a double integral of a function $\bar{f}$ of two variables.

Let $\mathbf{x}'$ be the image of $\mathbf{x}$ under the diagonal rotation (the one carrying $\mathbf{1}$ onto $\sqrt{n}\mathbf{e}_1$); and let f^* be the function related to f by

$$f^*(\mathbf{x}') = f(\mathbf{x}).$$

(Such a function was defined in Theorem 5-3, and also in the proofs of Theorems 4-4 and 5-4). The coordinates of $\mathbf{x}$ and $\mathbf{x}'$ are related by (5-8):

$$\sqrt{n}\bar{x} = x_1', \qquad \sum_{i=1}^{n} (x_i - \bar{x})^2 = \sum_{i=2}^{n} x_i'^2.$$

For a fixed value x_1' of the first coordinate of $\mathbf{x}'$, f^* is a continuous function of $x_2', \ldots, x_n'$. Let $\bar{f}(x_1', r^2)$ be the spherical average over the $(n - 1)$-dimensional sphere $\left\{(x_2', \ldots, x_n'): \sum_{i=2}^{n} x_i'^2 = r^2\right\}$, of radius r. Replace r^2 by $\sum_{i=2}^{n} x_i'^2$: $\bar{f}\left(x_1', \sum_{i=2}^{n} x_i'^2\right)$; then replace the arguments by their equivalents in terms of the coordinates of $\mathbf{x}$: by the equations above, we have

$$\bar{f}\left(x_1', \sum_{i=2}^{n} x_i'^2\right) = \bar{f}\left(\sqrt{n}\bar{x}, \sum_{i=1}^{n} (x_i - \bar{x})^2\right)$$
$$= \bar{f}(\sqrt{n}\bar{x}, (n - 1)s^2).$$

The last expression depends on $\mathbf{x}$ through the pair of functions $\bar{x}$ and s^2; it is called the "conditional expectation of f given $\sqrt{n}\bar{x}$ and $(n - 1)s^2$." Note that the likelihood function is not involved in this definition: $\bar{f}$ is simply a spherical average over certain variables. For the purpose of reference, we formally restate the definition of $\bar{f}$.

DEFINITION 6-1

For a continuous function f on R^n, the conditional expectation of f given $\sqrt{n}\bar{x}$ and $(n - 1)s^2$ is the function $\bar{f} = \bar{f}(\sqrt{n}\bar{x}, (n - 1)s^2)$ derived from f by

(a) defining f^* as $f^*(\mathbf{x}') = f(\mathbf{x})$, where $\mathbf{x}'$ is the image of $\mathbf{x}$ under the diagonal rotation; and

(b) averaging f^*, as a function of $x_2', \ldots, x_n'$ over the sphere of radius $\sum_{i=2}^{n} x_i'^2$; and

(c) replacing x_1' and $\sum_{i=2}^{n} x_i'^2$ by their equivalents, $\sqrt{n}\bar{x}$ and $(n - 1)s^2$, respectively.

The association between a continuous function and its conditional expectation is exploited through the following theorem; its proof requires almost all of the important results derived in the previous chapters.

THEOREM 6-1
The integral (6-9) is unchanged when $\bar{f}(\sqrt{n}\,\bar{x}, (n-1)s^2)$ is substituted for f: it is equal to

$$\int_{R^n}\cdots\int \bar{f}(\sqrt{n}\,\bar{x}, (n-1)s^2)L(\mathbf{x}; \mu, \sigma)\,d\mathbf{x}; \qquad (6\text{-}10)$$

furthermore, the latter is equal to

$$\int_0^{\infty}\int_{-\infty}^{\infty} \bar{f}(u, v)\,\frac{1}{\sigma}\,\phi\left(\frac{u-\sqrt{n}\mu}{\sigma}\right)\frac{1}{\sigma^2}\,\psi_{n-1}\left(\frac{v}{\sigma^2}\right)du\,dv. \qquad (6\text{-}11)$$

This is valid for all μ and $\sigma > 0$.

Proof: Recall that

$$L(\mathbf{x}; \mu, \sigma) = (2\pi\sigma^2)^{-n/2}\exp\left[(-1/2\sigma^2)\sum_{i=1}^{n}(x_i-\mu)^2\right].$$

Apply the sum of squares decomposition (4-13) to the exponent, which may then be written as

$$(1/2\sigma^2)\left[\sum_{i=1}^{n}(x_i-\bar{x})^2 + n(\bar{x}-\mu)^2\right].$$

By equations (5-8) relating the coordinates of $\mathbf{x}$ and $\mathbf{x}'$, the expression above is

$$(1/2\sigma^2)\left[\sum_{i=2}^{n}x_i'^2 + (x_1' - \sqrt{n}\mu)^2\right].$$

Substitute this in the integrand in (6-9), and replace $f(\mathbf{x})$ by $f^*(\mathbf{x}')$:

$$\int_{R^n}\cdots\int f^*(\mathbf{x}')(2\pi\sigma^2)^{-n/2}\exp\left[(-1/2\sigma^2)\sum_{i=2}^{n}x_i'^2\right]$$
$$\cdot\exp\left[(-1/2\sigma^2)(x_1'-\sqrt{n}\mu)^2\right]d\mathbf{x}.$$

The integral is unchanged when $\mathbf{x}$ is substituted for $\mathbf{x}'$ in the integrand:

$$\int \cdots \int_{R^n} f^*(\mathbf{x})(2\pi\sigma^2)^{-n/2} \exp\left[(-1/2\sigma^2)\sum_{i=2}^{n} x_i^2\right]$$

$$\cdot \exp\left[(-1/2\sigma^2)(x_1 - \sqrt{n}\mu)^2\right] d\mathbf{x};$$

this is a consequence of the invariance of integration under rotation (Theorem 3-4). Fix x_1, integrate over $x_2, \ldots, x_n$, and apply Eq. 6-5: the exponential is constant over spheres and so the integral is equal to

$$\int \cdots \int_{R^n} \bar{f}(x_1, x_2^2 + \cdots + x_n^2)(2\pi\sigma^2)^{-n/2}$$

$$\cdot \exp\left[(-1/2\sigma^2)\left(\sum_{i=2}^{n} x_i^2 + (x_1 - \sqrt{n}\mu)^2\right)\right] d\mathbf{x}. \qquad (6\text{-}12)$$

Again, by Theorem 3-4, we may replace $\mathbf{x}$ by $\mathbf{x}'$ as the variable in the integrand. Reversing the substitutions above and replacing the coordinates of $\mathbf{x}'$ by their equivalents in terms of $\mathbf{x}$, we find that $\bar{f}$ becomes $\bar{f}(\sqrt{n}\bar{x}, (n-1)s^2)$ and the integral (6-12) is equal to (6-10). This proves the equality of the integrals (6-9) and (6-10).

By Theorem 5-4, $\sqrt{n}(\bar{x} - \mu)/\sigma$ and $(n-1)s^2/\sigma^2$ are independent and have normal and chi-square distributions, respectively. By the remarks following the proof of that theorem, and Exercises 2 and 3, Sec. 5-2, it follows that $\sqrt{n}\bar{x}$ and $(n-1)s^2$ are also independent, the former having a normal distribution with mean $\sqrt{n}\mu$ and standard deviation σ, and the latter the density $(1/\sigma^2)\psi_{n-1} \cdot (v/\sigma^2)$. This and Theorem 5-1 (on the reduction of the integral of a composite function) now imply that the multiple integral (6-10) is equal to the double integral (6-11). ■

Theorem 6-1 is the precise statement of the statistical principle that the Sample mean and variance are "sufficient statistics": For any function f on the Sample space, there is a function $\bar{f}$ depending on $\mathbf{x}$ through $\bar{x}$ and s^2 alone such that the integral of $f \cdot L$ is equal to that of $\bar{f} \cdot L$, and this is true for all values of μ and $\sigma > 0$. The significance of this is: In maximizing or minimizing the integral (6-9) with respect to a class of continuous functions f, we can restrict f to the subclass of functions depending only on $\bar{x}$ and s^2; furthermore, the integral assumes the simpler form (6-11).

A major point of Theorem 6-1 is that $\bar{f}$ does not depend on the parameter values μ and σ. There is a variant of the conditional expectation and of Theorem 6-1 which is useful in the situation in which σ is known to be a certain fixed number but μ is unknown. Define

$$\bar{f}(u;\sigma) = \int_0^\infty \bar{f}(u,y)\,\psi_{n-1}\left(\frac{y}{\sigma^2}\right)\frac{dy}{\sigma^2}; \tag{6-13}$$

we call $\bar{f}(\sqrt{n}\bar{x};\sigma)$ the "conditional expectation of f, given $\sqrt{n}\bar{x}$, when σ is known." The following extension of Theorem 6-1 holds. For any continuous function f on R^n,

$$\underset{R^n}{\int\cdots\int} f(\mathbf{x})\,L(\mathbf{x};\mu,\sigma)\,d\mathbf{x} = \underset{R^n}{\int\cdots\int} \bar{f}(\sqrt{n}\bar{x};\sigma)\,L(\mathbf{x};\mu,\sigma)\,d\mathbf{x}$$

$$= \int_{-\infty}^{\infty} \bar{f}(x;\sigma)\,\frac{1}{\sigma}\,\phi\left(\frac{x-\sqrt{n}\mu}{\sigma}\right)dx; \tag{6-14}$$

this is a consequence of Theorem 6-1 (Exercise 4). The significance of (6-14) is that for any function f there is one $\bar{f}$ depending on $\mathbf{x}$ only through $\sqrt{n}\bar{x}$ such that the integral of $f\cdot L$ is equal to that of $\bar{f}\cdot L$; however, $\bar{f}$ depends on σ. This means that "$\sqrt{n}\bar{x}$ is a sufficient statistic when σ is known."

EXERCISES

1. Why is the relation between a function and its conditional expectation linear? (See Exercise 6, Sec. 6-1.)

2. Let g be a function on R^n depending on $\mathbf{x}$ only through the two functions $\sqrt{n}\bar{x}$ and $(n-1)s^2$. *Prove:* $\bar{g} = g$.

3. Let g be a function like the one in Exercise 2, and $f(\mathbf{x})$ an arbitrary function. *Prove:* The conditional expectation of the product $f(\mathbf{x})g(\mathbf{x})$ is $\bar{f}\cdot g$.

4. Prove the validity of (6-14).

6-3. SQUARE-INTEGRABLE FUNCTIONS

Let $f(\mathbf{x})$ be a continuous function on R^n. We say that f is square-integrable if

$$\int \cdots \int_{R^n} |f(\mathbf{x})|^2 L(\mathbf{x}; \mu, \sigma)\, d\mathbf{x} < \infty \tag{6-15}$$

for all μ and $\sigma > 0$. When convenient we shall write this integral in the concise form

$$\int |f|^2 L\, d\mathbf{x}.$$

This class of functions has a central position in the theory of "efficient estimation" to be discussed in Chapter 7. Square-integrable functions are dominant in many areas of analysis—for example, Fourier analysis. In this section we shall state and prove the well-known Cauchy-Schwarz inequality and its converse, and then prove that $\bar{f}$ is square-integrable if f is.

The following is the "continuous" analogue of the inequality proved in Exercise 9, Sec. 3-1:

THEOREM 6-2 (Cauchy-Schwarz Inequality)
Let f and g be square-integrable functions; then

$$\int \cdots \int_{R^n} |f(\mathbf{x})g(\mathbf{x})|\, L(\mathbf{x}; \mu, \sigma) d\mathbf{x} < \infty$$

and

$$\left(\int fgL\, d\mathbf{x}\right)^2 \leq \left(\int |f|^2 L\, d\mathbf{x}\right)\left(\int |g|^2 L\, d\mathbf{x}\right). \tag{6-16}$$

Proof: If f and g are square-integrable then so are $|f|$ and $|g|$; therefore, the second inequality implies the first; hence, it suffices to prove (6-16).

If f is square integrable, then the conditions

$$\int |f|^2 L\, d\mathbf{x} > 0 \tag{6-17}$$

and

$$f(\mathbf{x}) \neq 0 \qquad \text{for some } \mathbf{x} \tag{6-18}$$

are equivalent. On the one hand, if (6-18) does not hold—that is, f is identically equal to 0—then the integral in (6-17) must also be 0; on the other hand, if (6-18) is valid, then, as a continuous function, $|f(\mathbf{x})|^2$ is bounded away from 0 in some ball containing

$\mathbf{x}$, and so $|f|^2L$ has the same property and its integral is positive.

If f is identically equal to 0, then the inequality (6-16) is trivially true; thus we shall assume that it is not so—that (6-17) holds. For any real number λ, the integral

$$\int_{-\infty}^{\infty} (\lambda f + g)^2 L \, d\mathbf{x} \tag{6-19}$$

is nonnegative because the integrand is; thus the quadratic form

$$\lambda^2 \int |f|^2 L \, d\mathbf{x} + 2\lambda \int fgL \, d\mathbf{x} + \int |g|^2 L \, d\mathbf{x}$$
$$= \lambda^2 A + 2\lambda B + C \tag{6-20}$$

is nonnegative; therefore, the equation

$$\lambda^2 A + 2\lambda B + C = 0 \tag{6-21}$$

has at most one real solution. There are no real solutions if and only if

$$B^2 < AC$$

which is equivalent to the strict inequality form of (6-16). Equation 6-21 has one real solution if and only if

$$B^2 = AC; \tag{6-22}$$

in this case the solution is

$$\lambda_0 = -B/2A, \tag{6-23}$$

which is well defined because

$$A = \int |f|^2 L \, d\mathbf{x} > 0$$

by (6-17). Equation 6-22 signifies strict equality in (6-16); thus the latter holds in any case. ■

Here is a partial converse:

THEOREM 6-3

If f and g are square-integrable, and (6-17) holds, then there is strict equality in (6-16) if and only if there is a real number λ_0 such that

$$\lambda_0 f(\mathbf{x}) + g(\mathbf{x}) = 0 \qquad \textit{for all } \mathbf{x}. \tag{6-24}$$

Proof: By the proof of Theorem 6-2, there is strict equality in (6-16) if and only if Eq. 6-22 holds, in which case (6-21) has the unique solution λ_0, defined in (6-23); thus the quadratic form (6-20) vanishes for $\lambda = \lambda_0$. This means that the integral (6-19) is equal to 0 if $\lambda = \lambda_0$. Now apply the equivalence of the conditions (6-17) and (6-18), with $\lambda_0 f + g$ in place of f: the integral (6-19) vanishes for $\lambda = \lambda_0$ if and only if (6-24) holds. ■

A simple consequence of (6-16) is

$$\left(\int |f| \; L \, d\mathbf{x}\right)^2 \leq \int |f|^2 L \, d\mathbf{x}; \tag{6-25}$$

indeed, put $|f|$ and 1 in place of f and g, respectively. An important result is: If f is square-integrable, then the integral

$$\int fL \, d\mathbf{x}$$

is well defined and finite because the integral of $|f| \cdot L$ is. (See the remarks following Definition 4-3.)

There is a connection between square-integrability and spherical averages:

THEOREM 6-4

If f is a continuous function and $\bar{f}$ is its spherical average, then

$$(\bar{f})^2 \leq \overline{f^2}; \tag{6-26}$$

that is, the square of the spherical average is not more than the spherical average of the square of f.

Proof: The proof is based on a variation of the Cauchy-Schwarz inequality (6-16):

$$\left(\int_M \cdots \int f(\mathbf{x}) g(\mathbf{x}) \, d\mathbf{x}\right)^2 \leq \left(\int_M \cdots \int f^2(\mathbf{x}) \, d\mathbf{x}\right)\left(\int_M \cdots \int g^2(\mathbf{x}) \, d\mathbf{x}\right) \tag{6-27}$$

for any regular set M. The proof of this inequality is the same as that of (6-16) except that R^n and L are replaced by M and 1, respectively. In (6-27) take g to be the function identically equal to 1, and M the shell region $\{\mathbf{x}: r < \|\mathbf{x}\| \le r + h\}$, $r > 0$, $h > 0$. Divide each side of (6-27) by

$$\left(\int_M \cdots \int g^2(\mathbf{x})\, d\mathbf{x}\right)^2 = \left(\int_M \cdots \int d\mathbf{x}\right)^2 = (\text{volume of } M)^2,$$

and let $h \to 0$: the left-hand side of (6-27) converges to the square of the spherical average of f, and the right-hand side to the spherical average of $f^2(\mathbf{x})$. ■

THEOREM 6-5

If f is square-integrable, then so is $\bar{f}$, its conditional expectation, and

$$\int |\bar{f}|^2 L\, d\mathbf{x} \le \int |f|^2 L\, d\mathbf{x}.$$

Proof: We give the proof for the conditional expectation given $\sqrt{n}\bar{x}$ and $(n-1)s^2$ (Definition 6-1). Since the latter is a spherical average, Theorem 6-4 implies that $|\bar{f}|^2 \le (\overline{f^2})$. From this and the order preserving property of integration (3-16) we obtain:

$$\int |\bar{f}|^2 L\, d\mathbf{x} \le \int (\overline{f^2})\, L\, dx.$$

By Theorem 6-1 the integral on the right-hand side is unchanged if the "bar" is removed from f^2.

The proof for the conditional expectation given $\sqrt{n}\bar{x}$ is left to the reader (Exercises 9, 10). ■

The final result of this section will be applied to unbiased estimation (Chapter 7) and unbiased testing (Chapter 9).

THEOREM 6-6

Let $f(x)$ be a square-integrable function on R^1; then the integral

$$\int_{-\infty}^{\infty} f(x)\, \frac{1}{\sigma}\, \phi\left(\frac{x-\mu}{\sigma}\right) dx \tag{6-28}$$

is differentiable as a function of the parameter μ. The derivative is obtained by differentiating under the sign of integration, and is equal to

$$\int_{-\infty}^{\infty} f(x)\left(\frac{1}{\sigma}\right)^{3}(x-\mu)\phi\left(\frac{x-\mu}{\sigma}\, dx.\right) \tag{6-29}$$

Proof: It is sufficient to consider the case $\sigma = 1$ (Exercise 7). For $h \neq 0$, the difference quotient of (6-28) is

$$h^{-1}\int_{-\infty}^{\infty} f(x)\,\phi(x-\mu)[e^{h(x-\mu)-(1/2)h^2} - 1]\, dx, \tag{6-30}$$

by virtue of the relation

$$\phi(x+h) = \phi(x)\, e^{hx-(1/2)h^2} \tag{6-31}$$

(Exercise 2). The difference between (6-30) and (6-29) is

$$\int_{-\infty}^{\infty} f(x)\,\phi(x-\mu)[h^{-1}(e^{h(x-\mu)-(1/2)h^2} - 1) - (x-\mu)]\, dx,$$

which, by a change of variable, is

$$\int_{-\infty}^{\infty} f(x+\mu)\,\phi(x)[h^{-1}(e^{hx-(1/2)h^2} - 1) - x]\, dx. \tag{6-32}$$

By application of the formula (2-8) it is found that

$$\int_{-\infty}^{\infty} \phi(x)[h^{-1}(e^{hx-(1/2)h^2} - 1) - x]^2\, dx = h^{-2}(e^{h^2} - 1) - 1 \tag{6-33}$$

(Exercise 4). Now apply the Cauchy-Schwarz inequality (6-16) to the integral (6-32) with the identifications:

$$f(x+\mu) \to f(\mathbf{x}), \qquad \phi(x) \to L,$$
$$[h^{-1}(e^{hx-(1/2)h^2} - 1) - x] \to g;$$

then, the absolute value of (6-32) is at most equal to the square root of the product of

$$\int_{-\infty}^{\infty} |f(x)|^2\,\phi(x-\mu)\, dx$$

and (6-33). The latter converges to 0 as $h \to 0$; therefore, the difference (6-32) also does. ■

EXERCISES

1. *Prove:* If $f(\mathbf{x})$ is square-integrable, then so is $f(a\mathbf{x} + b\mathbf{1})$ for any real numbers a and b.

2. Verify (6-31).

3. *Prove:*

$$\int_{-\infty}^{\infty} xe^{tx}\phi(x)\,dx = te^{t^2/2}.$$

(*Hint:* Integrate by parts in Eq. 2-9.)

4. Verify (6-33) using the previous exercise and Eq. 2-9.

5. Using the proof of Theorem 6-6 as a model, prove that (6-28) is also differentiable with respect to σ^2 under the sign of integration. (Write σ as $\sqrt{\sigma^2}$.)

6. Prove the "discrete" version of (6-26): For any real numbers $x_1, \ldots, x_n$,

$$|\bar{x}|^2 \leq \frac{1}{n}(|x_1|^2 + \cdots + |x_n|^2).$$

7. By changing the variable of integration in (6-28) show that it is sufficient to prove Theorem 6-6 just for the case $\sigma = 1$.

8. *Prove:* If $\bar{f}$ is the conditional expectation of f given $\sqrt{n}\bar{x}$ and $(n-1)s^2$, then for any real number c,

$$\int |\bar{f} - c|^2 L\,d\mathbf{x} \leq \int |f - c|^2 L\,d\mathbf{x}.$$

(*Hint:* Use Theorem 6-4 and Exercise 1, Sec. 6-2.)

9. Let $\bar{f}(\sqrt{n}\bar{x};\sigma)$ be the conditional expectation of f given $\sqrt{n}\bar{x}$ when σ is known, defined by (6-13). Show that

$$|\bar{f}(u;\sigma)|^2 \leq \int_0^{\infty} (\bar{f}(u,y))^2 \psi_{n-1}(y/\sigma^2)\,dy/\sigma^2.$$

(*Hint:* Use an appropriate variation of the Cauchy-Schwarz inequality.)

10. Using the result of Exercise 9, complete the proof of Theorem 6-5 for the latter kind of conditional expectation.

11. Extend the result of Exercise 8 to the latter kind of conditional expectation.

6-4. SUFFICIENCY AND SQUARE-INTEGRABLE FUNCTIONS FOR MORE GENERAL SAMPLE SPACES (OPTIONAL)

The theory of sufficient statistics in sampling from a general family of populations can be completely and rigorously described only with either measure theory or at least advanced multidimensional integration theory; by contrast, the theory for the normal space in the previous sections demanded only spherical averages and invariance of integration under rotation. We shall now briefly give the definition and fundamental theory for the general case by indicating the necessary extensions of the reasoning in the normal case.

Let a population have a density $h(x;\theta)$ where θ is a parameter of one or more dimensions; put

$$L(\mathbf{x};\theta) = \prod_{i=1}^{n} h(x_i;\theta). \tag{6-34}$$

The function $\lambda(\mathbf{x})$ is said to be a sufficient statistic for the parameter θ if $L(\mathbf{x};\theta)$ can be factored into a function of $\mathbf{x}$ and a function of $\lambda(\mathbf{x})$ and θ:

$$L(\mathbf{x};\theta) = M(\mathbf{x})\,K(\lambda(\mathbf{x}),\theta), \tag{6-35}$$

where K depends on $\mathbf{x}$ only through $\lambda(\mathbf{x})$. (Sufficiency is usually defined by means of conditional distributions; however, the Fisher-Neyman criterion (see Bibliography) states that the equation above characterizes sufficiency.) If there is a set of functions $\lambda_1, \ldots, \lambda_k$ such that $K = K(\lambda_1(\mathbf{x}), \ldots, \lambda_k(\mathbf{x}), \theta)$, then $\{\lambda_1(\mathbf{x}), \ldots, \lambda_k(\mathbf{x})\}$ is called a *set* of sufficient statistics.

Example 6-1. Suppose $\theta = \mu$ and

$$h(x;\mu) = \phi(x - \mu).$$

The likelihood function is (as in the proof of Theorem 6-1)

$$(2\pi)^{-n/2} \exp\left(-\tfrac{1}{2}\|\mathbf{x}\|^2\right) \exp\left(n\bar{x}^2 - \tfrac{1}{2}n\mu^2\right).$$

Put $\lambda(\mathbf{x}) = \bar{x}$; then the likelihood factors as in (6-35) so that $\bar{x}$ is a sufficient statistic. If $\theta = \sigma$ and

$$h(x;\sigma) = \frac{1}{\sigma}\,\phi\left(\frac{x}{\sigma}\right),$$

then

$$L = (2\pi\sigma^2)^{-n/2} \exp(-\|\mathbf{x}\|^2/2\sigma^2)$$

so that (6-35) holds with $\lambda(\mathbf{x}) = \|\mathbf{x}\|^2$. If $\theta = (\mu, \sigma)$ and

$$h(x;\theta) = \frac{1}{\sigma}\phi\left(\frac{x-\mu}{\sigma}\right),$$

then

$$L = (2\pi\sigma^2)^{-n/2} \exp\left[-\frac{1}{2\sigma^2}\left(\sum_{i=1}^{n}(x_i - \bar{x})^2 + n(\bar{x} - \mu)^2\right)\right]$$

so that $\bar{x}$ and s^2 are sufficient. ■

Example 6-2. If $h(x;\theta)$ is the exponential density with mean θ, then

$$L(\mathbf{x};\theta) = 0 \qquad \text{if} \qquad \min x_i \le 0$$

$$= \theta^{-n} \exp\left(-\sum_{i=1}^{n} \frac{x_i}{\theta}\right), \quad \text{otherwise}$$

so that $\sum_{i=1}^{n} x_i$ is a sufficient statistic. ■

Example 6-3. Let $h(x;\theta)$ be the uniform density on $[0,\theta]$; then

$$L(\mathbf{x};\theta) = \theta^{-n} \quad \text{if} \quad \min x_i > 0,\ \max x_i < \theta$$

$$= 0, \quad \text{elsewhere.}$$

To show that $\max x_i$ is sufficient, let us represent L in the form (6-35). Let $D(x)$ be the function

$$D(x) = 0, \qquad x \le 0$$

$$= 1, \qquad x > 0;$$

then

$$L(\mathbf{x};\theta) = \theta^{-n} D(\min x_i)\, D(\theta - \max x_i).$$

Put $M(\mathbf{x}) = D(\min x_i)$ and $K(\max x_i, \theta) = \theta^{-n} D(\theta - \max x_i)$. ■

For each real y the set of points in R^n,

$$\{\mathbf{x}: \lambda(\mathbf{x}) = y\}, \tag{6-36}$$

is called *a surface*. The weighted average of a continuous function over this surface is defined as

$$\bar{f}(y) = \frac{\displaystyle\int_{\{\mathbf{x}:\,\lambda(\mathbf{x})=y\}} f(\mathbf{x})\,M(\mathbf{x})\,d\mathbf{x}}{\displaystyle\int_{\{\mathbf{x}:\,\lambda(\mathbf{x})=y\}} M(\mathbf{x})\,d\mathbf{x}}. \tag{6-37}$$

The integrals on the right-hand side are "surface integrals" in R^n. In the particular case $M \equiv 1$ and $\lambda(\mathbf{x}) = \|\mathbf{x}\|^2$, this is the spherical average defined in Sec. 6-1. The function $\bar{f}(\lambda(\mathbf{x}))$ is called the "conditional expectation of f given $\lambda(\mathbf{x})$."

The following generalization of Theorem 6-1 holds:

$$\int \cdots \int_{R^n} f(\mathbf{x})\,L(\mathbf{x};\theta)\,d\mathbf{x} = \int \cdots \int_{R^n} \bar{f}(\lambda(\mathbf{x}))\,L(\mathbf{x};\theta)\,d\mathbf{x} \tag{6-38}$$

for every continuous f and every θ. (It is assumed that both integrals are well defined.) While the proof of Theorem 6-1 involves the rotation, the proof of Eq. 6-38 requires a very general transformation of R^n (see Bibliography). The main ideas of the proof are

(i) Since $K(\lambda(\mathbf{x}),\theta)$ is constant on the surface (6-36), we may multiply the numerator and denominator in (6-37) by this factor and bring it under the signs of integration; thus, by (6-35):

$$\bar{f}(y) = \frac{\displaystyle\int_{\{\mathbf{x}:\,\lambda(\mathbf{x})=y\}} f(\mathbf{x})\,L(\mathbf{x};\theta)\,d\mathbf{x}}{\displaystyle\int_{\{\mathbf{x}:\,\lambda(\mathbf{x})=y\}} L(\mathbf{x};\theta)\,d\mathbf{x}}.$$

(ii) Put $y = \lambda(\mathbf{x}')$ in the equation above, multiply $f(\lambda(\mathbf{x}'))$ by $L(\mathbf{x}';\theta)$, and integrate over R^n. By a multidimensional change of variables it can be shown that (6-38) holds.

All of the results of Section 6-3—with the possible exception of (the special) Theorem 6-6—remain valid when the likelihood (6-34) takes the place of the normal one, and the weighted average (6-37) the place of the spherical average.

EXERCISES

1. Prove directly from the definition that any one-to-one transformation of a sufficient statistic is also a sufficient statistic—e.g., if λ is sufficient, so is λ^3.

2. Put

$$h(x;\theta) = 0, \quad x \leq 0$$
$$= (x/\theta)\exp(-x^2/2\theta^2), \quad x > 0.$$

Show that $x_1^2 + \cdots + x_n^2$ is sufficient.

3. Put $h(x;\theta) = (1/\theta)\exp(-|x|/\theta)$. Show that $|x_1| + \cdots + |x_n|$ is sufficient.

4. Put

$$h(x;\theta_1,\theta_2) = 1/(\theta_2 - \theta_1), \theta_1 < x < \theta_2,$$
$$= 0, \quad \text{elsewhere.}$$

Show that the two functions min x_i and max x_i form a set of sufficient statistics.

BIBLIOGRAPHY

The concept of the average value of a function over a circle arises in the theory of *harmonic* functions, which are studied in potential theory; see O. D. Kellog, *Foundations of Potential Theory* (Berlin: Springer, 1929).

The Fisher-Neyman criterion—that the "conditional distribution" definition of sufficiency is equivalent to the "factorization" definition—is stated and proved in Harald Cramér, *Mathematical Methods of Statistics* (Princeton, N.J.: Princeton U. P., 1946).

The topic of square-integrable functions is usually introduced in the context of measurable functions, which are more general than continuous ones. Although our definition includes continuity, our proof of the Cauchy-Schwarz inequality and its converse is almost the same as the one usually given (cf. Cramér's book).

CHAPTER 7

Estimation of Unknown Parameters

7-1. STATISTICAL BACKGROUND

Consider a normal population with mean μ and standard deviation σ. The quantities μ and σ completely specify the population: if μ and σ are given, then for every interval I, the proportion of the population whose numerical labels fall in I is determined. In a typical statistical problem either or both of these parameters are partially or completely unknown, and some decision has to be made about the population—a decision whose consequences depend on the imperfectly known values of the parameters. In order to get information about the parameters and to make a reasonable decision, the statistician observes the values of a Sample from the population, and makes an inference about the parameters; then the decision is made.

Here is an example. A manufacturer of men's shoes plans production. The shoes are to be produced in various sizes; the manufacturer must decide what proportion should be produced in each. For simplicity let us consider just the length size. If the distribution of the foot lengths in the population is known, then the manufacturer can choose the proportions (to be produced) in accordance with the distribution in the population. Suppose the distribution is known by experience to be normal; however, the mean μ and standard deviation σ are unknown. If μ and σ can be *estimated*, then the population distribution can be estimated, and size production planned accordingly. This *planning* is the decision that has to be made. The consequences of this decision—the economic consequences to the manufacturer—depend on the values of the unknown parameters and on their estimated values: while production is planned in accordance with the latter, the actual consumption is determined by the former. The loss to the manufacturer due to overproduction or underproduction in various sizes depends on the accuracy of the estimates of the unknown parameters; thus we are introduced to the statistical problem of "accurate" estimation of parameters.

We shall describe the usual method of estimating μ and σ (cf. Sec. 4-1). A Sample **X** is taken from the population and the values are noted. The Sample mean $\bar{x} = (x_1 + \cdots + x_n)/n$ is taken as the estimator of μ. The standard deviation is estimated by first estimating the variance σ^2, and then computing its square root. In the estimation of σ^2, there are two cases to consider. If μ is known, the estimator of the variance is the average sum of squares of the observations about μ:

$$\overline{(x - \mu)^2} = \frac{1}{n} \sum_{i=1}^{n} (x_i - \mu)^2.$$

If μ is unknown, the estimator is the Sample variance,

$$s^2 = \frac{1}{n - 1} \sum_{i=1}^{n} (x_i - \bar{x})^2.$$

One may ask: What justifies these procedures? Are these the best ones? This chapter is devoted to proving an affirmative answer. We shall specify—in mathematical terms—what we mean by a "best procedure"; then we shall prove that the ones described above are best.

We begin with the general theory of estimation in the case of the parameter μ; the discussion for σ^2 will be given later. Let $f(\mathbf{x})$ be a square-integrable function on R^n (Sec. 6-3). An *estimation procedure* based on f is: If **X** is the Sample, and **x** is the observed value of the Sample point, then the number $f(\mathbf{x})$ is the estimated value of μ; in other words, as in Sec. 4-1, $f(\mathbf{X})$ is the function of the Sample point which estimates μ. We call f an Estimator (capital "e") of μ. Note that the *procedure* is specified by f but the actual estimated value $f(\mathbf{X})$ depends also on the Sample. The classical Estimator of μ is the function $f(\mathbf{x}) = \bar{x}$; it is evidently square-integrable and so is an Estimator in our sense (Exercise 1).

Every Estimator is not necessarily or not always a good one because there is no apparent connection between f and the population. Consider the Estimator $f(\mathbf{x})$ identically equal to 1. The procedure specified by this function is to take 1 as the estimated value of μ for any value of the Sample. This will accurately estimate μ only if the latter is very nearly equal to 1; however, if it is significantly different from 1, the estimated value is far from the true one. In searching for a best Estimator, we shall ignore such Estimators as this, and restrict ourselves to those which vary in a specific way with the unknown value of μ. This property, known as *unbiasedness*, is defined as follows:

DEFINITION 7-1

Let $\sigma > 0$ be arbitrary but fixed. An Estimator f of μ is unbiased if the equation

$$\int_{R^n}\cdots\int f(\mathbf{x})L(\mathbf{x};\mu,\sigma)dx = \mu \qquad (7\text{-}1)$$

holds for all μ.

We note that the integral in (7-1) is well defined because f is square-integrable. [See the remarks following (6-25).] The meaning of unbiasedness is that the weighted average of the values of the Estimator, where each value is weighted by the corresponding value of L, is equal to the true though unknown value of μ. Unbiasedness is a mark of the validity of the estimation procedure. This is well illustrated in the case where the Estimator f is a statistic with a continuous density F'. By Theorem 4-1, with $g(y) = y$, the integral on the left-hand side of (7-1) reduces to a single integral, and the equation takes the form

$$\int_{-\infty}^{\infty} yF'(y)\,dy = \mu. \qquad (7\text{-}2)$$

This means that the unknown parameter value is the "center of gravity" of the distribution of the Estimator (cf. Sec. 2-2). The Sample mean is an unbiased Estimator of μ; indeed, by Corollary 1 of Theorem 4-5, $\bar{x}$ has a normal distribution with mean μ, so that Eq. 7-2 follows from Eq. 2-13. The Estimator $f(\mathbf{x}) \equiv 1$ is not unbiased because the integral in (7-1) is identically equal to 1.

The class of unbiased Estimators has many members; several examples are furnished in the exercises below. Not all are of the same quality: some are more "reliable" than others. We shall put the concept of reliability in a precise form. If f is an unbiased Estimator of μ, its variance is defined as

$$\int_{R^n}\cdots\int (f(\mathbf{x}) - \mu)^2 L(\mathbf{x}; \mu, \sigma)\, d\mathbf{x}. \tag{7-3}$$

The finiteness of this integral is a consequence of the square-integrability of f. The interpretation of variance of an Estimator is similar to the interpretation of the variance of the normal population, as defined by the parameter σ^2 (Sec. 2-2). The integral (7-3) is a weighted average of the squared differences between the values of the Estimator and the parameter. The variance is a measure of the reliability of the Estimator. This is most easily seen in the case when the Estimator f is a statistic with a continuous density F'. By Theorem 4-1, with $g(y) = (y - \mu)^2$, the integral (7-3) reduces to

$$\int_{-\infty}^{\infty} (y - \mu)^2 F'(y)\, dy. \tag{7-4}$$

This is analogous to the integral (2-14) characterizing the variance of the normal distribution. If an unbiased Estimator has a small variance, then the values y for which $(y - \mu)^2$ is large have small probability (Exercise 5); therefore, the Estimator is "reliable." We use the variance (7-4) as a numerical criterion for comparing unbiased Estimators: of two such Estimators, the one with the smaller variance is considered the better. For the Sample mean $\bar{x}$, the variance is σ^2/n; this follows from the form of the distribution of $\bar{x}$ (Corollary 1 of Theorem 4-5), and from the integral representation (2-14) of the variance of the normal distribution.

DEFINITION 7-2

An unbiased Estimator f is called efficient if its variance does not exceed that of any other unbiased Estimator for all values of μ.

An efficient Estimator is considered a best one. Later we shall prove that the Sample mean is an efficient Estimator, and that there exists not more than one efficient Estimator. As the unique efficient Estimator, the Sample mean is best.

EXERCISES

1. Prove that $\bar{x}$ is square-integrable.

2. A Sample of three observations is taken from a normal population. Consider these three Estimators of μ:

$$f_1(\mathbf{x}) = x_1, \quad f_2(\mathbf{x}) = \tfrac{1}{2}(x_1 + x_2), \quad f_3(\mathbf{x}) = \tfrac{1}{3}(x_1 + x_2 + x_3).$$

Show that these are unbiased, and find their variances. Which is the best?

3. A Sample of two observations is taken from a normal population. Let α and β be constants, and consider the Estimator $f(x) = \alpha x_1 + \beta x_2$. Under what conditions on α and β is f unbiased? In that case, what is the variance of f? What choice of α and β minimizes the variance?

4. Extend the previous exercise to n observations. Let $\alpha_1, \ldots, \alpha_n$ be constants, and $f(\mathbf{x}) = \sum_{i=1}^{n} \alpha_i x_i$. When is f unbiased? If so, what is the variance, and when is it minimized? [*Hint:* Substitute in the multiple integral (7-3), expand the square of the sum, and integrate term by term.]

5. Generalize the result of Exercise 4, Sec. 2-2. *Prove:* If f is a statistic, and its variance is given by (7-4), then the probability that $f(\mathbf{X})$ falls within t units of μ is at least equal to

$$1 - \frac{(\text{variance of } f)}{t^2}.$$

7-2. FORMULATION OF THE ESTIMATION PROBLEM AND REDUCTION BY SUFFICIENCY

The problem of finding an efficient Estimator may be put in this precise form: For fixed $\sigma > 0$, consider the class of square-integrable functions f such that Eq. 7-1 holds for all μ; then determine the function f—if it exists—for which the integral (7-3) is minimized for every value of μ. It happens that $\bar{x}$ is such a function; furthermore, since it does not depend on σ, it is "uniformly" efficient for all σ.

Before solving the minimization problem, we prove that there is never more than one solution: if f_1 and f_2 are two square-integrable functions satisfying (7-1) and minimizing (7-3), then f_1 and f_2 are identical.

THEOREM 7-1
There is at most one square-integrable function f satisfying (7-1) for all μ and minimizing (7-3).

Proof: Let f_1 and f_2 be two such functions; then, for every λ, $0 \leq \lambda \leq 1$, the function

$$h(\mathbf{x}) = \lambda f_1(\mathbf{x}) + (1 - \lambda) f_2(\mathbf{x})$$

is square-integrable and also satisfies (7-1) (Exercise 1). Let V stand for the minimum value of (7-3): it is the value of (7-3) for $f = f_i$, $i = 1, 2$. It follows from the role of V as a minimum that the integral (7-3)—with $\lambda f_1 + (1 - \lambda) f_2$ in place of f—is at least equal to V:

$$\int_{R^n} \cdots \int [\lambda(f_1(\mathbf{x}) - \mu) + (1 - \lambda)(f_2(\mathbf{x}) - \mu)]^2 L\, d\mathbf{x} \geq V.$$

Expand the square and integrate termwise; after some cancellation the above inequality becomes

$$V \leq \int_{R^n} \cdots \int (f_1 - \mu)(f_2 - \mu) L\, d\mathbf{x};$$

here we use the fact that

$$\int_{R^n} \cdots \int (f_i - \mu)^2 L\, d\mathbf{x} = V, \qquad i = 1, 2.$$

The Cauchy-Schwarz inequality (6-16) and the latter equation imply that

$$\int_{R^n} \cdots \int (f_1 - \mu)(f_2 - \mu) L\, d\mathbf{x} \leq V;$$

thus, combining the last two inequalities, we obtain

$$\int_{R^n} \cdots \int (f_1 - \mu)(f_2 - \mu) L\, d\mathbf{x} = V.$$

The function $f_1 - \mu$ is not identically equal to 0 for any value of μ because f_1 satisfies (7-1) for *all* μ; therefore, by the converse to the Cauchy-Schwarz inequality (Theorem 6-3), the above equation implies that for every μ, there exists a real number $\lambda(\mu)$ such that

$$\lambda(\mu)(f_1(\mathbf{x}) - \mu) + (f_2(\mathbf{x}) - \mu) = 0, \quad \text{for all } \mathbf{x}.$$

For any $\mu' \neq \mu$, multiply both sides of this equation by $L(\mathbf{x}; \mu', \sigma)$, and integrate over R^n: by Eq. 7-1 with μ' in place of μ, we obtain

$$\lambda(\mu)(\mu' - \mu) + (\mu' - \mu) = 0;$$

therefore $\lambda(\mu) \equiv -1$, and the previous equation involving f_1 and f_2 reduces to

$$f_1(\mathbf{x}) = f_2(\mathbf{x}), \quad \text{for all } \mathbf{x}. \quad \blacksquare$$

The first step in finding f minimizing (7-3) under the condition (7-1) is the reduction of the multiple integrals by the method of sufficiency. Let $\bar{f}(\sqrt{n}\bar{x}; \sigma)$ be the conditional expectation of f given $\sqrt{n}\bar{x}$, as defined by Eq. 6-13. By Theorem 6-5, $\bar{f}$ is square-integrable. By Eq. 6-14, the condition (7-1) takes the form

$$\int_{-\infty}^{\infty} \bar{f}(\sqrt{n}x; \sigma) \frac{\sqrt{n}}{\sigma} \phi\left(\frac{\sqrt{n}(x - \mu)}{\sigma}\right) dx = \mu; \tag{7-5}$$

here we have altered the integral (6-14) by a change of variable. By the result of Exercises 8 and 11 of Sec. 6-3, the integral (7-3) is decreased (or at least not increased) when f is replaced by $\bar{f}$; furthermore, by Theorem 4-1, the integral (7-3) with $\bar{f}$ in place of f is equal to

$$\int_{-\infty}^{\infty} (\bar{f}(\sqrt{n}x; \sigma) - \mu)^2 \frac{\sqrt{n}}{\sigma} \phi\left(\frac{\sqrt{n}(x - \mu)}{\sigma}\right) dx \tag{7-6}$$

because the conditional expectation is a function of $\sqrt{n}\bar{x}$. We infer: For any square-integrable function f satisfying (7-1), there is such a function $\bar{f}$ with the same properties and, in addition, depends on $\mathbf{x}$ only through $\bar{x}$, such that the variance of $\bar{f}$ is less than or equal to the variance of f. This means that in searching for a function f satisfying (7-1) and minimizing (7-3), we may restrict our quest to the subclass of those functions which depend on $\mathbf{x}$ only through $\bar{x}$. By virtue of the reduction of the multiple integrals to single integrals, the problem of finding f on R^n satisfying (7-1) and minimizing (7-3) is replaced by the simpler problem of finding a function $\bar{f}$ of a single variable satisfying (7-5) for all μ and minimizing the single integral (7-6). This is the significance of the sufficiency of the Sample mean in the estimation of the population mean.

We sketch the problem of the estimation of σ^2 when it is assumed that μ is known; for simplicity we shall put $\mu = 0$. An Estimator f of σ^2 is unbiased if the equation

$$\int_{R^n}\cdots\int f(\mathbf{x})\, L(\mathbf{x}; 0, \sigma)\, d\mathbf{x} = \sigma^2 \tag{7-7}$$

holds for all $\sigma > 0$. The variance of f is

$$\int_{R^n}\cdots\int (f(\mathbf{x}) - \sigma^2)^2 L(\mathbf{x}; 0, \sigma)\, d\mathbf{x}. \tag{7-8}$$

If f is also a statistic with a continuous density F', then, by Theorem 4-1, the equation (7-7) and the integral (7-8) assume the forms

$$\int_{-\infty}^{\infty} yF'(y)\, dy = \sigma^2 \tag{7-9}$$

and

$$\int_{-\infty}^{\infty} (y - \sigma^2)^2 F'(y)\, dy, \tag{7-10}$$

respectively. The integrals (7-7) and (7-8) can be reduced by means of the sufficient statistic $\|\mathbf{x}\|^2$. Let $\bar{f}(r^2)$ be the spherical average of f; then, by the equations (6-8), the condition (7-7) is equivalent to

$$\int_0^{\infty} \bar{f}(u)\, \psi_n\!\left(\frac{u}{\sigma^2}\right)\frac{du}{\sigma^2} = \sigma^2. \tag{7-11}$$

The variance of $\bar{f}(\|\mathbf{x}\|^2)$ is not greater than that of f, and is, by Theorem 4-1, equal to

$$\int_0^{\infty} (\bar{f}(u) - \sigma^2)^2\, \psi_n\!\left(\frac{u}{\sigma^2}\right)\frac{du}{\sigma^2}. \tag{7-12}$$

As in the case of μ the minimization of (7-8) subject to (7-7) is reduced to the minimization of the integral (7-12) subject to the condition (7-11). Theorem 7-1, stating the uniqueness of the solution of the minimization, is valid also for the estimation of σ^2. The only changes in the proof are substitution of σ^2 for μ, and Eqs. 7-7 and 7-8 for 7-1 and 7-3, respectively.

Up to now we have assumed that μ is known. The determination of the efficient Estimator of σ^2 when μ is unknown is much more complicated, and will be considered in Sec. 7-4.

EXERCISE

1. *Prove:* If f_1 and f_2 are square-integrable and satisfy (7-1), then the same is true of the function $\lambda f_1 + (1 - \lambda) f_2$.

7-3. SOLUTION OF THE MINIMIZATION PROBLEM: THE CRAMÉR-RAO LOWER BOUND

The estimation of μ is considered first. We shall derive a lower bound for the integral (7-6) defining the variance of an unbiased Estimator $\bar{f}$ of μ. This lower bound will be seen to coincide with the variance of the Estimator $\bar{x}$. Invoking Theorem 7-1, we conclude that $\bar{x}$ is the unique efficient Estimator of μ for all $\sigma > 0$.

The following statement is a special case of a general inequality known as the Cramér-Rao lower bound.

THEOREM 7-2

If $f(x)$ is a square-integrable function of one variable satisfying (7-5) for all μ:

$$\int_{-\infty}^{\infty} f(x) \frac{\sqrt{n}}{\sigma} \phi \left(\frac{\sqrt{n}(x - \mu)}{\sigma}\right) dx = \mu, \tag{7-13}$$

then the variance integral (7-6),

$$\int_{-\infty}^{\infty} (f(x) - \mu)^2 \frac{\sqrt{n}}{\sigma} \phi \left(\frac{\sqrt{n}(x - \mu)}{\sigma}\right) dx, \tag{7-14}$$

is at least equal to σ^2/n.

Proof: Differentiate each side of (7-13) with respect to μ: by Theorem 6-6, the differentiation may be done under the sign of integration, and the equation becomes

$$\int_{-\infty}^{\infty} f(x) \frac{n^{3/2}}{\sigma^3} (x - \mu) \phi \left(\frac{\sqrt{n}(x - \mu)}{\sigma}\right) dx = 1.$$

By Eq. 2-13, the integral above is unchanged when $f - \mu$ is substituted for f:

$$\int_{-\infty}^{\infty} (f(x) - \mu) \frac{n^{3/2}}{\sigma^3} (x - \mu) \phi \left(\frac{\sqrt{n}(x - \mu)}{\sigma}\right) dx = 1.$$

Square both sides of this equation, and apply the Cauchy-Schwarz inequality (6-16), with the correspondence

$$f \to f(x) - \mu, \qquad g \to \frac{n}{\sigma^2}(x - \mu);$$

it follows that

$$1 \le \int_{-\infty}^{\infty} (f(x) - \mu)^2 \frac{\sqrt{n}}{\sigma} \phi\left(\frac{\sqrt{n}(x - \mu)}{\sigma}\right) dx$$

$$\cdot n^2\sigma^{-4} \int_{-\infty}^{\infty} (x - \mu)^2 \frac{\sqrt{n}}{\sigma} \phi\left(\frac{\sqrt{n}(x - \mu)}{\sigma}\right) dx.$$

By Eq. 2-14 the last integral is equal to σ^2/n. The inequality above states that the integral (7-14) is at least equal to σ^2/n. ■

As noted in Sec. 7-1, $\bar{x}$ is unbiased and its variance is equal to σ^2/n. By Theorem 7-2 this is not more than the variance of any unbiased Estimator; therefore, $\bar{x}$ is efficient, and, by Theorem 7-1, it is the unique efficient Estimator.

Now we consider the estimation of σ^2 when $\mu = 0$. The Estimator $\|\mathbf{x}\|^2/n$ is unbiased and its variance is equal to $2\sigma^4/n$; the reader can verify this by noting that $\|\mathbf{x}\|^2/\sigma^2$ has a chi-square distribution with n degrees of freedom (extension of Theorem 4-2), and then evaluating the integrals in (7-9) and (7-10) by means of the gamma function (Exercises 1, 2). Now we show that the variance of $\|\mathbf{x}\|^2/n$ is smaller than that of any other unbiased Estimator of σ^2.

THEOREM 7-3

If $f(u)$ is a continuous function such that

$$\int_0^{\infty} |f(u)|^2 \psi_n\left(\frac{u}{\sigma^2}\right) du < \infty$$

and which satisfies (7-11) for all σ:

$$\int_0^{\infty} f(u) \psi_n\left(\frac{u}{\sigma^2}\right)\frac{du}{\sigma^2} = \sigma^2, \tag{7-15}$$

then the variance integral (7-12),

$$\int_0^{\infty} (f(u) - \sigma^2)^2 \psi_n\left(\frac{u}{\sigma^2}\right)\frac{du}{\sigma^2}, \tag{7-16}$$

is at least equal to $2\sigma^4/n$.

Proof: The explicit form of Eq. 7-15 is

$$\int_0^\infty f(y)\,K_n\,y^{(n/2)-1}e^{-y/2\sigma^2}\,dy = (\sigma^2)^{(n/2)+1}, \tag{7-17}$$

where $K_n = 1/2^{n/2}\,\Gamma(n/2)$. Differentiate each side with respect to σ^2; this may be done under the sign of integration:

$$\int_0^\infty f(y)\,K_n\,y^{(n/2)-1}\,\frac{y}{2\sigma^4}\,e^{-y/2\sigma^2}\,dy = \left(\frac{n}{2}+1\right)(\sigma^2)^{n/2}. \tag{7-18}$$

For f identically equal to 1 the integral on the left-hand side of (7-17) is equal to σ^n because

$$\int_0^\infty \psi_n(y)\,dy = 1.$$

Differentiation of the resulting equation leads to

$$\int_0^\infty K_n\,y^{(n/2)-1}\,\frac{y}{2\sigma^4}\,e^{-y/2\sigma^2}\,dy = \left(\frac{n}{2}\right)(\sigma^2)^{(n/2)-1}. \tag{7-19}$$

Multiply both sides of (7-19) by σ^2; and subtract corresponding members of the resulting equation from those of (7-18):

$$\int_0^\infty (f(y)-\sigma^2)\,\frac{y}{\sigma^4}\,\psi_n\!\left(\frac{y}{\sigma^2}\right)dy = \sigma^2.$$

Square both sides:

$$\left(\int_0^\infty (f(y)-\sigma^2)\,\frac{y}{\sigma^4}\,\psi_n\!\left(\frac{y}{\sigma^2}\right)dy\right)^2 = \sigma^4, \tag{7-20}$$

and apply the Cauchy-Schwarz inequality with the correspondence

$$f \to (f(y)-\sigma^2), \qquad g \to \frac{y}{\sigma^2}, \qquad L \to \frac{1}{\sigma^2}\,\psi_n\!\left(\frac{y}{\sigma^2}\right);$$

from (7-20) we obtain:

$$\sigma^4 \le \int_0^\infty (f(y)-\sigma^2)^2\psi_n\!\left(\frac{y}{\sigma^2}\right)\frac{dy}{\sigma^2}\cdot\int_0^\infty \left(\frac{y}{\sigma^2}\right)^2\psi_n\!\left(\frac{y}{\sigma^2}\right)\frac{dy}{\sigma^2}. \tag{7-21}$$

The first integral on the right-hand side of (7-21) is the variance (7-16). The second integral, after the change of variable from y to y/σ^2, is found to be $n/2$ (Exercise 4). From the inequality (7-21) we conclude that the variance (7-16) is at least equal to $2\sigma^4/n$. ■

As noted before the statement of the theorem $\| \mathbf{x} \|^2/n$ is unbiased and its variance is $2\sigma^4/n$; thus, by Theorem 7-3, it is efficient. If μ is not equal to 0, then the efficient Estimator of σ^2 is the average sum of squares about μ, defined in Sec. 7-1; this can be easily verified by transforming the variables of integration from x_i to $x_i - \mu$ (Exercise 5).

EXERCISES

1. Show that $\| \mathbf{x} \|^2/n$ is an unbiased Estimator of σ^2 when $\mu = 0$; use the method indicated before Theorem 7-3.
2. Show that the variance of $\| \mathbf{x} \|^2/n$ is $2\sigma^4/n$.
3. Following the proof of Theorem 6-6, show that the integral in (7-17) may be differentiated with respect to σ^2 under the sign of integration.
4. Show that the second integral in (7-21) is equal to $n/2$; use the definition and recursive relation for the gamma function.
5. Show that the results for the estimation of σ^2 remain valid if μ is not equal to 0, and the unbiased Estimator $(1/n) \sum_{i=1}^{n} (x_i - \mu)^2$ is used in place of $\| \mathbf{x} \|^2/n$.

7-4. ESTIMATION OF σ^2 WHEN μ IS UNKNOWN (OPTIONAL)

It was shown in the previous sections that the problem of finding an efficient Estimator of μ for a fixed value of σ has a solution independent of σ: the Sample mean is unbiased for all σ, and is efficient for all σ. The estimation of σ^2 is complicated by the fact that the efficient Estimator of σ^2 for one value of μ is not even unbiased for other values of μ: the Estimator $\| \mathbf{x} \|^2/n$ is *not* unbiased for $\mu \neq 0$ (Exercise 1). As noted in Sec. 7-1, the usual Estimator of σ^2 when μ is unknown is s^2; it is unbiased and its variance is $2\sigma^4/(n - 1)$. This can be verified by recalling that $(n - 1)s^2/\sigma^2$ has the chi-square distribution with $n - 1$ degrees of freedom (Exercises 2, 3). In this section we shall show that among all Estimators of σ^2 which are unbiased for all values of μ, s^2 has minimum variance for all μ and σ. The minimization problem assumes the form: Among all square-integrable functions $f(\mathbf{x})$ for which

$$\int_{R^n}\cdots\int f(\mathbf{x})\, L(\mathbf{x}; \mu, \sigma)\, dx = \sigma^2 \quad \textit{for all } \mu,\ \sigma > 0, \tag{7-22}$$

find the one minimizing the variance

$$\int_{R^n}\cdots\int (f(\mathbf{x}) - \sigma^2)^2 L(\mathbf{x}; \mu, \sigma)\, d\mathbf{x} \tag{7-23}$$

for all μ and $\sigma > 0$. We reduce these multiple integrals by sufficiency, as in Sec. 7-2. By Theorem 6-1, the minimization problem can be reduced to the following simpler one. Put

$$L_1(x; \mu, \sigma) = \frac{\sqrt{n}}{\sigma}\, \phi\left(\frac{\sqrt{n}(x - \mu)}{\sigma}\right), \quad L_2(y; \sigma) = \frac{1}{\sigma^2}\, \psi_{n-1}\left(\frac{y}{\sigma^2}\right);$$

then among all continuous functions $f(x, y)$ of two variables such that

$$\int_0^\infty \int_{-\infty}^\infty |f(x, y)|^2\, L_1(x; \mu, \sigma)\, L_2(y; \sigma)\, dx\, dy < \infty \tag{7-24}$$

and which satisfy

$$\int_0^\infty \int_{-\infty}^\infty f(x, y)\, L_1(x; \mu, \sigma)\, L_2(y; \sigma)\, dx\, dy = \sigma^2 \tag{7-25}$$

for all μ and $\sigma > 0$, find the one minimizing

$$\int_0^\infty \int_{-\infty}^\infty (f(x, y) - \sigma^2)^2\, L_1(x; \mu, \sigma)\, L_2(y; \sigma)\, dx\, dy \tag{7-26}$$

for all μ and $\sigma > 0$.

THEOREM 7-4

If $f(x,y)$ is a continuous function satisfying (7-24) and (7-25), then the integral (7-26) is at least equal to $2\sigma^4/(n - 1)$.

Proof: We combine the methods of proof of Theorems 7-2 and 7-3, differentiating each side of (7-25) with respect to μ and with respect of σ^2. To simplify the notation we put

$$L = L_1 \cdot L_2, \qquad f = f(x, y).$$

An extension of the reasoning in the proof of Theorem 6-6 indicates that the integral in (7-25) is *twice* differentiable with respect to μ

under the sign of integration:

$$\int_0^\infty \int_{-\infty}^\infty f\left(\frac{\partial^2 L}{\partial \mu^2}\bigg/L\right) L\,dx\,dy = 0.$$

Now put $f \equiv 1$, multiply the integral by σ^2, and subtract from the left hand member of the equation above:

$$\int_0^\infty \int_{-\infty}^\infty (f - \sigma^2)\left(\frac{\partial^2 L}{\partial \mu^2}\bigg/L\right) L\,dx\,dy = 0. \qquad (7\text{-}27)$$

Now differentiate (7-25) with respect to σ^2:

$$\int_0^\infty \int_{-\infty}^\infty f\left(\frac{\partial L}{\partial \sigma^2}\bigg/L\right) L\,dx\,dy = 1. \qquad (7\text{-}28)$$

Differentiate each member of the equation

$$\int_0^\infty \int_{-\infty}^\infty L\,dx\,dy = 1$$

with respect to σ^2:

$$\int_0^\infty \int_{-\infty}^\infty \frac{\partial L}{\partial \sigma^2}\,dx\,dy = 0.$$

Multiply the left-hand member of this equation by σ^2, and subtract it from the corresponding member of (7-28):

$$\int_0^\infty \int_{-\infty}^\infty (f - \sigma^2)\left(\frac{\partial L}{\partial \sigma^2}\bigg/L\right) L\,dx\,dy = 1. \qquad (7\text{-}29)$$

Multiply the left-hand member of (7-27) by an arbitrary number t, and add it to the corresponding member of (7-29):

$$\int_0^\infty \int_{-\infty}^\infty (f - \sigma^2)\left[\frac{\partial L}{\partial \sigma^2} \cdot \frac{1}{L} + t\,\frac{\partial^2 L}{\partial \mu^2} \cdot \frac{1}{L}\right] L\,dx\,dy = 1. \quad (7\text{-}30)$$

Apply the Cauchy-Schwarz inequality (6-16) with the correspondence

$$f \rightarrow f - \sigma^2, \qquad g \rightarrow \frac{\partial L}{\partial \sigma^2} \cdot \frac{1}{L} + t\,\frac{\partial^2 L}{\partial \mu^2} \cdot \frac{1}{L}.$$

As in the proofs of Theorems 7-2 and 7-3, we find the lower bound for the variance of f to be the reciprocal of

$$\int_0^\infty \int_{-\infty}^\infty \left[\frac{\partial L}{\partial \sigma^2} \cdot \frac{1}{L} + t\,\frac{\partial^2 L}{\partial \mu^2} \cdot \frac{1}{L}\right]^2 L\,dx\,dy. \qquad (7\text{-}31)$$

As a quadratic function of t, this is minimized at

$$t_0 = -\frac{\int_0^\infty \int_{-\infty}^\infty \left(\frac{\partial L}{\partial \sigma^2}\Big/L\right)\cdot\left(\frac{\partial^2 L}{\partial \mu^2}\Big/L\right) L\,dx\,dy}{\int_0^\infty \int_{-\infty}^\infty \left(\frac{\partial^2 L}{\partial \mu^2}\Big/L\right)^2 L\,dx\,dy},$$

where it has the value

$$\int_0^\infty \int_{-\infty}^\infty \left(\frac{\partial L}{\partial \sigma^2}\Big/L\right)^2 L\,dx\,dy - t_0^2\int_0^\infty \int_{-\infty}^\infty \left(\frac{\partial^2 L}{\partial \mu^2}\Big/L\right)^2 L\,dx\,dy; \qquad (7\text{-}32)$$

the reader can verify this by expanding the square in (7-31), integrating termwise, and expressing the sum of the integrals as a quadratic function $At^2 + Bt + C$ (Exercise 9). The smallest value of (7-31) yields the greatest—the best—lower bound.

The rest of the proof involves the computation of (7-32), which is found to be equal to $(n - 1)/2\sigma^4$. Since the integration involves no new ideas, we leave it to the reader to fill in the details of the steps outlined in Exercises 4–8 below. ■

It follows from the remarks preceding the theorem that s^2 is an efficient Estimator of σ^2, and from Theorem 7-1 that it is the unique one. This justifies the use of s^2 when μ is unknown.

EXERCISES

1. By a direct calculation of the multiple integral show that

$$\int \cdots \int_{R^n} \|\mathbf{x}\|^2 L(\mathbf{x}; \mu, \sigma)\,d\mathbf{x} \neq n\sigma^2$$

if $\mu \neq 0$.

2. Show that s^2 is unbiased.

3. Show that the variance of s^2 is $2\sigma^4/(n - 1)$.

4. *Verify:*

$$\frac{\partial L}{\partial \sigma^2}\Big/ L = \frac{y + n(x - \mu)^2}{2\sigma^4} - \frac{n}{2\sigma^2}$$

$$\frac{\partial^2 L}{\partial \mu^2}\Big/ L = \frac{n^2(x - \mu)^2}{\sigma^4} - \frac{n}{\sigma^2}.$$

5. *Verify:*

$$\int_0^\infty \int_{-\infty}^\infty \left(\frac{\partial L}{\partial \sigma^2}\Big/ \mathrm{L}\right)^2 L\,dx\,dy = \frac{n}{2\sigma^4}.$$

[*Hint:* $(\partial L/\partial \sigma^2)/L = \partial \log L/\partial \sigma^2$.]

6. *Verify:*

$$\int_0^\infty \int_{-\infty}^\infty \left(\frac{\partial^2 L}{\partial \mu^2}\Big/ \mathrm{L}\right)^2 L\,dx\,dy = \frac{2n^2}{\sigma^4}.$$

7. *Verify:*

$$\int_0^\infty \int_{-\infty}^\infty \left(\frac{\partial L}{\partial \sigma^2}\Big/ \mathrm{L}\right)\left(\frac{\partial^2 L}{\partial \mu^2}\Big/ \mathrm{L}\right) L\,dx\,dy = \frac{n}{\sigma^4}.$$

8. Use the results of Exercises 4–7 to show that the integral (7-32) is equal to $(n - 1)/2\sigma^4$.

9. Show that the integral (7-31) is minimized at $t = t_0$.

7-5. ESTIMATION OF PARAMETERS FOR A GENERAL POPULATION (OPTIONAL)

We generalize the foregoing theory of estimation to a population with density function $h(x;\theta)$, where θ is the unknown parameter to be estimated. As before, we put

$$L(\mathbf{x};\theta) = \prod_{i=1}^{n} h(x_i, \theta). \tag{7-33}$$

An Estimator $f(\mathbf{x})$ is a square-integrable function on the Sample space, that is, it is continuous and satisfies

$$\int_{R^n}\cdots\int f^2(\mathbf{x})\,L(\mathbf{x};\theta)\,d\mathbf{x} < \infty, \qquad \textit{for all } \theta.$$

It is called unbiased if

$$\int_{R^n} \cdots \int f(\mathbf{x})\, L(\mathbf{x};\theta)\, d\mathbf{x} = \theta, \qquad \textit{for all } \theta. \tag{7-34}$$

As in the normal case, this signifies that the mean of the distribution of f is equal to the unknown parameter. The variance of such an Estimator is

$$\int_{R^n} \cdots \int (f(\mathbf{x}) - \theta)^2 L(\mathbf{x};\theta)\, d\mathbf{x}. \tag{7-35}$$

An unbiased Estimator is called efficient if (7-35) is minimized for all θ; as in Sec. 7-2, an efficient Estimator is unique.

The Cramér-Rao lower bound has the following more general version:

THEOREM 7-5

Under the regularity conditions on h implied in the proof below, the variance of any unbiased Estimator of θ is at least equal to

$$\frac{1}{n \int_{-\infty}^{\infty} \left(\frac{\partial \log h(x;\theta)}{\partial \theta} \right)^2 h(x;\theta)\, dx}. \tag{7-36}$$

Proof: We follow the proof of Theorem 7-2. It is assumed that all indicated derivatives exist and all differentiation operations are justified. From (7-33) it follows that

$$\frac{\partial}{\partial \theta} \log L(\mathbf{x};\theta) = \sum_{i=1}^{n} \frac{\partial}{\partial \theta} \log h(x_i;\theta),$$

so that

$$\frac{\partial L}{\partial \theta} = L \cdot \sum_{i=1}^{n} \frac{\partial}{\partial \theta} \log h(x_i;\theta). \tag{7-37}$$

Differentiate both sides of (7-34), and take the derivative under the sign of integration:

$$\int_{R^n} \cdots \int f(\mathbf{x})\, L(\mathbf{x};\theta) \left[\sum_{i=1}^{n} \frac{\partial}{\partial \theta} \log h(x_i;\theta) \right] d\mathbf{x} = 1. \tag{7-38}$$

Perform the same differentiation on the identity

$$\int_{R^n}\cdots\int L(\mathbf{x};\theta)\,d\mathbf{x} = 1, \qquad \textit{for all } \theta,$$

and obtain:

$$\int_{R^n}\cdots\int L(\mathbf{x};\theta)\left[\sum_{i=1}^{n}\frac{\partial}{\partial\theta}\log h(x_i,\theta)\right]d\mathbf{x} = 0.$$

Multiply both sides of this equation by θ, and subtract corresponding members from (7-38):

$$\int_{R^n}\cdots\int [f(\mathbf{x}) - \theta]\left[\sum_{i=1}^{n}\frac{\partial}{\partial\theta}\log h(x_i;\theta)\right]L(\mathbf{x};\theta)\,d\mathbf{x} = 1.$$

Apply the Cauchy-Schwarz inequality:

$$\int_{R^n}\cdots\int [f(\mathbf{x}) - \theta]^2 L\,d\mathbf{x} \geq \frac{1}{\int_{R^n}\cdots\int\left[\sum_{i=1}^{n}\frac{\partial}{\partial\theta}\log h(x_i;\theta)\right]^2 L\,d\mathbf{x}}. \tag{7-39}$$

By expansion of the square in the denominator, and the relation

$$\int_{R^n}\cdots\int\left[\frac{\partial}{\partial\theta}\log h(x_i;\theta)\right]\left[\frac{\partial}{\partial\theta}\log h(x_j;\theta)\right]L\,d\mathbf{x} = 0, \qquad i \neq j, \tag{7-40}$$

(Exercise 1), we obtain the conclusion of the theorem from (7-39) (Exercise 2). ■

Example 7-1. If $\theta = \mu$ and $h(x;\mu) = (1/\sigma)\,\phi((x - \mu)/\sigma)$, where σ is known, then

$$\frac{\partial}{\partial\mu}\log\phi(x - \mu) = (x - \mu)/\sigma^2$$

and the lower bound (7-36) is equal to σ^2/n. This coincides with the lower bound in Theorem 7-2. An analogous result holds when $\theta = \sigma^2$ and $\mu = 0$: the lower bound (7-36) is equal to $2\sigma^4/n$ (Exercise 3). ■

Example 7-2. If

$$h(x;\theta) = 0, \qquad x \le 0$$
$$= \theta^{-1} \exp(-x/\theta), \qquad x > 0,$$

then it can be shown that the conditions of Theorem 7-5 are satisfied and that the lower bound (7-36) is equal to θ^2/n (Exercise 4). Since $\bar{x}$ is unbiased and has this variance (Exercise 5), it follows that $\bar{x}$ is efficient. ■

Although Theorem 7-5 gives a sufficient condition for an unbiased Estimator to be efficient, it does not furnish a method of constructing such an Estimator. A useful method is that of "maximum likelihood," popularized by R. A. Fisher in the 1920's. As noted in Sec. 3-6, $L(\mathbf{x};\theta)$ is approximately proportional to the probability that the Sample point $\mathbf{X}$ falls in a small rectangle containing $\mathbf{x}$. The maximum likelihood principle is that for a given Sample point $\mathbf{X}$ the value of θ selected as the estimate is the one which maximizes $L(\mathbf{X};\theta)$; in other words, we choose the value of θ which renders the observed Sample "most probable." The value of θ which maximizes $L(\mathbf{x};\theta)$ for a given $\mathbf{x}$ depends, of course, on $\mathbf{x}$. It is denoted $\hat{\theta}$ so that

$$L(\mathbf{x};\theta) \le L(\mathbf{x};\hat{\theta}), \qquad \textit{for all } \theta.$$

For certain density functions h the maximum likelihood function $\hat{\theta}$ can be found by solving the equation $\partial L/\partial\theta = 0$. Since log L is a monotone function of L, $\hat{\theta}$ also maximizes log L; thus, it suffices to solve the equation

$$\frac{\partial \log L}{\partial \theta} = 0,$$

or, equivalently,

$$\sum_{i=1}^{n} \frac{\partial}{\partial \theta} \log h(x_i;\theta) = 0. \tag{7-41}$$

This is known as the "likelihood equation."

A maximum likelihood Estimator is not necessarily unbiased; however, a small modification often makes it unbiased. Under certain regularity conditions these Estimators are also efficient; furthermore, in large samples, they are usually the "best" ones to use. A precise statement and proof of these results are found in the references in the Bibliography.

Example 7-3. For the exponential density in Example 7-2, Eq. 7-41 assumes the form

$$\sum_{i=1}^{n} (x_i - \hat{\theta}) = 0,$$

so that $\theta = \bar{x}$, which is the efficient Estimator. ■

There is a shortcoming in the use of Theorem 7-5. The lower bound is valid only when the density $h(x;\theta)$ satisfies certain regularity conditions. We shall give an example in which the regularity conditions are not satisfied, and produce an Estimator with a variance of order smaller than $1/n$.

Example 7-4. Put

$$h(x;\theta) = 1/\theta, \quad 0 < x < \theta$$
$$= 0, \quad \text{elsewhere.}$$

As shown in Example 3-9:

$$L(\mathbf{x};\theta) = \theta^{-n}, \quad \text{if} \quad \min x_i > 0, \quad \max x_i < \theta$$
$$= 0, \quad \text{elsewhere.}$$

L is maximized with respect to θ for a given $\mathbf{x}$ if θ is made as small as possible subject to the condition $\theta > \max x_i$; therefore

$$\hat{\theta} = \max x_i.$$

While $\hat{\theta}$ is not unbiased, the corrected Estimator $\hat{\theta}(n + 1)/n$ is unbiased (Exercise 6). The variance is equal to $\theta^2 n/(n + 2)(n + 1)^2$, which is of the order n^{-2}. The explanation for this is that L does not satisfy the regularity conditions under which Theorem 7-5 was proved. ■

We conclude this section with some remarks on the estimation of the mean

$$\mu = \int_{-\infty}^{\infty} xh(x;\theta)\,dx$$

and variance

$$\sigma^2 = \int_{-\infty}^{\infty} (x - \mu)^2 h(x;\theta)\,dx$$

of a population with density $h(x;\theta)$. In the case of a normal population it was shown (Sec. 7-3) that $\bar{x}$ is unbiased with variance σ^2/n and is efficient; and that $\overline{(x - \mu)^2}$ and s^2 are unbiased (and efficient) Estimators of σ^2 when μ is known and unknown, respectively (Sec. 7-3 and 7-4). In general, the equations above for μ and σ^2 indicate that these are functions of θ. It is still true here that $\bar{x}$ is an unbiased Estimator of μ and that its variance is σ^2/n:

$$\int_{R^n} \cdots \int \bar{x} L(\mathbf{x};\theta)\, d\mathbf{x} = \mu \qquad \text{(Exercise 10)} \tag{7-42}$$

$$\int_{R^n} \cdots \int (\bar{x} - \mu)^2 L(\mathbf{x};\theta)\, d\mathbf{x} = \sigma^2/n \qquad \text{(Exercise 11);} \tag{7-43}$$

and that the Estimators of σ^2 are unbiased:

$$\int_{R^n} \cdots \int \overline{(x - \mu)^2} L(\mathbf{x};\theta)\, d\mathbf{x} = \sigma^2 \qquad \text{(Exercise 12)} \tag{7-44}$$

$$\int_{R^n} \cdots \int s^2 L(\mathbf{x};\theta)\, d\mathbf{x} = \sigma^2 \qquad \text{(Exercise 13).} \tag{7-45}$$

While the distribution of $\bar{x}$ is normal if and only if the population density is normal, the central limit theorem implies that $\bar{x}$ has an approximately normal distribution if n is large (Sec. 4-6).

EXERCISES

1. Verify Eq. 7-40 using the argument following (7-38).
2. Show how the conclusion of Theorem 7-5 follows from (7-39) and (7-40).
3. Verify the assertions in Example 7-1.
4. Show that the lower bound of unbiased estimation for the exponential density (Example 7-2) is θ^2/n.
5. (Continuation of Exercise 4.) Prove that $\bar{x}$ is unbiased and has the variance θ^2/n.
6. Show that $\hat{\theta}(n + 1)/n$ is an unbiased Estimator of θ in sampling from the uniform population in Example 7-4. [*Hint:* Use the general

formula for the distribution of max x_i (Sec. 4-6, Exercise 2), and apply Theorem 4-1.]

7. (Continuation of Exercise 6.) Derive the expression for the variance of $\hat{\theta}(n + 1)/n$.

8. Show that $\bar{x}$ is the maximum likelihood Estimator of μ in sampling from a normal population. [*Hint:* Solve (7-41).]

9. Show that if μ is known, then $\overline{(x - \mu)^2}$ is the maximum likelihood Estimator of σ^2.

10. Verify (7-42) by substituting the expression (7-33) for L, writing $\bar{x}$ as a linear combination, and integrating termwise.

11. Verify (7-43) by putting

$$(\bar{x} - \mu)^2 = \frac{1}{n^2}\left[\sum_{i=1}^{n} (x_i - \mu)^2 + \sum_{i \neq j} (x_i - \mu)(x_j - \mu)\right],$$

substituting (7-33) for L, and integrating termwise.

12. Verify (7-44) using a method similar to that in Exercise 11.

13. Verify (7-45) as in Exercise 12.

14. The following result illustrates the importance of the sufficient statistic in estimation: *If f is an unbiased Estimator then there exists one depending on* **x** *only through the sufficient statistic, and having variance not greater than f.* This result is known as the Rao-Blackwell theorem. Prove this by means of an appropriate extension of the argument in Sec. 7-2.

7-6. NUMERICAL ILLUSTRATIONS

We shall illustrate the behavior of several Estimators by computing their values in Samples from a population with known parameters. Twenty-five Samples consisting of five observations each were taken from a standard normal population, that is, from Table VI, Appendix, of random normal deviates; here it is known that $\mu = 0$ and $\sigma = 1$.

An estimator of μ that is often used in the place of $\bar{x}$ is the "Sample median," which is a value such that equal numbers of observations fall above and below it; for example, the median of the Sample

$$-.202, \quad .420, \quad 2.417, \quad .260, \quad -.353$$

is .260. Its advantage is that it is easier to calculate than is the Sample mean; it is often used in practice as a quick, crude Estimator of μ. If the number of observations is even, the median may not be unique.

The *range* of the Sample is defined as the difference between the largest and smallest observations. When the range is divided by a certain known constant (depending on n) it is an unbiased Estimator of σ; for example, for $n = 5$, the range of the Sample just given is $2.417 - (-.353) = 2.770$, the constant is known to be 2.326, so that the Estimator of σ has the value $2.770 \div 2.326 = 1.191$.

In Table 7-1 are given the values of Sample mean and median,

TABLE 7-1
Values of Estimators in Samples of 5 from a Standard Normal Population.

Sample	$\overline{x}$	Median	s^2	$\overline{x^2}$	Range/2.326
1	−.734	−.157	1.596	1.818	1.229
2	−.650	−.845	.183	.569	.426
3	−.035	−.034	.245	.198	.582
4	.159	−.090	1.354	1.108	1.301
5	−.075	−.069	.385	.312	.678
6	.427	.760	.486	.571	.696
7	−.380	−.377	.310	.392	.615
8	−.102	.016	1.093	.885	1.163
9	−.677	−.551	.789	1.089	1.004
10	.239	.472	.985	.845	.978
11	.508	.260	1.240	1.250	1.191
12	−.391	−.580	.976	.934	1.068
13	−.334	−.168	.690	.671	.919
14	.091	−.140	.906	.734	.932
15	−.098	−.018	.396	.326	.728
16	−.210	−.482	1.858	1.531	1.467
17	.100	−.061	.437	.360	.712
18	−.004	−.412	1.056	.844	1.046
19	.187	.195	1.821	1.491	1.615
20	.822	.481	2.677	2.818	1.788
21	−1.023	−1.264	.736	1.635	.907
22	.709	.533	.352	.784	.626
23	.339	.558	1.830	1.579	1.358
24	−.206	.032	.474	.422	.715
25	−.415	−.401	.416	.505	.739

s^2, $\overline{x^2}$, and range/2.326 in twenty-five Samples of five from the first two pages of Table VI, Appendix. The statistic s^2 is computed from the formula

$$\frac{n \sum_{i=1}^{n} x_i^2 - \left(\sum_{i=1}^{n} x_i \right)^2}{n(n-1)} \tag{7-46}$$

in order to avoid an excessive rounding error. One notices that in this particular illustration there are few significant differences, between the mean and median, or between s^2 and $\overline{x^2}$. In this Sample, the range Estimator *appears* to be more accurate than s^2 and $\overline{x^2}$, which are known to be the best ones. The reason for this is that the range Estimator is estimating the *square root* of the variance, σ, while s^2 and $\overline{x^2}$ are estimating σ^2 itself. It would be more appropriate to compare the latter Estimators to the *square* of the range Estimator. Notice that the square of the range Estimator is very close to the other two (Exercise 3).

EXERCISES

1. Show that (7-46) is equivalent to the expression for s^2 given in Sec. 7-1.

2. (*Optional*) Construct a table like Table 7-1 for fifteen Samples of nine observations each. The constant in the denominator of the range Estimator is 2.970.

3. (*Optional*) Compute the squares of the range Estimator in Table 7-1.

BIBLIOGRAPHY

The general theory of efficient estimation, and properties of maximum likelihood Estimators are discussed in detail in Harald Cramér, *Mathematical Methods of Statistics* (Princeton, N.J.: Princeton U. P., 1946).

A comprehensive exposition of many results in modern estimation theory is available in "Notes on the Theory of Estimation," in *Lectures by E. L. Lehmann*, recorded by Colin Blyth (Stanford: U. of California Press, 1950), Chapters 1–5.

CHAPTER 8

Confidence Intervals

8-1. CONFIDENCE INTERVAL FOR μ WHEN σ IS KNOWN

We are given a normal population with unknown mean μ and known standard deviation σ, and wish to estimate μ on the basis of a Sample of n observations. The results of estimation theory (Chapter 7) indicate that (even when σ is unknown) the Sample mean is the best Estimator: however, we also want to know "how good" it is in each situation.

Example 8-1. A Sample of n observations is taken from a population with standard deviation σ. Suppose that the observed value of $\bar{x}$ is 4.8: this is the estimated value of μ. As mentioned at the beginning of Chapter 7, important consequences may depend on the accuracy of the Estimator. Our first question is: Is it "likely" that the unknown value μ is close to 4.8?

In order to answer this, let us choose a particular value for σ; for example, take $\sigma = 2$. Let us now compare two situations. Suppose in one case that the Sample mean is based on four observations ($n = 4$) and, in the second case, on 100 observations ($n = 100$). It is natural to think that the Estimator in the second case is more reliable than in the first; as a matter of fact, while both Estimators are unbiased, the first has variance $\sigma^2/4 = 1$, and the second variance .04. This means that the distribution of the Estimator in the second case is much more concentrated about the parameter μ; however, the Estimator in the first case is still the best one available on the basis of four observations.

Let us compare another two cases. Suppose in one that $\sigma = 2$ and, in the other, $\sigma = 10$; and $n = 4$ in both cases. Here the numbers of observations are equal but not the population standard deviations. While the Estimators are efficient in their respective situations, the variance is 1 in the first and 25 in the second.

For the particular observed value $\bar{x} = 4.8$ the question of how likely it is for μ to be near 4.8 has no obvious meaning because μ is a fixed number; for example, if $\mu = 3$ there is no content in "the probability that 3 is near 4.8." We put the question in a meaningful form in this way: What is the probability that the Estimator will yield an estimated value near the true but unknown mean? This is determined on the basis of the knowledge of σ and n—before the Sample is taken. Suppose $\sigma = 2$ and $n = 16$. Let us determine the probability that $\bar{x}$ differs from μ by less than 1 unit. We employ Corollary 1 of Theorem 4-5: $\bar{x}$ has a normal distribution with mean μ and standard deviation $\sigma/\sqrt{n} = 1/2$; thus, the probability that $\bar{x}$ differs from μ by less than one unit is $\Phi(2) - \Phi(-2)$ which, from Table I, Appendix, is equal to .9544. Another way of stating this is: The probability that the interval $\bar{x} \pm 1$ contains μ is equal to .9544. The probability .9544 is a measure of our *confidence* that the interval $\bar{x} \pm 1$ contains μ, and is called the *confidence coefficient* of the *confidence interval* $\bar{x} \pm 1$. In the case $\bar{x} = 4.8$, the confidence interval is (3.8, 5.8). Although we do not *know* whether μ is contained in the confidence interval, we have much confidence that it is because the interval was constructed in such a way that its probability of containing μ is .9544. ■

This example illustrates a general procedure of constructing confidence intervals with given confidence coefficients. For every real number α, $0 < \alpha < 1$, let z_α stand for the number z such that the integral of the standard normal density from z to $+\infty$ is α; in other words, z_α is the unique solution of the equation

$$\alpha = 1 - \Phi(z). \tag{8-1}$$

The number z_α is called the upper α-percentile of the standard normal distribution. By the symmetry of the distribution z_α also satisfies

$$\Phi(-z_\alpha) = \alpha.$$

THEOREM 8-1

In sampling from a normal population with standard deviation σ, the interval

$$\bar{x} \pm \frac{z_{\alpha/2}\sigma}{\sqrt{n}} \tag{8-2}$$

contains, with probability $1 - \alpha$, *the value of the (unknown) parameter* μ; *in other words, the interval (8-2) is a confidence interval of confidence coefficient* $1 - \alpha$.

Proof: $\bar{x}$ falls within $z_{\alpha/2}\sigma/\sqrt{n}$ units of μ if and only if $\sqrt{n}(\bar{x} - \mu)/\sigma$ falls between $-z_{\alpha/2}$ and $z_{\alpha/2}$. By Theorem 4-5 this statistic has a standard normal distribution; thus, it follows from the definition (8-1) of z_α (with $\alpha/2$ in place of α) that the probability that $\sqrt{n}(\bar{x} - \mu)/\sigma$ falls between $-z_{\alpha/2}$ and $z_{\alpha/2}$ is equal to

$$\Phi(z_{\alpha/2}) - \Phi(-z_{\alpha/2}) = \alpha. \quad \blacksquare$$

The length of the confidence interval, $2z_{\alpha/2}\sigma/\sqrt{n}$, is directly proportional to σ and inversely proportional to the square root of the number of observations. The confidence coefficient $1 - \alpha$, though arbitrary, is commonly chosen as one of the numbers .95, .98 or .99. The corresponding values of $\alpha/2$ are .025, .01 and .005, and those of $z_{\alpha/2}$ are found in Table I, Appendix:

$1 - \alpha$	.95	.98	.99
$z_{\alpha/2}$	1.96	2.33	2.58

Example 8-2. A Sample of 25 observations is taken from a normal population with $\sigma = .5$. A confidence interval of confidence coefficient .95 is of length

$$2(.5)(1.96)/5 = .39,$$

and is centered at the computed value of $\bar{x}$. If the coefficient is raised to .98, the length of the interval is increased to

$$2(.5)(2.33)/5 = .47.$$

For confidence coefficient .99 it is .51. ■

EXERCISES

1. What happens to the confidence interval as n increases?

2. Derive a formula for the minimum Sample size n necessary to obtain a confidence interval of length at most 2δ. If $\sigma = 7$ what number n is necessary to get a confidence interval of length 1 and of coefficient .98?

In the following exercises confidence intervals for σ^2 are derived.

3. Let h_α and k_α be positive numbers defined by

$$\int_0^{h_\alpha} \psi_n(y)\,dy = \int_{k_\alpha}^{\infty} \psi_n(y)\,dy = \alpha, \qquad 0 < \alpha < 1.$$

Suppose μ is known. Consider the Estimator of σ^2, $\overline{(x - \mu)^2}$, defined in Sec. 7-1. *Prove:* The probability that it falls between

$$\frac{\sigma^2 h_{\alpha/2}}{n} \qquad \text{and} \qquad \frac{\sigma^2 k_{\alpha/2}}{n}$$

is equal to $1 - \alpha$.

4. *Prove:* The probability that the interval from

$$\frac{n\overline{(x - \mu)^2}}{k_{\alpha/2}} \qquad \text{to} \qquad \frac{n\overline{(x - \mu)^2}}{h_{\alpha/2}}$$

contains the true value of σ^2 is equal to $1 - \alpha$. Use the result of Exercise 3.

5. Values of h_α and k_α for selected values of α and n are given in Table II, Appendix. Find the explicit confidence interval for σ^2 where the confidence coefficient is .95, $n = 6$, and $\overline{(x - \mu)^2} = 5$; use the general formula in Exercise 4.

6. If μ is unknown, the Estimator s^2 is used to form a confidence interval for σ^2. Show that when s^2 is used in place of $\overline{(x - \mu)^2}$, the only change in the formulas in Exercises 3 and 4 is that n is replaced by $n - 1$.

7. Repeat Exercise 5 for the case in which μ is unknown and the observed value of s^2 is 5.

8. Repeat Exercises 5 and 7 for 10 observations; then with confidence coefficient .98 (and 5 and 10 observations).

8-2. CONFIDENCE INTERVAL FOR μ WHEN σ IS UNKNOWN

The confidence interval (8-2) is expressed in terms of the known value of σ. If the latter is unknown, but s, the square root of the efficient Estimator of σ^2, is substituted for σ, then the derivation of the confidence interval (8-2) is no longer valid; indeed, the ratio

$$t = \sqrt{n}(\bar{x} - \mu)/s \tag{8-3}$$

does not have a normal distribution. The function (8-3) is known as the *t-statistic*. First we shall show that it is a statistic in the technical sense of Definition 4-1. Subtract the constant μ from each coordinate of $\mathbf{x}$ in (8-3); then the ratio becomes

$$\sqrt{n}\bar{x}/s. \tag{8-4}$$

Under the diagonal rotation, the function (8-4) is transformed into

$$\frac{x_1}{\left[\frac{1}{n-1}(x_2^2 + \cdots + x_n^2)\right]^{1/2}}. \tag{8-5}$$

The property of being a statistic is unchanged by these two operations on the function (8-3). It was indicated in Example 5-4 that (8-5) is a statistic; thus, (8-3) also is. Now we show that (8-3) has the t-distribution with $n - 1$ degrees of freedom. It follows from the transformation of the variables of integration from x_i to $(x_i - \mu)/\sigma$ in the multiple integral defining the distribution of t that the latter is the same for all μ and $\sigma > 0$; in particular, it is the same as that of (8-4) in sampling from a standard normal population. By the invariance of the distribution under rotation of the Sample space (Theorem 4-4), the distributions of (8-4) and (8-5) are the same; thus, by the result of Example 5-4, they have a common t-distribution with $n - 1$ degrees of freedom.

The remarkable property of the t-statistic is that neither it nor its distribution depends on σ; this enables us to construct a confidence interval for the unknown mean even when the standard deviation is unknown. The density function (5-21) of the t-distribution is symmetric about the origin, and has a unique maximum at $y = 0$. For $|y| \to \infty$ the density tends to 0 at the same asymptotic rate as $|y|^{-n}$. For large values of n the density resembles that of the standard normal (Exercise 8, Sec. 5-3). In Fig. 8-1 the graph of the t-density with 4 degrees of freedom is compared to the graph of the standard normal. By analogy to the definition of the percentile of the normal density we define the corresponding percentile $t_{n,\alpha}$ for the t-distribution with n degrees of freedom:

$$\int_{t_{n,\alpha}}^{\infty} s_n(y)\,dy = \alpha. \tag{8-6}$$

As in the normal case, symmetry implies:

$$\int_{-\infty}^{-t_{n,\alpha}} s_n(y)\,dy = \alpha.$$

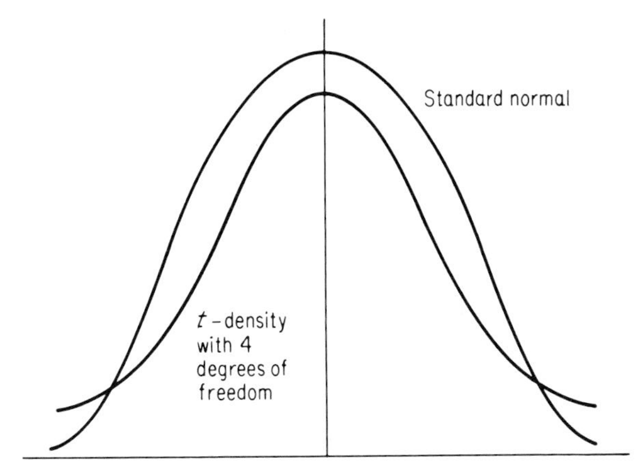

Fig. 8-1. Comparison of t and standard normal.

Values of $t_{n,\alpha}$ for various values of α and for $n = 1, \ldots, 45$ are given in Table III, Appendix. It is interesting to compare these to the corresponding percentiles of the normal distribution when the number of degrees of freedom is large; for example, for $n = 45$, we have these percentiles for the standard normal and t-distribution, respectively:

α	.025	.01	.005
z_α	1.96	2.33	2.58
$t_{45,\alpha}$	2.0141	2.4121	2.6896

The procedure for the construction of a confidence interval based on the t-distribution is embodied in

THEOREM 8-2

In sampling from a normal population with any (unknown) standard deviation, the interval

$$\bar{x} \pm \frac{t_{n-1,\alpha/2}s}{\sqrt{n}} \tag{8-7}$$

contains, with probability $1 - \alpha$, *the value of the unknown parameter* μ; *thus, (8-7) is a confidence interval of confidence coefficient* $1 - \alpha$.

Proof: The proof is similar to that of Theorem 8-1: s and t replace σ and z, respectively, and the t-distribution replaces the normal. ■

Example 8-3. As in Example 8-2 we consider a Sample of 25 observations. Suppose that σ is unknown, and the observed value of s is .5. A confidence interval of coefficient .95 is of length

$$2(.5)(2.0639)/5 = .4128,$$

and is centered at $\bar{x}$. This is slightly larger than the length of the corresponding interval for the case when σ is known to be .5. If the coefficient is raised to .98, the length is increased to

$$2(.5)(2.4922)/5 = .4984.$$

For confidence coefficient .99 the length is

$$2(.5)(2.7969)/5 = .5594. \quad \blacksquare$$

After comparing the previous two examples one may ask if there is any major difference between the confidence intervals based on the normal and t-distributions. Is the confidence interval constructed without the knowledge of σ inferior to the one constructed with the value of σ? There *is* a significant difference between the confidence intervals (8-2) and (8-7): The former has a fixed length $2z\sigma/\sqrt{n}$ which can be prescribed before the Sample is taken; on the other hand, the latter interval has the length $2st_{n-1,\alpha/2}/\sqrt{n}$ which depends on the value of s computed from the Sample. In the latter case the length of the interval cannot be prescribed before the Sample is observed: it varies with the random value of s. In the next section it will be shown how this disadvantage can be overcome by means of a "two-Sample" procedure.

EXERCISES

1. The following Sample of five observations is from a normal population with unknown standard deviation: -1, $+1$, 3, 2, 1. Construct a confidence interval with confidence coefficient .95. In computing s^2 use the formula $(n - 1)s^2 = \Sigma x_i^2 - n\bar{x}^2$ (Eq. 4-12) to avoid an excessive rounding error.

2. Repeat Exercise 1 with these nine observations: 1.5, 2.0, 5.3, 4.7, 3.6, 2.2, 4.0, 3.2, 3.9.

3. For $\sigma = s$, find the ratio of the length of the interval (8-2) to the length of (8-7) for $n = 5, 10, 15, 20, 25, 40$, and for $\alpha = .05, .01$.

4. As indicated in Sec. 7-4 the variance of s^2 is $2\sigma^4/(n-1)$. What does this imply about the distribution of s^2 when n is very large? (cf. Exercise 5, Sec. 7-1.)

5. Use the result of the previous exercise and the result of Exercise 8, Sec. 5-3 to show that there is little difference between the confidence intervals (8-2) and (8-7) when n is large.

8-3. STEIN'S TWO-SAMPLE PROCEDURE (OPTIONAL)

As shown at the end of the previous section the confidence interval for μ based on the t-distribution has length proportional to s; thus, it is a random quantity and so cannot be prescribed before the Sample is taken. The problem of constructing a confidence interval of prescribed length when σ is unknown was solved by C. Stein (1946) by the use of a two-Sample procedure. The idea of the method is to take an initial Sample of a fixed number of observations from the population; estimate σ^2 on the basis of these; finally, take an additional set of observations and form a confidence interval using the estimated value of σ^2 from the first Sample. The unique feature of the second set of observations—the second Sample—is that the number of observations is not fixed beforehand but varies in accordance with the estimated value of σ^2 obtained from the first Sample. This is one of the first examples of a *sequential* statistical procedure—one in which the number of observations taken varies with the values of previous observations. The two-Sample procedure is not clearly superior to the one in Theorem 8-2: The gain of a fixed-length confidence interval is paid for with the loss of a fixed Sample size.

In contrast to the elementary distribution theory behind the earlier confidence intervals, the theory for the two-Sample procedure is complicated. Let n stand for the number of observations in the initial Sample; it is an arbitrary positive integer. Let $\bar{x}$ and s^2 be the Sample mean and variance, respectively, computed from the Sample $\mathbf{X}$ of the first n observations. Having fixed n, we prescribe the length L of the confidence interval; then we define the number $c > 0$, depending on n, L, and α, by means of the relation

$$L = 2t_{n-1,\alpha/2}\sqrt{c}, \tag{8-8}$$

where $t_{n-1,\alpha/2}$ is defined as in (8-7). The procedure is completely specified by the choice of n, L, and α.

The number of observations in the second Sample is now defined as an integer-valued function on R^n. Let $m = m(\mathbf{x})$ be the function:

$$m(\mathbf{x}) = \max(n, [s^2/c] + 1),$$

where $[s^2/c]$ stands for the largest integer less than or equal to s^2/c; in other words,

$$\begin{aligned} m(\mathbf{x}) &= n && \text{if } s^2 < cn \\ &= [s^2/c] + 1 && \text{if } s^2 \geq cn. \end{aligned} \tag{8-9}$$

Here is the statistical procedure for the construction of a confidence interval of length L and of coefficient $1 - \alpha$:

Stein's Two-Sample Procedure

Take n observations and compute s^2:

1. If $s^2 < cn$, take no more observations and form the confidence interval based on the t-distribution with $n - 1$ degrees of freedom:

$$\bar{x} \pm \sqrt{c}\, t_{n-1,\alpha/2}. \tag{8-10}$$

2. If $s^2 \geq cn$, then, by (8-9), $m > n$. Take $m - n$ observations as a second Sample, and form the confidence interval centered at the mean of *all* m observations

$$\left(\sum_{i=1}^{m} x_i/m\right) \pm \sqrt{c}\, t_{n-1,\alpha/2}. \tag{8-11}$$

The form (8-10) is the special case of (8-11) for m $= n$. Eq. (8-8) implies that the length of the interval (8-11) is equal to L, as prescribed, We would like to prove:

THEOREM 8-3

The probability that the interval (8-11) contains μ is at least equal to $1 - \alpha$.

First we note that the confidence coefficient is not claimed to be exactly equal to $1 - \alpha$ but *at least* $1 - \alpha$. This does not tarnish the procedure because the confidence coefficient is *at least* as high as claimed. Secondly, we have to make some important qualifications in the statement of the theorem. The "probability" in

Theorem 8-3 is still undefined because the dimension of the Sample space has not been fixed: apparently it varies with the size of the second Sample. Probability is defined in Sec. 3-6 only for a fixed number n of observations but not a random number m; hence the probability distribution of the average of m observations, used in (8-11), is not well defined within our scope. (It is defined only in the context of an infinite-dimensional Sample space considered in advanced theory in a measure-theoretic setting.)

The probability statement in Theorem 8-3 is given precision in the following way. For each integer $N \geq n$, let T_N be the set of points in R^N:

$$T_N = \{\mathbf{x}: \mathbf{x} \in R^N, s^2 < cN\}, \qquad (8\text{-}12)$$

where s^2 is defined, as before, in terms of the first n coordinates of $\mathbf{x}$. Note that the membership of a point in T_N is determined by its first n coordinates. T_N is a regular subset of R^N. Its probability $P(T_N)$ is equal to the distribution function of s^2 at the point cN:

$$[(n - 1)/\sigma^2] \int_0^{cN} \psi_{n-1}(y(n - 1)/\sigma^2)\, dy;$$

thus $P(T_N) \to 1$ for $N \to \infty$. The importance of the set T_N is that the average of all m coordinates, $\sum_{i=1}^{m} x_i/m$, can be defined on the *subset* T_N of R^N: If $\mathbf{x} \in T_N$, then $s^2 < cN$ so that, by (8-9), $m \leq N$; thus the average of m coordinates is well defined because m is defined in terms of the first n coordinates, and $n \leq N$.

The set T_N engulfs all of R^N as $N \to \infty$ in the sense that $P(T_N) \to 1$. Theorem 8-3 embodies a probability statement about $\sum_{i=1}^{m} x_i/m$. This probability is defined as the limit for $N \to \infty$ of the probability of the intersection of T_N with the subset of R^N upon which (8-11) is well defined; more precisely, let P_N be the probability of the intersection of T_N and

$$\left\{\mathbf{x}: \mathbf{x} \in R^N, \left| \sum_{i=1}^{m} x_i/m - \mu \right| \leq \sqrt{c}\, t_{n-1,\alpha/2}\right\}. \qquad (8\text{-}13)$$

We shall prove this qualified version of Theorem 8-3:

THEOREM 8-3′

$$\liminf_{N\to\infty} P_N \geq 1 - \alpha. \qquad (8\text{-}14)$$

The meaning of (8-14) is: If N is very large, the probability—in terms of the Sample space R^N—that

$$s^2 < cN \quad \text{and} \quad \left| \sum_{i=1}^{m} x_i/m - \mu \right| \leq \sqrt{c}\, t_{n-1,\alpha/2}$$

is not appreciably less than $1 - \alpha$. Since $P(T_N) \to 1$, for large N it is very probable that $s^2 < cN$; thus the probability that the second inequality above holds by itself is not much less than $1 - \alpha$. This means that the probability that the interval (8-11) contains μ is not much less than $1 - \alpha$ for large N; furthermore, the limit inferior for $N \to \infty$ is at least $1 - \alpha$.

Proof of the Theorem 8-3': The proof involves the computation of the probability P_N of the set (8-13). It is done by decomposing the set into pieces on which m is constant; and then calculating the probability in terms of the functions s^2 and $\sum_{i=1}^{m} x_i/m$ for fixed m.

For each integer k, $n < k \leq N$, the equation $m = k$ holds if and only if

$$k - 1 \leq s^2/c < k \tag{8-15}$$

and $m = n$ if and only if $s^2/c < n$. T_N is the union of the disjoint sets in R^N

$$\{\mathbf{x}: k - 1 \leq s^2/c < k\}, \qquad k = 1, \ldots, N. \tag{8-16}$$

Now the probability P_N has been defined as the probability of the intersection of (8-12) and (8-13); thus, by the decomposition of T_N into the sets (8-16), the probability is equal to the sum of the probabilities of the pieces

$$\left\{\mathbf{x}: k - 1 \leq s^2/c < k, \left| \frac{1}{m} \sum_{i=1}^{m} x_i - \mu \right| \leq \sqrt{c}\, t \right\}, \qquad k = 1, \ldots, N \tag{8-17}$$

[additivity of probability—(3-33)]. Here and in the remainder of the proof we suppress the subscripts of $t = t_{n-1,\alpha/2}$. If $k \leq n$, the set (8-17) above takes on the special form

$$\{\mathbf{x}: k - 1 \leq s^2/c < k, \; |\bar{x} - \mu| \leq \sqrt{c}\, t\}; \tag{8-18}$$

if $n < k \leq N$, then (8-17) is equivalent to

$$\{\mathbf{x}: k - 1 \leq s^2/c < k, \left|\frac{1}{k}\sum_{i=1}^{k} x_i - \mu\right| \leq \sqrt{c}t\} \tag{8-19}$$

because (8-15) holds if and only if $m = k$.

The integrals of L over the sets (8-18) and (8-19) respectively are the same for all μ because we may change the variables of integration from x_i to $x_i - u$; thus, without loss of generality, we take $\mu = 0$. We now employ Theorem 4-4, which permits us to replace the sets (8-18) and (8-19) by their images under rotation; recall that the rotation-invariance holds for all $\sigma > 0$ as long as $\mu = 0$ (Exercise 1, Sec. 4-5). Apply the diagonal rotation to the first n coordinates: by Eqs. 5-8, the sets (8-18) and (8-19) are transformed into

$$\left\{\mathbf{x}: c(n - 1)(k - 1) \leq \sum_{i=2}^{n} x_i^2 < c(n - 1)k, \ |x_1| \leq \sqrt{cn}t\right\}, \quad k \leq n, \tag{8-20}$$

and

$$\left\{\mathbf{x}: c(n - 1)(k - 1) \leq \sum_{i=2}^{n} x_i^2 < c(n - 1)k, \quad \frac{\sqrt{n}x_1}{k} + \sum_{i=n+1}^{k} \frac{x_i}{k} \leq \sqrt{c}t\right\}, \quad n < k \leq N, \tag{8-21}$$

respectively. By means of two more rotations we further simplify the set (8-21). The first is the diagonal rotation in the variables $n + 1, \ldots, k$. The sum

$$\sum_{i=n+1}^{k} x_i$$

is transformed into $\sqrt{k - n}x_{n+1}$; thus the inequality describing the second condition in (8-21) becomes

$$|\sqrt{n}x_1 + \sqrt{k - n}x_{n+1}| \leq k\sqrt{c}t, \tag{8-22}$$

and the first inequality in (8-21) is unchanged because the rotation does not affect the variables of indices up to and including n. The final rotation is a plane rotation in the pair x_1, x_{n+1}:

$$x_1' = \sqrt{n/k}x_1 + \sqrt{(n - k)/k}x_{n+1}$$

$$x_{n+1}' = \sqrt{(n - k)/k}x_1 - \sqrt{n/k}x_{n+1}.$$

This rotation transforms the condition (8-22) into $|x_1| \le \sqrt{ckt}$ and leaves the first inequality in (8-21) unchanged; thus, under the succession of the three rotations the region (8-21) is transformed into

$$\left\{\mathbf{x}: c(k-1)(n-1) \le \sum_{i=2}^{n} x_i^2 < ck(n-1), \ |x_1| \le \sqrt{ckt}\right\}, \quad n < k \le N. \tag{8-23}$$

For each k, $k \le N$, the set

$$\left\{\mathbf{x}: |x_1| \le t\left(\sum_{i=2}^{n} x_i^2/(n-1)\right)^{1/2}, \quad c(k-1)(n-1) \le \sum_{i=2}^{n} x_i^2 < ck(n-1)\right\} \tag{8-24}$$

is contained in (8-20) for $k \le n$ and in (8-23) for $n < k \le N$; therefore, the probabilities of (8-20) and (8-23) are at least equal to that of (8-24) for $k \le n$ and for $n < k \le N$, respectively [property (3-32)]. The sets (8-20) and (8-23) are all disjoint so that the probability of their union over all indices k is the sum of the probabilities; and the same is true for the sets (8-24); therefore, the probability of the union of all sets in (8-20) and (8-23) is at least equal to the probability of the union of the sets (8-24). Now the union of the sets (8-24) over $1 \le k \le N$ is

$$\left\{\mathbf{x}: |x_1| \le t\left(\sum_{i=2}^{n} x_i^2/(n-1)\right)^{1/2}, \sum_{i=2}^{n} x_i^2 < cN(n-1)\right\}. \tag{8-25}$$

Even though this set is defined as a subset of a space of dimension N, it is specified by conditions on just the first n coordinates; thus the probability is computed as if the set (8-25) were a subset of n-dimensional space. [The point is that even though N appears in the inequality defining the domain of integration (8-25), the space may be considered to be of fixed dimension n.]

The probability of (8-25) is equal to value of the joint distribution of the statistics

$$\frac{|x_1|}{\left(\sum_{i=2}^{n} x_i^2/(n-1)\right)^{1/2}} \tag{8-26}$$

and

$$\sum_{i=2}^{n} x_i^2 \tag{8-27}$$

at the point $(t, cN(n - 1))$; therefore, as $N \to \infty$, it converges to the marginal distribution of (8-26) at t (Sec. 5-1)—that is, to the probability that (8-26) does not exceed t. This is equal to $1 - \alpha$; indeed, by the result of Example 5-4, the statistic

$$\frac{x_1}{\left(\sum_{i=2}^{n} x_i^2/(n - 1)\right)^{1/2}}$$

has the t-distribution with $n - 1$ degrees of freedom, and so the assertion follows from the definition of the $\alpha/2$-percentile.

We have shown that P_N is at least equal to the probability of the set (8-25), whose limit is $1 - \alpha$; thus, the relation (8-14) is established. ■

Here is a numerical example illustrating the two-Sample method.

Example 8-4. Suppose we want a confidence interval of length $L = 2$ and coefficient .98; and that there are $n = 10$ observations in the first Sample. From Table III, Appendix, we find that

$$t_{9,.01} = 2.8214.$$

Suppose that s^2 is found in the first Sample to have the value 5.2; then from Eq. 8-8 we have

$$s^2/c = s^2(2t_{9,.01}/L)^2 = (5.2)(2.8214)^2 = 41.5,$$

so $[s^2/c] + 1 = 42$—that is, $m = 42$. The second Sample consists of 32 observations. The confidence interval is centered at the average of all 42 observations and is of length 2. ■

8-4. SOME REMARKS ON CONFIDENCE INTERVALS FOR PARAMETERS OF OTHER DENSITIES (OPTIONAL)

In contrast to the complete theory of confidence intervals for normal populations the theory for general populations contains only of a few concrete results. Let us generalize the technique used above in the normal case. Suppose $f(\mathbf{x})$ is an Estimator of the parameter θ based on a Sample of n observations from the population with density $h(x;\theta)$; and suppose also that there is a function $g(f;\theta)$ of two variables such that the statistic

$$g(f(\mathbf{x});\theta)$$

has, for each θ, a distribution which does not depend on the particular value of θ; for example, in sampling from a normal population with mean μ and standard deviation σ, the statistic $\sqrt{n}(\bar{x} - \mu)/\sigma$ has a standard normal distribution, which is independent of the parameter values μ and σ. A confidence interval can sometimes be constructed from the distribution of $g(f(\mathbf{x});\theta)$ by inverting a probability statement about this statistic; for example, if, for $0 < \alpha < 1$, there exist numbers z_1 and z_2 such that the probability of the set

$$\{\mathbf{x}\colon z_1 \leq g(f(\mathbf{x});\theta) \leq z_2\}$$

is equal to $1 - \alpha$ for all θ, and if the double inequality defining this set can be inverted to a double inequality on θ:

$$\{x\colon g^*(z_1, f(\mathbf{x})) \leq \theta \leq g^*(z_2; f(\mathbf{x}))\}$$

then the interval $[g^*(z_1; f(\mathbf{x})),\ g^*(z_2; f(\mathbf{x}))]$ is a confidence interval for θ of coefficient $1 - \alpha$. The main difficulties in the application of this method are (i) The function g may not exist, or, if it does, may not be easily invertible; and, (ii) The distribution of g may be hard to calculate.

We conclude this section by showing that the confidence interval (8-7) is "robust" if the sample size n is large. Let $h(x;\theta)$ be a density function such that

$$\int_{-\infty}^{\infty} x^4 h(x;\theta)\,dx < \infty, \qquad \textit{for all } \theta. \tag{8-28}$$

By means of the Cauchy-Schwarz inequality it can then be shown that μ and σ^2 are well defined and finite (Exercise 1). In accordance with Sec. 7-5, $\bar{x}$ and s^2 are unbiased Estimators of μ and σ^2, respectively. In the normal case the variance of s^2 is $2\sigma^4/(n - 1)$ (Sec. 7-4); furthermore, in the general case under the condition (8-28) it can be shown that

$$n \cdot \mathrm{Var}\,(s^2)$$

is bounded in n, so that $\mathrm{Var}(s^2)$ tends to 0 as $n \to \infty$ (Exercise 2). This implies that the distribution of s^2 "clusters" about the expected value σ^2 as n increases (Sec. 7-1, Exercise 5, and Sec. 8-2, Exercise 4), so that s^2/σ^2 "tends" to be close to 1. It follows that the ratios

$$\sqrt{n}(\bar{x} - \mu)/s \quad \text{and} \quad \sqrt{n}(\bar{x} - \mu)/\sigma$$

will probably differ by very little. Now the central limit theorem (Sec. 4-6) implies that the second of these statistics has a nearly normal distribution; furthermore, for large n, the normal and t-distribution with n degrees of freedom are nearly the same (Sec. 5-3, Exercise 8). We conclude that Theorem 8-2 remains valid for an arbitrary population satisfying (8-28) if the probability statement is changed to "approximately $1 - \alpha$," and n is large.

EXERCISES

1. Show that (8-28) implies the finiteness of

$$\int_{\infty}^{\infty} x^2 h(x;\theta)\,dx.$$

(*Hint:* Apply the Cauchy-Schwarz inequality with $f \to x^2$, $g \to 1$, $L \to h$.) This implies the existence and finiteness of μ and σ^2 (Sec. 2-8).

2. Show, directly from the definition of s^2 that n Var (s^2) is bounded in n under the assumption (8-28).

8-5. NUMERICAL COMPARISON OF CONFIDENCE INTERVALS BASED ON NORMAL DEVIATES

In order to compare the confidence intervals based on the normal and t-distribution, we shall use the 25 values of $\bar{x}$ and s^2 in Table 7-1, and construct 25 corresponding pairs of confidence intervals for $\mu = 0$; here $\sigma = 1$. The confidence intervals are given in Table 8-1. In each case a check indicates that the confidence interval contains the value $\mu = 0$. The confidence coefficient is .95, and $n = 5$; thus:

$$z_{\alpha/2} = z_{.025} = 1.960; \qquad t_{n-1,\alpha/2} = t_{4,.025} = 2.776,$$

and so

$$z_{\alpha/2}\sigma/\sqrt{n} = .877; \qquad t_{n-1,\alpha/2}s/\sqrt{n} = 2.776\sqrt{s^2/5}.$$

It is interesting that 15 of the 25 intervals based on the t-distribution are longer than the one based on the normal distribution; and that 24 of the 25 intervals—or 96 percent—contain the "true" parameter value $\mu = 0$.

In Table 8-2 we also compute the number of necessary observations for the two-sample confidence interval of the fixed length 2(.877), comparable to the one given in Table 8-1. The

TABLE 8-1
Confidence Intervals of Coefficient .95 for Samples of 5 When $\mu = 0$ and $\sigma = 1$.

$\bar{x} \pm .877$	Contains 0	$\bar{x} \pm 2.776s/\sqrt{5}$	Contains 0
$-.734 \pm .877$	√	$-.734 \pm 1.568$	√
$-.650 \pm .877$	√	$-.650 \pm .534$	
$-.035 \pm .877$	√	$-.035 \pm .615$	√
$.159 \pm .877$	√	$.159 \pm 1.445$	√
$-.075 \pm .877$	√	$-.075 \pm .770$	√
$.427 \pm .877$	√	$.427 \pm .864$	√
$-.380 \pm .877$	√	$-.380 \pm .691$	√
$-.102 \pm .877$	√	$-.102 \pm 1.299$	√
$-.677 \pm .877$	√	$-.677 \pm 1.103$	√
$.239 \pm .877$	√	$.239 \pm 1.232$	√
$.508 \pm .877$	√	$.508 \pm 1.382$	√
$-.391 \pm .877$	√	$-.391 \pm 1.225$	√
$-.334 \pm .877$	√	$-.334 \pm 1.031$	√
$.091 \pm .877$	√	$.091 \pm 1.181$	√
$-.098 \pm .877$	√	$-.098 \pm .780$	√
$-.210 \pm .877$	√	$-.210 \pm 1.693$	√
$.100 \pm .877$	√	$.100 \pm .819$	√
$-.004 \pm .877$	√	$-.004 \pm 1.275$	√
$.187 \pm .877$	√	$.187 \pm 1.675$	√
$.822 \pm .877$	√	$.822 \pm 2.030$	√
$-1.023 \pm .877$		-1.023 ± 1.064	√
$.709 \pm .877$	√	$.709 \pm .735$	√
$.339 \pm .877$	√	$.339 \pm 1.679$	√
$-.206 \pm .877$	√	$-.206 \pm .856$	√
$-.415 \pm .877$	√	$-.415 \pm .800$	√

initial Sample is taken to consist of five observations; then $c = (.877/2.776)^2$; thus, to three significant digits, $c^{-1} = 10.0$. The total number of observations is given by (8-9) under the column headed m. Note that the average value of m is approximately 10, so that the average number of additional observations needed to get an interval of fixed length is 5.

TABLE 8-2
Number of Observations in Two-Sample Confidence Intervals for $L/2 = .877, \alpha = .05, \mu = 0, \sigma = 1.$

s^2	$s^2/c = s^2(10.0)$	$m = \max(5, [s^2/c] + 1)$
1.596	15.96	16
.183	1.83	5
.245	2.45	5
1.354	13.54	14
.385	3.85	5
.486	4.86	5
.310	3.10	5
1.093	10.93	11
.789	7.89	8
.985	9.85	10
1.240	12.40	13
.976	9.76	10
.690	6.90	7
.906	9.06	10
.396	3.96	5
1.858	18.58	19
.437	4.37	5
1.056	10.56	11
1.821	18.21	19
2.677	26.77	27
.736	7.36	8
.352	3.52	5
1.830	18.30	19
.474	4.74	5
.416	4.16	5. $\overline{m} = 10$

EXERCISE

1. (*Optional*) Using Table VI, Appendix, construct a version of Table 8-2 for ten Samples of $n = 4$ observations each, and with $\alpha = .05$ and $L/2 = .877$.

BIBLIOGRAPHY

Confidence intervals based on the normal and t-distributions are given in all introductory books on statistics. Procedures for general parametric families are given in Cramér's book, which has been mentioned at the end of several chapters. The two-Sample confidence interval is given in "Notes on the Theory of Estimation," in *Lectures by E. L. Lehmann* (recorded by Colin Blyth) (Stanford: U. of California Press, 1950), Chapters 1 to 5, and in Samuel S. Wilks *Mathematical Statistics* (New York: Wiley, 1963).

CHAPTER 9

Testing Hypotheses

9-1. SIMPLE HYPOTHESES

Consider a normal population with given standard deviation σ. Suppose that μ is only partially known: it is known that μ is equal to one of two specified values μ_0 and μ_1 but not which of the two it is. We identify μ_0 with the "hypothesis that the true mean μ is equal to μ_0"; and μ_1 with the "hypothesis that $\mu = \mu_1$." These are referred to as the "simple hypothesis μ_0" and the "simple hypothesis μ_1," respectively. By convention, one of these is called the *null* hypothesis and the other the *alternative* hypothesis; thus we shall call μ_0 and μ_1 the null and alternative simple hypotheses, respectively. As in the problem of estimation (Sec. 7-1) we suppose that some decision has to be made about the population, and that the consequences depend not only on the decision but also on the imperfectly known value of μ.

The statistician must first decide whether μ_0 or μ_1 is the true value of μ. His choice is liable to error: he might decide that the alternative hypothesis is true when the null hypothesis really is; or, that the null hypothesis is true when the alternative really is. These two types of error are called "error of the first kind" and "error of the second kind," respectively. In order to get information about the truth of the competing hypotheses, the statistician takes a sample from the population, and infers from the observed values. Let us illustrate this by means of a simple example. The null and alternative hypotheses are $\mu_0 = 5$ and $\mu_1 = -5$, respectively, and $\sigma = 1$ is given. The density functions are drawn on a single graph in Fig. 9-1.

A Sample of $n = 3$ observations is taken, and the observed values are 5.3, 4.6, and 4.7. In choosing between -5 and $+5$ as the true mean, we make a reasonable decision in taking $\mu = 5$; indeed, it is much more likely (in a nontechnical sense) that the values

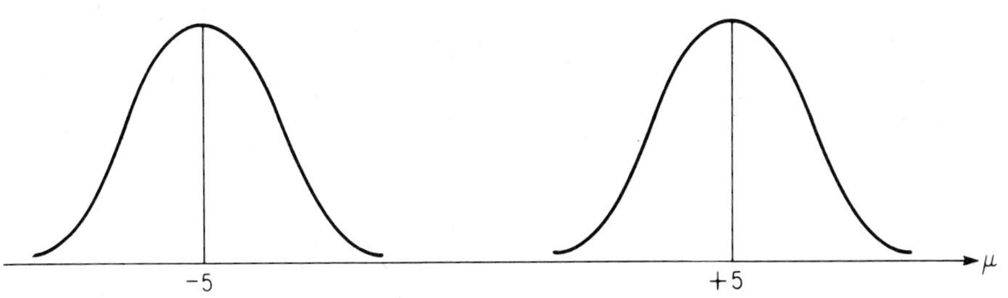

Fig. 9-1. Densities under the null and alternative hypotheses $\mu = \pm 5$ when $\sigma = 1$.

observed were from a normal population with mean 5 and standard deviation 1 than from one with mean -5. It would be harder to choose between the two if σ were large and the observed values not so closely clustered about μ_0; for example, if $\sigma = 10$ and the observed values 6, -4, and 0, it is not obvious which of the two means is likely to be true (see Fig. 9-2).

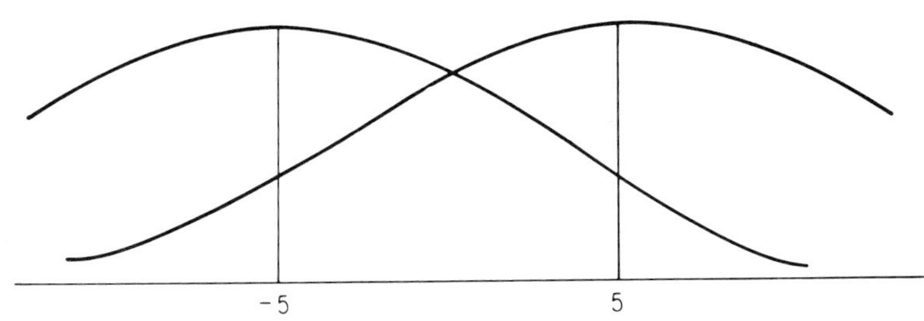

Fig. 9-2. Densities for $\mu = \pm 5$ and $\sigma = 10$.

Here is an outline of the general theory of testing simple hypotheses. Suppose that the decision μ_0 or μ_1 is to be made on the basis of a Sample of n observations from the normal population. The Sample space is decomposed into two nonoverlapping regions, the *acceptance* and *rejection* regions. The Sample point is observed. If it falls in the first region, the null hypothesis μ_0 is accepted as true; if it falls in the second region, the null hypothesis is rejected in favor of the alternative hypothesis μ_1. This procedure is called a *test of* μ_0 *against* μ_1; it is completely specified by the choice of a rejection region. The rejection region (and so its complement, the acceptance region) is always assumed to be a regular set. By definition the probability that **X** falls in the acceptance (or rejection) region is the intergral of L over the region. There are two likelihood functions L corresponding to the two hypotheses: when $\mu = \mu_0$, the likelihood function is $L_0 = L(\mathbf{x}; \mu_0, \sigma)$, and when $\mu = \mu_1$ it is $L_1 = L(\mathbf{x}; \mu_1, \sigma)$. The probability of an error of the first kind is the probability that

$\mathbf{X}$ falls in the rejection region, computed as the integral of L_0 over the region. The probability of an error of the second kind is, analogously, the integral of L_1 over the acceptance region. The probabilities of error of the first and second kind are denoted α and β, respectively; thus

$$\alpha = \underset{\text{[rejection region]}}{\int \cdots \int} L_0 \, d\mathbf{x}$$

$$\beta = \underset{\text{[acceptance region]}}{\int \cdots \int} L_1 \, d\mathbf{x}. \tag{9-1}$$

Example 9-1. We wish to test $\mu_0 = -2$ against $\mu_1 = +1$; σ is assumed to be 3. A Sample of one observation is taken; the Sample space is the real line. Decompose the line into acceptance and rejection regions; for example, take the half-line $(-\infty, 0]$ as the acceptance region and $(0, \infty)$ as the rejection region. This choice, though arbitrary, is based on the consideration that if μ_0 is true, then the Sample point X is likely to be closer to it (thus negative) while if μ_1 is true, X is more likely to be positive. The probabilities of error of the two kinds are computed as follows:

$$\alpha = \int_0^\infty \frac{1}{3} \phi\left(\frac{x+2}{3}\right) dx = 1 - \Phi\left(\frac{2}{3}\right) = .2514.$$

$$\beta = \int_{-\infty}^0 \frac{1}{3} \phi\left(\frac{x-1}{3}\right) dx = \Phi\left(\frac{-1}{3}\right) = .3707. \quad \blacksquare$$

The probability α is known as the *level of significance* or *size* of the test. The probability of rejecting μ_0 in favor of μ_1 when the latter is true is called the *power* of the test; it is equal to $1 - \beta$.

There are usually many tests (choices of rejection regions) available in a given situation; the corresponding values (α, β) may vary. Which is the best test? The usefulness of a test is determined not only by the error probabilities but also by the relative costs of the two kinds of errors. One kind may be much more serious than the other. Suppose that an engineer wants to improve the quality of a building material by means of a chemical additive. For simplicity, let us assume that the natural mean breaking strength of the material is a known number μ_0; and that the developer of the additive claims that it will raise the strength to the level $\mu_1 > \mu_0$. We shall also suppose that μ_0 is too low for the material to be used in certain structures, while μ_1 is sufficiently high for its use. The engineer tests the material after treatment with the additive. The two kinds of error have costs of different magnitudes. If μ_0 is accepted when μ_1 is really true—that is, the additive is rejected even though it is of real value, then the engineer unnecessarily loses at least some of the cost of developing the additive and may cancel or postpone some projects. If μ_1 is accepted when μ_0 is really true, that is, the material is considered to be safe when it really is not, then there may be large losses caused by the collapse of unsafe structures. The costs of these two errors are determined not by the statistician but by the engineer, who knows the economics of the industry.

Before World War II statistics was applied mainly to agricultural and low-cost scientific experiments, so that the concept of the cost of error was not important. Cost was introduced into the theory of statistics by Abraham Wald during the war when statistics was first used for large and expensive industrial and military experiments. We shall consider only the classical theory of hypothesis testing which does not involve cost; it is still very useful in many scientific and commercial problems. The reader interested in the more general theory can find it in the references.

Let (α, β) and (α', β') be the pairs of error probabilities for two different tests. The first test is considered better than the second if

$$\alpha \leq \alpha' \quad \text{and} \quad \beta \leq \beta'$$

with at least one of these inequalities strict. It is impossible to find a best test in the sense that α and β are simultaneously minimized. If such a test existed its error probabilities would be equal to 0 (that is, it would always yield the correct decision); indeed, the test whose rejection region is the whole space R^n has $\beta = 0$, and the test whose rejection region is empty has $\alpha = 0$. It can be shown that a test with $\alpha = \beta = 0$ does not exist. It has been the convention in classical statistical theory to fix the level of significance α and seek a best test among those of size α; among those of this size, we look for the one minimizing β. Since a test minimizing β maximizes the power $1 - \beta$, such a test is called a *most powerful test of size* α.

Example 9-2. Consider another test for the hypotheses in Example 9-1. It will be of the same approximate size as the one in the latter example but its power will be smaller. Put

$$\text{Acceptance region} = [-6.98, 0.5]$$

$$\text{Rejection region} = \text{union of } (-\infty, -6.98) \quad \text{and} \quad (0.5, +\infty);$$

then

$$\alpha = \left(\int_{-\infty}^{-6.98} + \int_{0.5}^{\infty}\right) \frac{1}{3}\,\phi\left(\frac{x+2}{3}\right) dx$$
$$= \Phi(-1.66) + 1 - \Phi(.83) = .2518,$$

and

$$\beta = \int_{-6.98}^{0.5} \frac{1}{3}\,\phi\left(\frac{x-1}{3}\right) dx = \Phi(-1.67) - \Phi(-2.66) = .4286.$$

The level of significance is slightly larger than that in Example 9-1 but the probability of error of the second kind is appreciably larger; thus the first test is better than this one. ■

Referring to the representation (9-1) of α and β, we put the problem of determining the most powerful test of size α in this precise form:

Among all regular sets M in the Sample space satisfying

$$\int_M \cdots \int L_0 \, d\mathbf{x} = \alpha, \tag{9-2}$$

find the one—if it exists—which maximizes the integral

$$\int_M \cdots \int L_1 \, d\mathbf{x}. \tag{9-3}$$

The latter represents the power of the test with the rejection region M because β is the integral of L_1 over the *complement* of M.

In order to solve this problem we reduce it to one involving the integrals of continuous functions over R^n instead of integrals over regular sets. Then we show that the solution of the reduced problem furnishes the solution of the original one. This is done by recalling that the integral of L over a regular set is approximable by an integral of $f \cdot L$ over the space, where f is continuous [properties (3-34) and (3-35)]. So we now replace the maximization problem (9-2) and (9-3) by the following one:

Considering all continuous functions f on R^n such that

$$0 \leq f(\mathbf{x}) \leq 1, \qquad \mathbf{x} \in R^n, \tag{9-4}$$

and which satisfy

$$\int_{R^n} \cdots \int f(\mathbf{x}) \, L_0 \, d\mathbf{x} = \alpha, \tag{9-5}$$

find an upper bound of the set of real numbers of the form

$$\int_{R^n} \cdots \int f(\mathbf{x}) \, L_1 \, d\mathbf{x}. \tag{9-6}$$

We do not attempt to prove that the solution of the latter problem *always* yields that of the former; however, this will be shown to be so in the cases considered.

The hypothesis testing problem is now the mathematical problem of maximizing the integral (9-6) under the conditions (9-4) and (9-5). Note that this is similar to the form of the problem of finding the efficient Estimator (Chapter 7); there we minimized the integral for the variance under the integral condition for unbiasedness. As in estimation theory, we first reduce the integrals (9-5) and (9-6) by the method of sufficiency. Let f be the conditional expectation of f given $\sqrt{n}\bar{x}$; then, as an average of f, the function f also satisfies (9-4), and the integrals (9-5) and (9-6) are unchanged if $f(\mathbf{x})$ is replaced by $f(\sqrt{n}\bar{x};)$ (Eq. 6-14). Invoking (6-14) once more, we replace the maximization problem involving (9-4), (9-5), and (9-6) by:

Considering all continuous functions f on the real line such that

$$0 \leq f(x) \leq 1, \tag{9-7}$$

and which satisfy

$$\int_{-\infty}^{\infty} f(x) \, \frac{\sqrt{n}}{\sigma} \, \phi\left(\frac{\sqrt{n}(x - \mu_0)}{\sigma}\right) dx = \alpha, \tag{9-8}$$

find an upper bound of

$$\int_{-\infty}^{\infty} f(x) \, \frac{\sqrt{n}}{\sigma} \, \phi\left(\frac{\sqrt{n}(x - \mu_1)}{\sigma}\right) dx. \tag{9-9}$$

This is solved in the next section.

EXERCISES

1. Consider testing $\mu_0 = -1$ against $\mu_1 = +1$ when $\sigma = 1$. A Sample of four observations is taken. The null hypothesis is accepted if the Sample mean is less than $-.2$, and otherwise rejected. Using the fact that $\bar{x}$ has a normal distribution (with mean μ and standard deviation $\sigma/\sqrt{n}$) find α and β.

2. Repeat Exercise 1 for $n = 9$.

3. In Exercise 1, what happens to α and β as $n \to \infty$?

4. Consider a normal population with $\mu = 0$ but σ unknown. We want to test the null hypothesis $\sigma = 1$ against the alternative $\sigma = 2$. One observation X is taken: the null hypothesis is accepted if $|X| \leq 1.5$ and otherwise rejected. Find α and β.

5. Repeat Exercise 4 for two observations: the null hypothesis is accepted if the Sample point falls inside the circle of radius $\sqrt{2}(1.5)$ centered at the origin, and otherwise rejected. Find α and β by using the distribution of the statistic $x_1^2 + x_2^2$ (Example 4-4 or Theorem 4-2).

6. If $\mu = 0$, and the null hypothesis $\sigma = \sigma_0$ is tested against the alternative $\sigma = \sigma_1$, the only change in the formulation of the maximization problem for (9-3) and (9-6) is that L_0 and L_1 are now the likelihood functions with $\mu = 0$, $\sigma = \sigma_0$ and $\mu = 0$, $\sigma = \sigma_1$, respectively. What changes are necessary in (9-8) and (9-9)?

9-2. THE NEYMAN-PEARSON LEMMA AND THE MOST POWERFUL TEST

The maximization of (9-9) is effected by means of the following proposition, which is a special case of a general theorem known as the *Fundamental Lemma of Neyman and Pearson*:

LEMMA 9-1 (Neyman-Pearson)
Let $p_0(x)$ and $p_1(x)$ be two continuous positive functions such that

$$\int_{-\infty}^{\infty} p_i(x)\,dx = 1, \qquad i = 0, 1,$$

and $p_1(x)/p_0(x)$ is nondecreasing in x. Let f be a continuous function on the line such that $0 \leq f(x) \leq 1$; put

$$\alpha = \int_{-\infty}^{\infty} f(x)\,p_0(x)\,dx,$$

so that $0 \leq \alpha \leq 1$. *If* $0 < \alpha < 1$, *let* x_α *be the real number satisfying*

$$\int_{x_\alpha}^{\infty} p_0(x)\,dx = \alpha;$$

and, if $\alpha = 0$ *or* 1, *then* $x_a = +\infty$ *or* $-\infty$, *respectively. Our conclusion is*

$$\int_{-\infty}^{\infty} f(x) p_1(x)\,dx \leq \int_{x_\alpha}^{\infty} p_1(x)\,dx. \tag{9-10}$$

Proof: Suppose first $0 < \alpha < 1$. The assumption on p_0 and the definition of x_α imply:

$$\int_{-\infty}^{x_\alpha} f(x)\, p_0(x)\,dx = \int_{x_\alpha}^{\infty} (1 - f(x))\, p_0(x)\,dx. \tag{9-11}$$

Split the integral on the left-hand side of (9-10) into integrals over $(-\infty, x_\alpha]$ and (x_α, ∞), respectively; then the inequality (9-10) is equivalent to

$$\int_{-\infty}^{x_\alpha} f(x)\, p_1(x)\,dx \leq \int_{x_\alpha}^{\infty} (1 - f(x))\, p_1(x)\,dx. \tag{9-12}$$

Since, by hypothesis, p_1/p_0 does not decrease as x increases, $p_1(x)/p_0(x) \leq p_1(x_\alpha)/p_0(x_\alpha)$ for $x \leq x_\alpha$, so that the integral on the left-hand side of (9-12) is at most equal to

$$[p_1(x_\alpha)/p_0(x_\alpha)] \int_{-\infty}^{x_\alpha} f(x)\, p_0(x)\,dx.$$

By (9-11) this is equal to

$$[p_1(x_\alpha)/p_0(x_\alpha)] \int_{x_\alpha}^{\infty} (1 - f(x))\, p_0(x)\,dx.$$

Again applying the monotone property of p_1/p_0, we find that

$$p_0(x) \leq p_1(x)\, p_0(x_\alpha)/p_1(x_\alpha), \qquad x \geq x_\alpha,$$

so that the expression above is at most equal to the integral on the right-hand side of (9-12). This completes the proof of (9-10) for $0 < \alpha < 1$.

Suppose $\alpha = 0$; then the right-hand side of (9-10) is equal to 0. We shall prove that the left-hand side is also 0:

$$\int_{-\infty}^{\infty} f(x)p_1(x)\,dx = \lim_{c\to\infty}\int_{-c}^{c} f(x)p_1(x)\,dx$$

$$\le \lim_{c\to\infty} (p_1(c)/p_0(c)) \int_{-c}^{c} f(x)p_0(x)\,dx$$

$$\le \lim_{c\to\infty} (p_1(c)/p_0(c)) \int_{-\infty}^{\infty} f(x)p_0(x)\,dx = 0.$$

Suppose $\alpha = 1$; then, the right-hand side of (9-10) is equal to 1, and the inequality is trivial because $f(x) \le 1$. ■

As a consequence of this lemma we get an upper bound for the integral (9-9) under the condition $\mu_0 < \mu_1$. There is an analogous result under the alternative condition $\mu_1 < \mu_0$.

COROLLARY

Suppose $\mu_0 < \mu_1$. Under the conditions (9-7) and (9-8) the integral (9-9) is at most equal to

$$1 - \Phi\left(z_\alpha + \frac{\sqrt{n}(\mu_0 - \mu_1)}{\sigma}\right) \qquad (9\text{-}13)$$

where z_α is the upper α-percentile of the standard normal distribution.

Proof: Put $x_\alpha = z_\alpha \sigma/\sqrt{n} + \mu_0$, and

$$p_i(x) = \frac{\sqrt{n}}{\sigma}\,\phi\left(\frac{\sqrt{n}(x - \mu_i)}{\sigma}\right), \qquad i = 0, 1;$$

then the ratio p_1/p_0 increases with x and the conditions of Lemma 9-1 are satisfied (Exercise 1). The expression (9-13) corresponds to the right-hand side of Eq. 9-10. ■

By means of the corollary, we solve the maximization of (9-6) and then of (9-3):

THEOREM 9-1

Suppose $\mu_0 < \mu_1$. Among all regular sets M satisfying (9-2), the one maximizing the integral (9-3) is

$$M_0 = \left\{\mathbf{x}: \bar{x} > \mu_0 + \frac{\sigma z_\alpha}{\sqrt{n}}\right\},$$

and the maximum value is given by (9-13). In statistical terms the most powerful test of size α has the rejection region M_0.

Proof: Since every integral of the form (9-6) is reducible to one like (9-9), we infer from the corollary that the integral (9-6) is bounded above by (9-13) under the conditions (9-4) and (9-5). Now we shall show that the integral (9-3) has the same bound.

Let M be a regular set satisfying (9-2) and f a function in $\mathfrak{L}(M)$. (We recall from Sec. 3-4 that f is continuous, assumes values from 0 to 1, and vanishes outside M.) It follows from the definition of the integral over regular sets that

$$\int_{R^n}\cdots\int fL_i\,d\mathbf{x} \le \int_M\cdots\int L_i\,d\mathbf{x}, \qquad i = 0, 1. \tag{9-14}$$

Put $\alpha' = \int_{R^n}\cdots\int fL_0\,d\mathbf{x}$; then, by the above inequality, we have $\alpha' \le \alpha$. In accordance with the opening statement of this proof—with α' in place of α—we have

$$\int_{R^n}\cdots\int fL_1\,dx \le 1 - \Phi\left(z_{\alpha'} + \frac{\sqrt{n}(\mu_0 - \mu_1)}{\sigma}\right).$$

Since $z_\alpha \le z_{\alpha'}$, and Φ is increasing, it follows that the inequality above is preserved when α' is replaced by α on the right-hand side:

$$\int_{R^n}\cdots\int fL_1\,dx \le 1 - \Phi\left(z_\alpha + \frac{\sqrt{n}(\mu_0 - \mu_1)}{\sigma}\right).$$

Since f is an arbitrary function in the class $\mathfrak{L}(M)$, the latter inequality is also preserved upon replacing the left-hand member by its least upper bound over all f in $\mathfrak{L}(M)$; thus, by property (3-34) of the integral, the inequality becomes

$$\int_M\cdots\int L_1\,d\mathbf{x} \le 1 - \Phi\left(z_\alpha + \frac{\sqrt{n}(\mu_0 - \mu_1)}{\sigma}\right). \tag{9-15}$$

This shows that the integral (9-3) has the bound (9-13); it remains for us to show that the bound is actually attained with the set M_0.

The statistic $\bar{x}$ has a normal distribution with mean μ_i and standard deviation $\sigma/\sqrt{n}$ with respect to the likelihood function L_i, $i = 0, 1$ (Theorem 4-5 and its corollaries); thus the integral of L_0 over M_0 is equal to α, and the integral of L_1 over M_0 is equal to the expression on the right-hand side of (9-15) (Exercise 2). ■

Example 9-3. Let us test $\mu_0 = -1$ against the alternative $\mu_1 = +2$ with $\sigma = 2$ and $\alpha = .05$. Suppose that $n = 16$. Since $z_{.05} = 1.64$, the rejection region of the most powerful test of size .05 is

$$\{\mathbf{x}: \bar{x} > -1 + (1/2)(1.64) = -.18\};$$

thus the null hypothesis is rejected if $\bar{x}$ exceeds $-.18$, and is otherwise accepted. ■

Example 9-4. Theorem 9-1 explains why the test in Example 9-1 is more powerful than the one in Example 9-2. The theorem asserts that the most powerful test accepts or rejects the null hypothesis accordingly as $\bar{x}$ does not or does exceed some fixed number. The test in Example 9-1 is of this form, whereas the one in Example 9-2 is not. ■

These results are suitably modified when $\mu_1 < \mu_0$. Lemma 9-1 is altered: the ratio p_1/p_0 is assumed to be nonincreasing, and x_α is defined on the lower tail of p_0 instead of the upper one; and the right-hand side of (9-10) has the range of integration from $-\infty$ to x_α. In Theorem 9-1 the rejection region is changed to $\{\mathbf{x}: \bar{x} < \mu_0 - \sigma z_\alpha/\sqrt{n}\}$. We leave the details as exercises for the reader.

We indicate the derivation of the most powerful test of size α for testing simple hypotheses about σ (see Sec. 9-1, Exercise 6). For simplicity take $\mu = 0$. Suppose the null and alternative hypotheses are $\sigma = \sigma_0$ and $\sigma = \sigma_1$, respectively, with $\sigma_0 < \sigma_1$. L_0 and L_1 are the corresponding likelihood functions. Formulas (9-2) through (9-5) are the same in this case, and (9-8) and (9-9) are changed to

$$\sigma_0^{-2} \int_0^\infty f(y)\, \psi_n\left(\frac{y}{\sigma_0^2}\right) dy = \alpha \tag{9-16}$$

and

$$\sigma_1^{-2} \int_0^\infty f(y)\, \psi_n\left(\frac{y}{\sigma_1^2}\right) dy \tag{9-17}$$

respectively; the latter follow from (6-8). Applying Lemma 9-1 with

$$p_i(x) = \sigma_i^{-2}\psi_n\left(\frac{x}{\sigma_i^2}\right), \qquad i = 0, 1,$$

we find that the integral (9-17) is at most equal to

$$\int_{(\sigma_0/\sigma_1)\chi^2_{n,\alpha}}^{\infty} \psi_n(y)\,dy, \tag{9-18}$$

where $\chi^2_{n,\alpha}$ is the upper α-percentile of the chi-square distribution with n degrees of freedom. The rejection region of the most powerful test of size α is

$$\{\mathbf{x}: \|\mathbf{x}\|^2 > \sigma_0^2\chi^2_{n,\alpha}\}.$$

EXERCISES

1. *Prove:* The ratio

$$\frac{\phi(\sqrt{n}(x - \mu_1)/\sigma)}{\phi(\sqrt{n}(x - \mu_0)/\sigma)}$$

increases with x if $\mu_0 < \mu_1$, and decreases if $\mu_1 < \mu_0$.

2. Show that the test with the rejection region M_0 has the level of significance α and that its power is the quantity (9-13).

3. Explicitly determine the rejection regions of the most powerful tests of size α in these particular cases:

(a) $\mu_0 = 0,\ \mu_1 = 5,\ n = 9,\ \sigma = 2,\ \alpha = .05$
(b) $\mu_0 = -1,\ \mu_1 = 2,\ n = 25,\ \sigma = 5,\ \alpha = .04$
(c) $\mu_0 = 1,\ \mu_1 = -2,\ n = 36,\ \sigma = 10,\ \alpha = .01$
(d) $\mu_0 = 4,\ \mu_1 = 2,\ n = 9,\ \sigma = 3,\ \alpha = .02.$

What is the role (if any) of μ_1 in the rejection region?

4. State and prove the version of Lemma 9-1 when p_1/p_0 is nonincreasing; define x_α through

$$\int_{-\infty}^{x_\alpha} p_0(x)\,dx = \alpha.$$

5. Modify the proof of Theorem 9-1 to fit the case $\mu_1 < \mu_0$.

6. Show that the ratio

$$\frac{\psi_n(x/\sigma_1^2)\,\sigma_1^{-2}}{\psi_n(x/\sigma_0^2)\,\sigma_0^{-2}}$$

increases if $\sigma_1 > \sigma_0$.

7. Prove that the integral (9-17) has the upper bound (9-18).

8. Verify that the rejection region following (9-18) is of size α, and that its power is equal to (9-18).

9. Modify the results in Exercises 6–8 for the case $\sigma_1 < \sigma_0$.

10. *Prove:* The power of the test with the rejection region M_0 converges to 1 as $n \to \infty$.

11. Suppose $\mu_0 < \mu_1$, and let $1 - \beta$ be a preassigned number. *Prove:* In order that the power of the most powerful test of size α be at least $1 - \beta$, the number n of observations must be at least equal to

$$\frac{\sigma^2(z_\alpha - z_{1-\beta})^2}{(\mu_1 - \mu_0)^2}$$

(The same is true if $\mu_1 < \mu_0$.)

12. What changes are necessary in testing σ when μ is known but not equal to 0?

13. Determine the rejection regions of the most powerful tests of size $\alpha = .05$ in each of these cases:
    (a) $\mu = 0,\ \sigma_0 = 1,\ \sigma_1 = 2,\ n = 6$
    (b) $\mu = 0,\ \sigma_0 = 1,\ \sigma_1 = 2,\ n = 12$
    (c) $\mu = 1,\ \sigma_0 = 2,\ \sigma_1 = 2.1,\ n = 10$.

14. Find the power of each test in Exercise 4.

9-3. COMPOSITE ALTERNATIVE HYPOTHESIS: UNBIASED TESTS (OPTIONAL)

The hypotheses μ_0 and μ_1 in Sec. 9-1 are called *simple* because there is one specified "parameter point" in each. A more general problem involves deciding not between two parameter values but whether the true value belongs to one or the other of two specified subsets of values; for example, for a given number μ_0, one may have to decide whether the true value μ is less than or equal to μ_0, or greater than μ_0. We now consider a slightly more special problem: the null hypothesis is simple, but the alternative is a subset of values. Such a problem is called that of testing a simple null hypothesis against a composite alternative.

Suppose there is a normal population with known standard deviation σ. We wish to test the simple null hypothesis $\mu = \mu_0$ against the composite alternative $\mu > \mu_0$, namely, that the value of

μ belongs to the subset (μ_0, ∞) of the real line. For any test (that is, any choice of a rejection region) the probability of error of the first kind is defined as before; however, the power, or alternatively, the probability of error of the second kind, depends on the particular value of the alternative hypothesis; for example, in testing $\mu_0 = 5$ against the alternative that $\mu > 5$, the probability of error of the second kind is different in the two cases $\mu = 6$ and $\mu = 8$. The power of a test against a composite alternative is a function $P = P(\mu)$ on the set of parameter points of the alternative hypothesis.

Example 9-5. Let us test the simple null hypothesis $\mu_0 = 0$ against the composite alternative that $\mu > 0$; the latter hypothesis is denoted simply as "$\mu_1 > 0$." We take $\sigma = 1$, $\alpha = .05$, and $n = 9$, and the rejection region $\{\mathbf{x}: \bar{x} > (1.64)/3 = .55\}$. It is directly seen that the level of significance of this test is .05, as claimed. Now we examine the probability of rejection of the null hypothesis under various alternatives. For any value $\mu > 0$ under the alternative hypothesis the power of the test is, in accordance with (9-13), equal to $1 - \Phi(1.64 - 3\mu)$; for example, for $\mu = .5$ and $\mu = 1$, the power is $1 - \Phi(.14) = .4443$ and $1 - \Phi(-1.36) = .9131$, respectively. It is evident from the form of the power function that it increases with μ: as the value of the alternative moves away from that of the null hypothesis, the probability of rejecting the latter increases. ■

A test of size α of a simple null hypothesis against a composite alternative is called *uniformly* most powerful if it is most powerful against every value of the alternative at the same time; in other words, the power function of the test exceeds that of all others at all values of the parameter in the set forming the alternative hypothesis. For testing the hypothesis $\mu = \mu_0$ against the composite alternative $\mu_1 > \mu_0$, the test with the rejection region M_0 (Theorem 9-1) is actually *uniformly* most powerful; indeed, the rejection region is the same for all alternatives $\mu_1 > \mu_0$, and Theorem 9-1, states that the power is maximized for every $\mu_1 > \mu_0$ because μ_1 is arbitrary. The power function is given by (9-13). The test in Example 9-5 is the most powerful against both alternatives $\mu = .5$ and $\mu = 1$ even though the power is different in the two cases.

For testing the simple null hypothesis $\mu = \mu_0$ against the composite alternative $\mu_1 < \mu_0$ the test with the rejection region

$$\left\{\mathbf{x}: \bar{x} < \mu_0 - \frac{\sigma z_\alpha}{\sqrt{n}}\right\}$$

is uniformly most powerful of size α (Sec. 9-2); the power function is

$$\Phi\left(-z_\alpha + \frac{\sqrt{n}(\mu_0 - \mu_1)}{\sigma}\right). \tag{9-19}$$

The two testing problems just mentioned are said to have "one-sided" composite alternatives: in the first, the alternative hypothesis specifies the parameter values greater than μ_0, and, in the second, values less than μ_0. As noted there is a uniformly most powerful test of size α for each. Now we consider testing the simple null hypothesis $\mu = \mu_0$ against the "two-sided" alternative that either $\mu_1 < \mu_0$ or $\mu_1 > \mu_0$. The alternative is represented as $\mu_1 \neq \mu_0$. Neither of the one-sided uniformly most powerful tests is uniformly most powerful against the two-sided alternative; indeed, the power function (9-13) of the test against $\mu_1 > \mu_0$ is smaller than the power (9-19) of the test against $\mu_1 < \mu_0$ for values $\mu_1 < \mu_0$:

$$1 - \Phi\left(z_\alpha + \frac{\sqrt{n}(\mu_0 - \mu_1)}{\sigma}\right) < \Phi\left(-z_\alpha + \frac{\sqrt{n}(\mu_0 - \mu_1)}{\sigma}\right), \quad \mu_1 < \mu_0,$$

(Exercise 2). By symmetry, the function (9-19) is smaller than (9-13) for $\mu_1 > \mu_0$. The two power functions are plotted in Fig. 9-3.

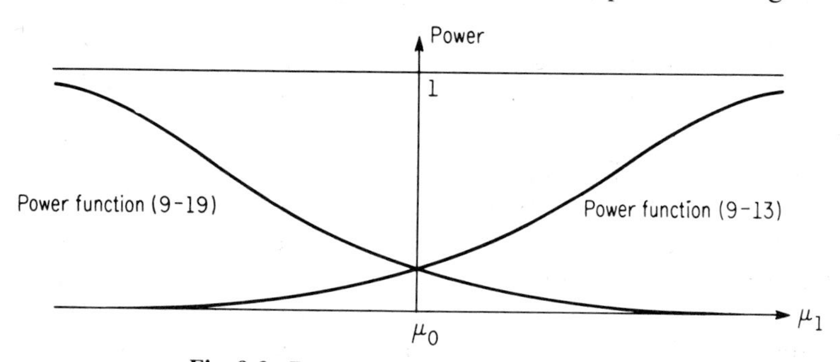

Fig. 9-3. Power functions of one-sided tests.

Does there exist a test of $\mu = \mu_0$ against the two-sided composite alternative $\mu_1 \neq \mu_0$ which is uniformly most powerful? It is conceivable that there exists a third test that does as well as each one-sided test against the corresponding one-sided alternatives.

We shall indicate why such a test does not exist, giving two independent arguments. The first, which we present without a complete proof, is based on the "essential uniqueness" of the most powerful test: using the advanced concept of a set of Lebesgue measure 0, one can show that the rejection region M_0 is the unique set—modulo a set of measure 0—maximizing the power against the simple alternative μ_1. It follows that the rejection region of any uniformly most powerful test against $\mu_1 \neq \mu_0$ would have to coincide with M_0; however, as shown above, the power function of this test falls below (9-19) for $\mu_1 < \mu_0$, so that the test is not uniformly most powerful.

Now we give a rigorous second argument. Suppose that there exists a uniformly most powerful test against $\mu_1 \neq \mu_0$; let $P(\mu)$ be its power function. The latter would have to be equal to the power function of the uniformly most powerful one-sided test in each case $\mu_1 > \mu_0$ and $\mu_1 < \mu_0$:

$$P(\mu) = 1 - \Phi\left(z_\alpha + \frac{\sqrt{n}(\mu_0 - \mu)}{\sigma}\right), \qquad \mu > \mu_0$$

$$= \Phi\left(-z_\alpha + \frac{\sqrt{n}(\mu_0 - \mu)}{\sigma}\right), \qquad \mu < \mu_0.$$

Such a function cannot be extended to one differentiable at $\mu = \mu_0$; as a matter of fact, the right-hand and left-hand derivatives are equal to $\pm\sqrt{n}\phi(z_\alpha)/\sigma$, respectively. (Note that ϕ is symmetric.) We shall show that this leads to a contradiction of the following proposition:

LEMMA 9-2

For a test with arbitrary rejection region M the power function

$$P(\mu) = \int_M \cdots \int L(\mathbf{x}; \mu, \sigma)\, d\mathbf{x}$$

is everywhere differentiable with respect to μ.

Proof: We shall show that the difference quotient converges to

$$\int_M \cdots \int \frac{n}{\sigma^2}(\bar{x} - \mu)\, L(\mathbf{x}; \mu, \sigma)\, d\mathbf{x}. \tag{9-20}$$

Noting that

$$L(\mathbf{x}; \mu + h, \sigma) = L(\mathbf{x}; \mu, \sigma) \exp\left(\frac{nh(\bar{x} - \mu)}{\sigma^2} - \frac{nh^2}{2\sigma^2}\right),$$

we find that the difference between $[P(\mu + h) - P(\mu)]/h$ and (9-20) is

$$\int_M \cdots \int \left\{h^{-1}\left[\exp\left(\frac{nh(\bar{x} - \mu)}{\sigma^2} - \frac{nh^2}{2\sigma^2}\right) - 1\right] - \frac{n}{\sigma^2}(\bar{x} - \mu)\right\} L d\mathbf{x}.$$

In order to show that this integral converges to 0 as $h \to 0$, we first replace the integrand by its absolute value, and then the domain M by the whole space R^n; these two alterations cannot but augment the value of the integral, which becomes

$$\int_{R^n} \cdots \int \left| h^{-1}\left[\exp\left(\frac{nh(\bar{x} - \mu)}{\sigma^2} - \frac{nh^2}{2\sigma^2}\right) - 1\right] - \frac{n}{\sigma^2}(\bar{x} - \mu)\right| L\, d\mathbf{x}.$$

The coefficient of L in the integrand above depends on $\mathbf{x}$ only through the statistic $\sqrt{n}(\bar{x} - \mu)/\sigma$ which has a standard normal distribution; thus, by Theorem 4-1, the integral reduces to a simple integral with respect to $\phi(x)$. The convergence of this integral to 0 follows from the method used in the proof of Theorem 6-6 (Exercise 3). ■

We have now finished the proof of the nonexistence of a uniformly most powerful test against the two-sided alternative. In order to determine a "best" test, we shall restrict the class of tests; this was done in determining a best Estimator, when the class of Estimators was limited to unbiased ones. By analogy, we now define an unbiased test of a simple null hypothesis against a composite alternative: it is a test in which the power function is nowhere less than the level of significance: $\alpha \leq 1 - \beta$; in other words, the probability of rejecting the null hypothesis is smallest when the null hypothesis is true.

We have to add a technical condition on the rejection region of the test to complete the definition of unbiasedness; this condition is not necessary in the measure-theoretic setting. The power function of a test with the rejection region M,

$$P(\mu) = \int_M \cdots \int L(\mathbf{x}; \mu, \sigma)\, d\mathbf{x},$$

is, by (3-34), the least upper bound of

$$\int_{R^n}\cdots\int f L(\mathbf{x}; \mu, \sigma)\, d\mathbf{x}, \qquad f \in \mathcal{L}(M). \tag{9-21}$$

Let $\mathcal{L}_0(M)$ be the subclass of $\mathcal{L}(M)$ consisting of functions f for which the integral (9-21) is a minimum at $\mu = \mu_0$; then the test is considered unbiased if not only $P(\mu)$ is minimized at μ_0 but, in addition, $P(\mu)$ is approximable by integrals (9-21) having the same property: the least upper bound of the integral (9-21) is the same over the subclass $\mathcal{L}_0(M)$ as over the whole class $\mathcal{L}(M)$. We record this condition as

$$P(\mu) = \operatorname*{lub}_{f \in \mathcal{L}_0(M)} \int_{R^n}\cdots\int f L(\mathbf{x}; \mu, \sigma)\, d\mathbf{x}. \tag{9-22}$$

Following the method of Secs. 9-1 and 9-2, we consider the following maximization problem:

LEMMA 9-3

Let $f(x)$ be a continuous function on the line such that $0 \le f(x) \le 1$ and such that the integral

$$\int_{-\infty}^{\infty} f(x)\, \frac{\sqrt{n}}{\sigma}\, \phi\left(\frac{\sqrt{n}(x-\mu)}{\sigma}\right) dx \tag{9-23}$$

is minimized at $\mu = \mu_0$, where it is equal to α; then, the integral is not more than

$$\Phi\left(-z_{\alpha/2} + \frac{\sqrt{n}(\mu_0 - \mu)}{\sigma}\right) + \Phi\left(-z_{\alpha/2} - \frac{\sqrt{n}(\mu_0 - \mu)}{\sigma}\right) \tag{9-24}$$

for all μ.

Proof: To simplify the notation we substitute the single factor σ for $\sigma/\sqrt{n}$ and put $\mu_0 = 0$; this changes the proof in no essential way (Exercise 8).

The first point of the proof is that *there exist constants k_1 and k_2 such that the inequality*

$$\frac{1}{\sigma}\,\phi\left(\frac{x-\mu}{\sigma}\right) \le k_1\,\frac{1}{\sigma}\,\phi\left(\frac{x}{\sigma}\right) + k_2\,\frac{x}{\sigma^2}\,\phi\left(\frac{x}{\sigma}\right) \tag{9-25}$$

holds for those and only those x for which the inequality

$$|x| \leq z_{\alpha/2}\sigma \tag{9-26}$$

holds. To prove this, divide each side of (9-25) by $(1/\sigma)\phi(x/\sigma)$; when simplified, the left hand member is an exponential function of a linear function of x, and the right hand side becomes the linear function $k_1 + k_2 x/\sigma$. Since the exponential function is convex, there is a linear function whose graph intersects that of the exponential at any two prescribed points $\pm x_0$, and is such that the linear function exceeds the exponential for $|x| < |x_0|$, and is exceeded by it for $|x| > |x_0|$ (Fig. 9-4).

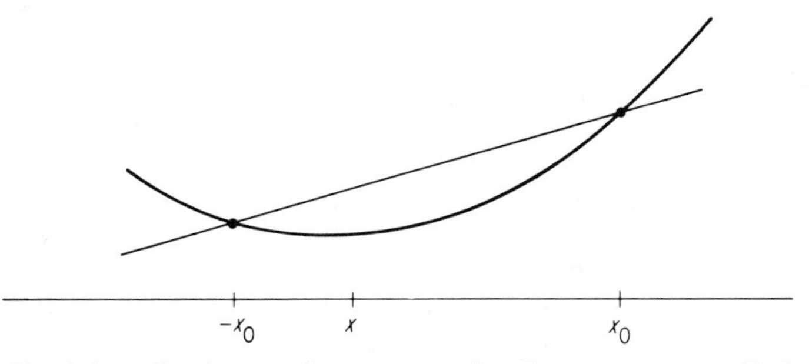

Fig. 9-4. A line intersecting a convex function at two prescribed points.

For the two points $\pm z_{\alpha/2}\sigma$, there accordingly exist constants k_1 and k_2 such that the linear function $k_1 + k_2 x/\sigma$ has the intersection property above; thus (9-25) holds if and only if (9-26) holds. The particular values of k_1 and k_2 are not of relevance to us.

We continue with the proof of the lemma. The hypothesis implies that

$$\int_{-\infty}^{\infty} f(x)\,\frac{1}{\sigma}\,\phi\left(\frac{x}{\sigma}\right) dx = \alpha,$$

which, as in the proof of Lemma 9-1, implies

$$\int_{|x| \leq z_{\alpha/2}\sigma} f(x)\,\frac{1}{\sigma}\,\phi\left(\frac{x}{\sigma}\right) dx = \int_{|x| > z_{\alpha/2}\sigma} (1 - f(x))\,\frac{1}{\sigma}\,\phi\left(\frac{x}{\sigma}\right) dx. \tag{9-27}$$

By Theorem 6-6 the integral (9-23) is differentiable with respect to μ; furthermore, by the method of proof of that theorem, it can be shown that the derivative is continuous (Exercise 4); therefore, the condition that (9-23) have a minimum at $\mu = 0$

implies that the derivative vanishes at that point:

$$\int_{-\infty}^{\infty} f(x)\,\frac{x}{\sigma^3}\,\phi\left(\frac{x}{\sigma}\right) dx = 0.$$

This implies

$$\int_{|x|\le z_{\alpha/2}\sigma} f(x)\,\frac{x}{\sigma^3}\,\phi\left(\frac{x}{\sigma}\right) dx = \int_{|x|>z_{\alpha/2}\sigma} (1 - f(x))\,\frac{x}{\sigma^3}\,\phi\left(\frac{x}{\sigma}\right) dx, \qquad (9\text{-}28)$$

because $(-x)\,\phi(-x) = -x\phi(x)$, and

$$\int_{|x|>z_{\alpha/2}\sigma} \frac{x}{\sigma^3}\,\phi\left(\frac{x}{\sigma}\right) dx = 0.$$

Suppose that the constants k_1 and k_2 have been chosen so that (9-25) and (9-26) hold simultaneously; then

$$\int_{|x|\le z_{\alpha/2}\sigma} f(x)\,\phi\left(\frac{x-\mu}{\sigma}\right)\frac{dx}{\sigma} \le k_1 \int_{|x|\le z_{\alpha/2}\sigma} f(x)\,\phi\left(\frac{x}{\sigma}\right)\frac{dx}{\sigma} + k_2 \int_{|x|\le z_{\alpha/2}\sigma} f(x)\,\frac{x}{\sigma}\,\phi\left(\frac{x}{\sigma}\right)\frac{dx}{\sigma}.$$

By (9-27) and (9-28) the right-hand member of this inequality is equal to

$$\int_{|x|>z_{\alpha/2}\sigma} (1 - f(x))\,\phi\left(\frac{x}{\sigma}\right)\left(k_1 + \frac{k_2 x}{\sigma}\right)\frac{dx}{\sigma}.$$

By the reversed inequalities (9-25) and (9-26), the latter integral is less than or equal to

$$\int_{|x|>z_{\alpha/2}\sigma} (1 - f(x))\,\frac{1}{\sigma}\,\phi\left(\frac{x-\mu}{\sigma}\right) dx.$$

From this and the previous inequality we conclude:

$$\left(\int_{|x|\le z_{\alpha/2}\sigma} + \int_{|x|>z_{\alpha/2}\sigma}\right) f(x)\,\phi\left(\frac{x-\mu}{\sigma}\right)\frac{dx}{\sigma} \le \int_{|x|>z_{\alpha/2}\sigma} \phi\left(\frac{x-\mu}{\sigma}\right)\frac{dx}{\sigma}. \quad \blacksquare$$

Now we derive a uniformly most powerful unbiased test of the two-sided alternative.

THEOREM 9-2
The test with the rejection region

$$M_l = \left\{\mathbf{x}: |\bar{x} - \mu_0| > \frac{z_{\alpha/2}\sigma}{\sqrt{n}}\right\}$$

is a uniformly most powerful unbiased test of size α of the null hypothesis $\mu = \mu_0$ against the composite alternative $\mu_l \neq \mu_0$.

Proof: The test with the rejection region M_1 is certainly of size α, and its power function is equal to (9-24) (Exercise 5). The power is minimized at $\mu = \mu_0$, where it has the value α; this is verified by the standard method of differentiation (Exercise 6). In order to complete the proof that the test is unbiased, we must show that M_1 satisfies the technical requirement (9-22). Let $f_m(x)$ be a continuous function of a real variable defined as

$$f_m(x) = 1, \qquad |x| > \frac{z_{\alpha/2}\sigma}{\sqrt{n}} + \frac{1}{m}$$

$$= 0, \qquad |x| \leq \frac{z_{\alpha/2}\sigma}{\sqrt{n}},$$

and varying continuously from 0 to 1 for $z_{\alpha/2}\sigma/\sqrt{n} \leq |x| < z_{\alpha/2}\sigma/\sqrt{n} + (1/m)$, $m = 1, 2, \ldots$; then the integral

$$\int_{R^n}\cdots\int f_m(\bar{x} - \mu_0)\, L(\mathbf{x}; \mu, \sigma)\, d\mathbf{x} \tag{9-29}$$

converges to the power function (9-24) for $m \to \infty$, and each function (9-29) is minimized at $\mu = \mu_0$ (Exercise 7). This shows that the conditions for unbiasedness are satisfied.

Now let M be the rejection region of any unbiased test of size α. Let f be an arbitrary function in $\mathfrak{L}_0(M)$ and $\bar{f}$ its conditional expectation given $\sqrt{n}\bar{x}$; then, by Eq. 6-14, the integrals

$$\int_{R^n}\cdots\int f\, L(\mathbf{x}; \mu, \sigma)\, d\mathbf{x} \quad \text{and} \quad \int_{-\infty}^{\infty} \bar{f}(\sqrt{n}x; \sigma)\frac{\sqrt{n}}{\sigma}\, \phi\left(\frac{x - \mu}{\sigma/\sqrt{n}}\right) dx$$

are equal for all μ, and the latter satisfies the conditions in the hypothesis of Lemma 9-3 with

$$\alpha' = \int_{-\infty}^{\infty} \bar{f}(\sqrt{n}x;\sigma)\frac{\sqrt{n}}{\sigma}\,\phi\left(\frac{\sqrt{n}(x-\mu_0)}{\sigma}\right)dx$$

in place of α. In accordance with the conclusion of Lemma 9-3, the function

$$\int_{-\infty}^{\infty} \bar{f}(\sqrt{n}x;\sigma)\,\frac{\sqrt{n}}{\sigma}\,\phi\left(\frac{\sqrt{n}(x-\mu)}{\sigma}\right)dx$$

is bounded above by (9-24) with α' in place of α. It follows from

$$\alpha' = \int_{R^n}\cdots\int f(\mathbf{x})\,L(\mathbf{x};\mu_0,\sigma)\,d\mathbf{x} \le \int_M\cdots\int L(\mathbf{x};\mu_0,\sigma)\,d\mathbf{x} = \alpha$$

that $z_{\alpha/2} \le z_{\alpha'/2}$; thus

$$\int_{R^n}\cdots\int f(\mathbf{x})\,L(\mathbf{x};\mu,\sigma)\,d\mathbf{x} = \int_{-\infty}^{\infty} \bar{f}(\sqrt{n}x;\sigma)\,\frac{\sqrt{n}}{\sigma}\,\phi\left(\frac{x-\mu}{\sigma/\sqrt{n}}\right)dx$$

$$\le \Phi\left(-z_{\alpha/2} + \frac{\sqrt{n}(\mu_0-\mu)}{\sigma}\right) + \Phi\left(-z_{\alpha/2} - \frac{\sqrt{n}(\mu_0-\mu)}{\sigma}\right).$$

This inequality is valid for any f in $\mathfrak{L}_0(M)$; therefore, it remains valid when the first member is replaced by its least upper bound over all f in $\mathfrak{L}_0(M)$. By (9-22), the power function of the test with the rejection region M is similarly bounded; thus, the power is less than or equal to that of the test with rejection region M_1. ■

Example 9-6. Test $\mu = 0$ against the two-sided composite alternative $\mu \neq 0$; take $\sigma = 2$, $\alpha = .05$ and $n = 16$. From the table of the normal distribution we find $z_{\alpha/2} = z_{.025} = 1.96$. The uniformly most powerful unbiased test is based on the procedure: accept or reject the null hypothesis accordingly as the Sample mean satisfies

$$|\bar{x}| \le 2(1.96)/4 = .98, \quad \text{or} \quad |\bar{x}| > .98. \quad ■$$

EXERCISES

1. Find the rejection regions and power functions of the uniformly most powerful unbiased tests of the following hypotheses against the corresponding two-sided alternatives:

(a) $\mu_0 = 1, n = 9, \sigma = 3, \alpha = .04$
(b) $\mu_0 = -1, n = 25, \sigma = 1, \alpha = .02$
(c) $\mu_0 = 0, n = 16, \sigma = .5, \alpha = .05$
(d) $\mu_0 = 2, n = 4, \sigma = 1, \alpha = .01$.

2. *Prove:* $1 - \Phi(z + \epsilon) < \Phi(-z + \epsilon)$, for every $z, \epsilon > 0$.

3. Complete the last step in the proof of Lemma 9-2.

4. Show that the derivative of (9-23) is continuous in μ; extend the method of proof of Theorem 6-6. (It can be shown that the derivative is even differentiable.)

5. Show that the level of significance of the test with the rejection region M_1 (Theorem 9-2) is α, and that the power function is given by (9-24).

6. Prove that the function (9-24) is minimized at $\mu = \mu_0$ where it is equal to α.

7. Carry out the details of the proof of the two assertions concerning (9-29).

8. Point out the alterations necessary in the given proof of Lemma 9-3 (for $\mu_0 = 0, n = 1$) to cover the case of general μ_0 and n.

9. The usual test of the simple null hypothesis $\sigma = \sigma_0$ against the two-sided alternative $\sigma \neq \sigma_0$ when $\mu = 0$ is as follows. Let h and k the lower and upper $\alpha/2$-percentiles of the chi-square distribution with n degrees of freedom. The rejection region is the union of the two sets $\{\mathbf{x}: \sigma_0^2 k < \| x \|^2\}$ and $\{\mathbf{x}: \| \mathbf{x} \|^2 < \sigma_0^2 h\}$. Show that the test is of size α. (There is no satisfactory theoretical justification for this test; however, its use is supported by intuitive reasons.)

9-4. COMPOSITE HYPOTHESES; INVARIANT TESTS (OPTIONAL)

We consider the problem of testing the null hypothesis that the mean of a normal population is μ_0 against the one-sided composite alternative that $\mu_1 > \mu_0$ *when σ is unknown.* The null hypothesis is really composite because σ may be any positive number. The uniformly most powerful test proposed in Sec. 9-2 is not usable because its rejection region M_0 depends on the value of σ. The level of significance of an arbitrary test is not necessarily fixed because the probability of an error of the first kind,

$$\int \cdots \int_M L(\mathbf{x}; \mu_0, \sigma)\, d\mathbf{x}$$

varies with σ. In this section we are going to restrict ourselves to a certain class of tests—invariant tests—whose level of significance is the same for all σ, and find the most powerful of these tests. We shall also indicate why it is reasonable to limit ourselves to this class of tests.

There is no loss of generality in assuming that $\mu_0 = 0$; indeed, the general case can be reduced to this one by subtracting μ_0 from the value of each member of the population. So we test the null hypothesis $\mu = 0$ against the alternative $\mu > 0$ when σ is unknown. This testing problem has an invariance property.

Imagine that the normal population is a population of balls with labels x. Suppose that the system of labels is transformed in this way. There is a constant $c > 0$ such that every label x is replaced by its multiple cx. The resulting population is called the *transformed population*; it is also normal but with mean $c\mu$ and standard deviation $c\sigma$. (See the definition of the general normal population in Sec. 2-2.)

If $\mathbf{X}$ is a Sample from the original population, then $c\mathbf{X}$ is the corresponding Sample from the transformed population; the latter is called the *transformed Sample*. Under the null hypothesis the means of both the original and transformed populations are equal to 0; under the alternative the means are both positive; this is the invariance of the testing problem. It suggests the use of an *invariant test*. Such a test has the property that the original Sample point $\mathbf{X}$ falls in the rejection region if and only if the transformed Sample point $c\mathbf{X}$ does, and for every $c > 0$; in other words, the decision to accept or reject the null hypothesis is insensitive to the introduction of constant multiples c in the observations. The rejection region of an invariant test is invariant under all scale changes: a point $\mathbf{x}$ belongs to the region if and only if $c\mathbf{x}$ does, for all $c > 0$.

For technical reasons we must put another condition on the rejection region to complete the definition of an invariant test; it is analogous to (9-22). Like the latter this condition is dispensable in the measure-theoretic setting. A function $f(\mathbf{x})$ on R^n is called *invariant* if $f(c\mathbf{x}) = f(\mathbf{x})$, $\mathbf{x} \in R^n$, for every $c > 0$. For any rejection region M, let $\mathfrak{L}_0(M)$ be the subclass of $\mathfrak{L}(M)$ consisting of the invariant functions in $\mathfrak{L}(M)$. A test with this rejection region is called invariant if not only it has the property

$$\mathbf{x} \in M \qquad \text{if and only if} \qquad c\mathbf{x} \in M, \qquad c > 0,$$

but, in addition, the power function

$$P(\mu) = \int_M \cdots \int L(\mathbf{x}; \mu, \sigma)\, dx$$

is approximable by the integrals (9-21) of *invariant* functions. This condition, with the appropriate definition of $\mathfrak{L}_0(M)$, is formally the same as the condition (9-22); hence, with this understanding, we shall refer to (9-22) throughout the rest of this section on invariance.

By sufficiency and invariance the integral of every continuous invariant function over the Sample space is reducible to a single integral:

LEMMA 9-4

Let f be a continuous invariant function on the Sample space for which the integral

$$\int_{R^n} \cdots \int f(\mathbf{x}) L\mathbf{x}; \mu, \sigma)\, d\mathbf{x}$$

is defined and finite for all μ and $\sigma > 0$. Then there exists a continuous function $\bar{f}(y)$ of a real variable y such that the integral above is equal to

$$\int_{R^n} \cdots \int \bar{f}\left(\frac{\sqrt{n}\,\bar{x}}{s}\right) L(\mathbf{x}; \mu, \sigma)\, d\mathbf{x} \tag{9-30}$$

for all μ and $\sigma > 0$, where $(n - 1)s^2 = \sum_{i=1}^{n} (x_i - x)^2$; furthermore, the latter integral is equal to

$$\int_{-\infty}^{\infty} \bar{f}(y) \int_0^{\infty}$$

$$\phi(u\sqrt{v} - \sqrt{n}\,\mu/\sigma)(n - 1)\psi_{n-1}((n - 1)v)\sqrt{v}\, dv\, dy. \tag{9-31}$$

Proof: By the assumed invariance of f, the first integral in the statement of the lemma is unchanged is $\mathbf{x}$ is replaced by $c\mathbf{x}$. By the fundamental result on the sufficiency of $\bar{x}$ and s^2 (Theorem 6-1), the integral of $f(c\mathbf{x})L$ is equal to that of $f(c\sqrt{n}\bar{x}, (n - 1)c^2s^2)L$, where f is the conditional expectation of f; moreover, f inherits the invariance of f in this way

$$\bar{f}(uc, vc^2) = \bar{f}(u, v), \quad \text{for all } u \text{ and } v > 0$$

(Exercise 5). A function satisfying this equation for all $c > 0$ necessarily depends on the variables u and v through the single function $u/\sqrt{v}$; in fact, putting c equal to the particular value $1/\sqrt{v}$ in the equation above, one finds

$$\bar{f}(u, v) = \bar{f}\left(\frac{u}{\sqrt{v}}, 1\right).$$

The expression on the right-hand side depends only on $u/\sqrt{v}$; thus, $\bar{f}(\sqrt{n}\bar{x}, (n - 1)\, s^2)$ depends only on $\sqrt{n}\,\bar{x}/s$; therefore, the integral of $f \cdot L$ is equal to that in (9-30).

In reducing (9-30) to (9-31) we invoke Theorems 5-1 and 5-4: by the former, the multiple integral is equal to the double integral of $\bar{f}(u/\sqrt{v})$ with respect to the joint density of $\sqrt{n}\,\bar{x}$ and s^2; and, by the latter theorem, the joint density of $\sqrt{n}\,\bar{x}/\sigma$ and s^2/σ^2 is the product of the normal density with mean $\sqrt{n}\,\mu/\sigma$ and standard deviation 1, and of the "scaled" chi-square density $(n - 1)\psi_{n-1}((n - 1)v)$ (Exercise 2). ■

COROLLARY

The probability of an error of the first kind for an invariant test of the hypothesis $\mu = 0$ is the same for all values of σ.

Proof: The probability is equal to

$$\int_M \cdots \int L(\mathbf{x}; 0, \sigma)\, d\mathbf{x}$$

where M is the rejection region. By condition (9-22), the integral is approximable by

$$\int_{R^n} \cdots \int f(\mathbf{x})\, L(\mathbf{x}; \mu, \sigma)\, d\mathbf{x}$$

where f is continuous and invariant. By Lemma 9-4, the integral is equivalent to the one in (9-31); the latter is independent of σ when $\mu = 0$ because it depends on μ/σ. ■

Next we prove an extension of the Neyman-Pearson Lemma (Lemma 9-1):

LEMMA 9-5

Let $p_0(x)$ and $p_i(x)$ be positive continuous functions such that

$$\int_{-\infty}^{\infty} p_i(x)\,dx = 1, \qquad i = 1,2$$

$$p_1(x)/p_0(x) \text{ is nondecreasing;}$$

then for any nondecreasing, nonnegative continuous function $G(x)$, the following inequality holds:

$$\int_{-\infty}^{\infty} p_0(x)\,G(x)\,dx \le \int_{-\infty}^{\infty} p_1(x)\,G(x)\,dx. \tag{9-32}$$

Proof: For fixed x, let $\alpha = \alpha(x)$ be the value of the integral

$$\alpha(x) = \int_x^{\infty} p_0(y)\,dy;$$

then $0 \le \alpha(x) \le 1$. Let $g(y)$ be the constant function identically equal to α; then, by Lemma 9-1, we have

$$\alpha = \int_{-\infty}^{\infty} g(y)\,p_1(y)\,dy \le \int_x^{\infty} p_1(y)\,dy;$$

therefore, by the definition of α:

$$\int_x^{\infty} p_0(y)\,dy \le \int_x^{\infty} p_1(y)\,dy.$$

Now let $G(x)$ be a nondecreasing, nonnegative continuous function. For every number y in the range of G, let $G^{-1}(y)$ be the smallest (glb) x satisfying the equation $G(x) = y$; then, by the previous inequality:

$$\int_{G^{-1}(y)}^{\infty} p_0(x)\,dx \le \int_{G^{-1}(y)}^{\infty} p_1(x)\,dx.$$

Integrate over y from $G(0)$ to $G(\infty)$:

$$\int_{G(0)}^{G(\infty)} \int_{G^{-1}(y)}^{\infty} p_0(x)\,dx\,dy \le \int_{G(0)}^{G(\infty)} \int_{G^{-1}(y)}^{\infty} p_1(x)\,dx\,dy.$$

The inequality (9-32) follows from this one; indeed:

$$\int_{G(0)}^{G(\infty)} \int_{G^{-1}(y)}^{\infty} p_i(x)\,dxdy = \iint\limits_{\{(x,y):\, G(0) \le y \le G(\infty),\, G^{-1}(y) \le x < \infty\}} p_i(x)\,dxdy$$

$$= \iint\limits_{\{(x,y):\, G(0) \le y < G(x)\}} p_i(x)\,dxdy = \int_{-\infty}^{\infty} [G(x) - G(0)]\,p_i(x)\,dx$$

$$= \int_{-\infty}^{\infty} G(x)\,p_i(x)\,dx - G(0), \qquad i = 1, 2. \quad \blacksquare$$

The inner integral in (9-31),

$$\int_0^{\infty} \phi(u\sqrt{v} - \sqrt{n}\,\mu/\sigma)(n - 1)\,\psi_{n-1}((n - 1)\,v)\,\sqrt{v}\,dv \tag{9-33}$$

is, as a function of u, the density of $\sqrt{n}\bar{x}/s$; this is evident from Theorems 5-4 and 5-5. The density depends on the unknown parameters μ and σ through the single function

$$\delta = \sqrt{n}\mu/\sigma.$$

When $\mu = 0$, the density is that of the t-distribution with $n - 1$ degrees of freedom. When $\mu \neq 0$ the distribution is called the "noncentral t-distribution with $n - 1$ degrees of freedom and noncentrality parameter δ." The density is not expressible in closed analytic form. Now we show that the t- and noncentral t-densities satisfy the conditions of the Neyman-Pearson Lemma.

LEMMA 9-6

Let $q_0(u)$ be the density (9-33) for $\mu = 0$, and $q_1(u)$ the density for a fixed value $\mu > 0$; then the ratio $q_1(u)/q_0(u)$ is nondecreasing.

Proof: Substitute δ and the explicit form of the normal density in (9-33), and change the variable of integration: $y = |u|\sqrt{v}$, for fixed u. The ratio q_1/q_0 assumes the form

$$e^{-\delta^2/2}\,\frac{\displaystyle\int_0^{\infty} e^{2\delta y\,\mathrm{sgn}\,u - y^2/2}\psi_{n-1}(y^2/u^2)\,y^2\,dy}{\displaystyle\int_0^{\infty} e^{-y^2/2}\psi_{n-1}(y^2/u^2)\,y^2\,dy},$$

where $\operatorname{sgn} u = 1$ or -1 accordingly as $u \geq 0$ or $u < 0$. Put

$$p(y; u) = \frac{e^{-y^2/2}\psi_{n-1}(y^2/u^2)\, y^2}{\int_0^\infty e^{-w^2/2}\psi_{n-1}(w^2/u^2)\, w^2\, dw},$$

and

$$G(y) = e^{2\delta y \operatorname{sgn} u};$$

then q_1/q_0 is equal to

$$e^{-\delta^2/2}\int_0^\infty G(y)\, p(y; u)\, dy.$$

To complete the proof we show that the integral

$$\int_0^\infty G(y)\, p(y; u)\, dy \tag{9-34}$$

is a nondecreasing function of u. Let u_0 and u_1 be values of u such that $u_0^2 < u_1^2$; and define

$$p_0(y) = p(y; u_0), \quad p_1(y) = p(y; u_1), \qquad y > 0.$$

We are now going to apply Lemma 9-5; note that the statements and proofs of Lemmas 9-1 and 9-5 remain the same when the variable x is restricted to positive values, as is the case with the above functions p_i. These satisfy the conditions of the latter lemma for $y > 0$ (Exercise 6). If $u > 0$, then $G(y)$ increases with y; thus the lemma implies that

$$\int_0^\infty G(y)\, p(y; u_0)\, dy \leq \int_0^\infty G(y)\, p(y; \mu_1)\, dy,$$

$$0 < u_0, \quad 0 < u_1, \quad u_0^2 < u_1^2.$$

If $u < 0$, then $1 - G(y)$ increases with y, and the lemma implies that

$$\int_0^\infty (1 - G(y))\, p(y; u_0)\, dy \leq \int_0^\infty (1 - G(y))\, p(y; u_1)\, dy,$$

or, equivalently,

$$\int_0^\infty G(y)\, p(y; u_1)\, dy \leq \int_0^\infty G(y)\, p(y; u_0)\, dy,$$

$$u_0 < 0, \quad u_1 < 0, \quad u_0^2 < u_1^2.$$

We conclude that the integral is a nondecreasing function of u^2 for $u > 0$, and a nonincreasing function of u^2 for $u < 0$; thus, it is nondecreasing for all u. ■

Having completed the preliminary lemmas we turn to the statement and proof of the main result, which is that a uniformly most powerful invariant test of size α has a one-sided rejection region based on the t-statistic $\sqrt{n}\bar{x}/s$.

THEOREM 9-3

For the test of the hypothesis $\mu = 0$ against the one-sided alternative $\mu > 0$ with σ unknown, a sufficient condition that an invariant test of size α be uniformly most powerful is that it have the rejection region

$$M_2 = \{\mathbf{x}: \sqrt{n}\bar{x}/s > t_{n-1,\alpha}\},$$

where $t_{n-1,\alpha}$ is the upper α-percentile of the t-distribution with $n - 1$ degrees of freedom.

Proof: Under the null hypothesis $\mu = 0$, the t-statistic has the t-distribution with $n - 1$ degrees of freedom (Sec. 8-2); hence, the level of significance is α, as claimed. The function $\sqrt{n}\bar{x}/s$ is invariant, so that M_2 is unchanged under scalar multiplication of its members. The proof of the technical condition (9-22) is carried out in the same way as for unbiased tests (Sec. 9-2). All these statements imply that the test with rejection region M_2 is invariant and of size α. It remains to be shown that it is uniformly most powerful.

The proof is similar in structure to that of Theorem 9-1 for testing the simple null hypothesis against the simple alternative: we find the invariant set M maximizing the integral

$$\int_M \cdots \int L(\mathbf{x}; \mu, \sigma)\, d\mathbf{x}$$

under the condition that the latter is equal to α when $\mu = 0$. First we find an upper bound on the integral of the invariant continuous function f assuming values between 0 and 1,

$$\int_{R^n} \cdots \int f(\mathbf{x})\, L(\mathbf{x}; \mu, \sigma)\, d\mathbf{x}$$

under the condition that the integral is equal to α at $\mu = 0$. By

Lemma 9-4, this multiple integral is representable as the simple integral (9-31); thus, it is sufficient to find an upper bound on

$$\int_{-\infty}^{\infty} \bar{f}(y)\, q_1(y)\, dy$$

under the condition

$$\int_{-\infty}^{\infty} \bar{f}(y)\, q_0(y)\, dy = \alpha,$$

where q_0 and q_1 are the densities in the statement of Lemma 9-6. By the latter lemma, q_1/q_0 is nondecreasing; therefore, by Lemma 9-1 and the fact that q_0 is the t-density we have

$$\int_{-\infty}^{\infty} \bar{f}(y)\, q_1(y)\, dy \leq \int_{t_{n-1,\alpha}}^{\infty} q_1(y)\, dy.$$

By the method of proof of Theorems 9-1 and 9-2—involving the approximation of the power function by integrals of continuous invariant functions [condition (9-22)]—it follows that

$$\underset{M}{\int \cdots \int} L(\mathbf{x}; \mu, \sigma)\, dx \leq \int_{t_{n-1,\alpha}}^{\infty} q_1(y)\, dy$$

for any invariant test of size α with rejection region M. The latter integral is the power of the test with the rejection region M_2 (Exercise 3); thus

$$\underset{M}{\int \cdots \int} L(\mathbf{x}; \mu, \sigma)\, d\mathbf{x} \leq \underset{M_2}{\int \cdots \int} L(\mathbf{x}; \mu, \sigma)\, d\mathbf{x}.$$

This inequality is also independent of the particular value $\mu > 0$. It follows that the power of any invariant test of size α is less than or equal to that of the test with rejection region M_2 for all alternatives $\mu > 0$. ■

Example 9-7. Let us test the null hypothesis $\mu = 0$ against the one-sided composite alternative $\mu > 0$ by means of the test with the rejection region M_2; suppose $\alpha = .05$ and $n = 25$ so that

$$t_{24,.05} = 1.711.$$

The null hypothesis is rejected if $\bar{x}/s$ exceeds .370, and otherwise accepted. ■

EXERCISES

1. Determine the rejection region M_2 in the following particular cases:

(a) $n = 9$, $\alpha = .05$ (d) $n = 9$, $\alpha = .01$
(b) $n = 25$, $\alpha = .05$ (e) $n = 25$, $\alpha = .01$
(c) $n = 16$, $\alpha = .05$ (f) $n = 16$, $\alpha = .01$.

2. Verify that the joint density of $\sqrt{n}\bar{x}/\sigma$ and s^2/σ^2 is the product $\phi(u - \sqrt{n}\mu/\sigma)(n-1)\psi_{n-1}((n-1)v)$ at the point (u, v); use Theorem 5-4.

3. Why is

$$\int_{t_{n-1,\alpha}}^{\infty} q_1(y)\,dy$$

the power of the test with the rejection region M_2?

4. At what points in the derivation of the uniformly most powerful test was the fact that $\mu > 0$ under the alternative used? Indicate the changes necessary to adapt the proofs to the one-sided alternative $\mu < 0$.

5. *Prove:* If f is invariant, then so is its conditional expectation given $\sqrt{n}\bar{x}$ and $(n-1)s^2$.

6. Show that the functions p_i defined in the proof of Lemma 9-6 satisfy the conditions of Lemma 9-1.

9-5. TESTING HYPOTHESES FOR A GENERAL POPULATION (OPTIONAL)

The main ideas and definitions of testing developed above for normal populations (mainly those in Sec. 9-1 and 9-2) extend to more general populations. Consider a population with the density function $h(x; \theta)$, where θ is a parameter. Suppose that θ is known to be equal to one of two values θ_0 and θ_1. We identify the first with the "simple null hypothesis that $\theta = \theta_0$" and the second with the "simple alternative hypothesis that $\theta = \theta_1$." A test based on n observations is a decomposition of the Sample space into acceptance and rejection regions. The probabilities of error, α and β, are defined as in Eq. 9-1, with the understanding that

$$L_0 = L(\mathbf{x}; \theta_0), \qquad L_1 = L(\mathbf{x}; \theta_1),$$

where $L(\mathbf{x}; \theta) = \prod_{i=1}^{n} h(x_i, \theta)$. The problem of finding a most powerful test of size α is that of determining, among all regular sets M satisfying (9-2), the one maximizing (9-3). The Neyman-Pearson Lemma, in its general form, provides the solution. The idea behind the Lemma is similar to that justifying the maximum likelihood method (Sec. 7-5): the rejection region is determined by the *relative likelihood* of points of the Sample space under the two competing hypotheses. The rejection region is the set of all points $\mathbf{x}$ for which the ratio $L(\mathbf{x}; \theta_0)/L(\mathbf{x}; \theta_1) \leq k$, where k is a constant. This means that θ_0 is rejected if the relative likelihood of the Sample point is "too small." The constant k, as will be shown, is determined by θ_0, θ_1, and α.

THEOREM 9-4 (General Neyman-Pearson Lemma)
If M is the rejection region of a test of size α, i.e.,

$$\int_M \cdots \int L(\mathbf{x}; \theta_0)\, d\mathbf{x} = \alpha \tag{9-35}$$

and if there exists a constant k such that

$$M = \left\{\mathbf{x}: \frac{L(\mathbf{x}; \theta_0)}{L(\mathbf{x}; \theta_1)} \leq k\right\}, \tag{9-36}$$

then the test determined by M is most powerful, that is, if M' is any regular set satisfying (9-35), then

$$\int_{M'} \cdots \int L(\mathbf{x}; \theta_1)\, d\mathbf{x} \leq \int_M \cdots \int L(\mathbf{x}; \theta_1)\, d\mathbf{x}. \tag{9-37}$$

The proof of this theorem is an extension of that of the special form, Lemma 9-1. The details are outlined in Exercise 3 below. ■

The idea of this theorem is that the decision to accept or reject the null hypothesis depends only on the value of the statistic $L(\mathbf{x}; \theta_0)/L(\mathbf{x}; \theta_1)$ at the observed Sample point. If there exists a sufficient statistic λ, then, by the defining equation of sufficiency, Eq. 6-35, the ratio depends on $\mathbf{x}$ only through the single function $\lambda(\mathbf{x})$. This signifies that a sufficient statistic summarizes all the necessary "information" in the Sample for the purpose of testing the hypotheses. This was illustrated above in the normal case where it was shown that the decision about μ depended just on the observed value of $\bar{x}$ (Sec. 9-2).

The ideas of Secs. 9-3 and 9-4—on composite hypotheses, unbiased tests and invariant tests—have been generalized; however, these require a measure-theoretic treatment and so we omit them.

Example 9-8. Let (X_1, X_2) be a Sample of two observations from an exponential population with mean θ. We test the null hypothesis $\theta = \theta_0$ against the alternative hypothesis $\theta = \theta_1$, where $\theta_0 < \theta_1$. The likelihood ratio is

$$(\theta_1/\theta_0) \exp \{(x_1 + x_2)(\theta_1^{-1} - \theta_0^{-1})\}. \tag{9-38}$$

Theorem 9-4 states that if there exists a constant k such that θ_0 is rejected whenever this ratio is too small, then the corresponding test is a most powerful one of those of the same level of significance. The likelihood ratio clearly depends on (x_1, x_2) only through $x_1 + x_2$; furthermore, the coefficient $(\theta_1^{-1} - \theta_0^{-1})$ is negative because $\theta_0 < \theta_1$; thus, the test is equivalent to one for which the null hypothesis is rejected if and only if the Sample point falls in the set $\{(x_1, x_2): x_1 + x_2 \geq c\}$, where c is a constant. For any α, $0 < \alpha < 1$, there exists a number $c = c(\alpha)$ such that

$$\iint\limits_{\{(x_1,x_2):x_1+x_2\geq c(\alpha)\}} \theta_0^{-2} \exp[-(x_1 + x_2)/\theta_0]\, dx_1\, dx_2 = \alpha. \tag{9-39}$$

It follows that the test for which the null hypothesis is rejected whenever $X_1 + X_2 \geq c(\alpha)$ is of size α and is a most powerful test of that size. The number $c(\alpha)$ itself can be obtained from the distribution of the statistic $x_1 + x_2$, which was calculated in Sec. 4-6 in the special case $\theta = 1$. Note that $c(\alpha)$ depends only on α and θ_0 but not on θ_1, so that the test is actually uniformly most powerful against all alternatives $\theta_1 > \theta_0$ (cf. Sec. 9-3). ■

EXERCISES

1. A Sample of one observation is taken from the exponential population with mean θ. Find a most powerful test of $\theta_0 = 1$ against $\theta_1 = 3$ at the level $\alpha = .05$.

2. What changes are necessary in the test in Example 9-8 if $\theta_1 < \theta_0$?

3. The proof of Theorem 9-4 can be obtained from that of Lemma 9-1 by means of the following identifications: Let L_0 and L_1 take the places

of p_0 and p_1, respectively, and the multiple integral over R^n the place of the single integral over R^l. *The semi-infinite interval* $[x_\alpha, \infty)$ is replaced by the set M. The function $f(x)$ on the line is replaced by the function on R^n equal to 1 at all points of M', and 0 elsewhere; consequently, $1 - f(\mathbf{x})$ is replaced by the function equal to 0 on M' and 1 elsewhere.

4. Verify that the tests on the mean and variance of the normal distribution in Sec. 9-2 can be shown to be most powerful on the basis of the general result Theorem 9-4.

9-6. NUMERICAL EXAMPLES

Let us illustrate the test of the simple hypothesis $\mu_0 = 0$ against the alternative $\mu_1 = 1$ for the mean of a normal distribution with $\sigma = 1$, and for a Sample of five observations. The most powerful test of size $\alpha = .05$ is to reject the null hypothesis if $\bar{x} > z_{.05}/\sqrt{5}$, where

$$z_{.05}/\sqrt{5} = .73.$$

The means of 25 Samples of size 5 from a standard normal population are given in Table 7-1; these represent values of $\bar{x}$ under the null hypothesis. Notice that only one of these 25 means, .822, exceeds .73; thus, the test leads to an error of the first kind in only one case out of 25, i.e., in four percent of the cases. This is very close to the five percent "size" of the test.

Now we examine the performance of the test when the value of μ is really 1. If X is a random variable from a normal population with mean 0, then $X + 1$ is a random variable from such a population with mean 1; therefore, by adding 1 to the Sample means in Table 7-1, we obtain representatives of Sample means from a normal population with mean 1 and standard deviation 1; in other words, $\bar{x} + 1$ is a Sample mean from such a population if $\bar{x}$ is a mean from Table 7-1. The null hypothesis $\mu = 0$ is accepted if $\bar{x} + 1 < .73$, i.e., if $\bar{x} < -.27$; this represents an error of the second kind. Observe that this is true for eight of the 25 means in Table 7-1; this proportion is very close to the value β for this test, which is $\Phi(1.64 - \sqrt{5}) = \Phi(-.60) = .2743$.

The data in Table 8-1 furnish a numerical example of a two-sided test of the mean of a normal population with $\sigma = 1$. If $\mu_0 = 0$ the null hypothesis is rejected if $|\bar{x}| > z_{\alpha/2}/\sqrt{n}$. If $\alpha = .05$, $n = 5$, then, as in Table 8-1, $z_{\alpha/2}/\sqrt{n} = .877$. The null hypothesis is rejected if and only if the confidence interval does not contain 0 because

$$|\bar{x}| < z/\sqrt{n} \quad \text{if and only if} \quad \bar{x} - z/\sqrt{n} < 0 < \bar{x} + z/\sqrt{n}.$$

In the same way, the confidence intervals based on the t-distribution represent tests of the two-sided hypothesis $\mu = 0$ when σ is unknown.

BIBLIOGRAPHY

The most complete and general survey of hypothesis testing is E. L. Lehmann, *Testing Statistical Hypotheses* (New York: Wiley, 1959). The introduction of cost in general statistical problems is described in Abraham Wald, *Statistical Decision Functions* (New York: Wiley, 1950). These two books are at a high mathematical level. A simpler book which considers the cost of decision is Lionel Weiss, *Statistical Decision Theory* (New York: McGraw-Hill, 1961).

CHAPTER 10

Sequential Testing (Optional)

10-1. FIXED SAMPLE SIZE AND SEQUENTIAL TESTS

In the test procedures described in Chapter 9, the number n of observations—the size of the Sample—is fixed before the Sample is observed. The statements about the performance of the test—that is, the error probabilities—depend on the given value of n: the power function (9-13) clearly depends on n. In many applications there is concern about the number of observations necessary for a test procedure because of the cost of each observation: we wish to minimize the number of observations. Here we introduce the theory of the "sequential" test which often permits significant reduction in the actual number of observations. This theory was developed by A. Wald in the decade 1940–50.

We are given a normal population with known standard deviation, and wish to test a simple null hypothesis μ_0 against a simple alternative hypothesis μ_1. There is no loss in assuming $\sigma = 1$, because for any $\sigma > 0$ the numerical labels of the population may be divided by σ, and the standard deviation is changed to 1. A second simplification we make is taking μ_0 to be equal in absolute value but opposite in sign to μ_1; this can always be effected by shifting the origin on the axis of the values of the population. Let μ be a known positive number, and consider the testing problem

Null hypothesis: $-\mu$
Alternative hypothesis: $+\mu$

For each n and significance level α, there is a most powerful test of size α, the one in which the null hypothesis is rejected if the Sample mean exceeds $-\mu + z_\alpha/\sqrt{n}$; this is an application of Theorem 9-1. By the latter theorem, the power of the test is

$$1 - \Phi(z_\alpha - 2\mu\sqrt{n}),$$

which increases with n. For a prescribed number β, $0 < \beta < 1$, one can determine the minimum value of n such that the power is at least $1 - \beta$: solve the equation

$$\Phi(z_\alpha - 2\mu\sqrt{n}) = \beta.$$

By the definition of the percentile $z_{1-\beta}$, we have

$$z_{1-\beta} = z_\alpha - 2\mu\sqrt{n}.$$

Noting that $z_{1-\beta} = -z_\beta$ (symmetry of the standard normal density), we conclude that

$$n = (z_\alpha + z_\beta)^2/4\mu^2. \tag{10-1}$$

The smallest integer at least equal to this value of n is the minimum Sample size necessary to achieve power at least $1 - \beta$.

Since $\bar{x}$ is a multiple of $x_1 + \cdots + x_n$, the above test procedure leads to rejection or acceptance of the null hypothesis accordingly as $x_1 + \cdots + x_n$ exceeds $-n\mu + \sqrt{n}\, z_\alpha$, or does not. The idea behind the sequential test is to stop taking observations as soon as the partial sum $x_1 + \cdots + x_k$, $k = 1, 2, \ldots$, gets either too large or too small and accordingly reject or accept the null hypothesis. We take n to be the *scheduled* number of observations, the maximum number allowed in the Sample. In our proofs n is assumed to be very large; however, the precise value of n hardly influences the results because the *actual* number of observations is almost always less than the scheduled number. Let $A > 0$ and $B < 0$ be prescribed real numbers, whose exact values will be fixed later. Sampling begins with X_1, the first observation: if it exceeds A, we take no more observations and reject the null hypothesis; if it falls below B, we also stop sampling but accept the null hypothesis; finally, if it falls between A and B inclusive we postpone the decision and take the next observation X_2. The same procedure is repeated for $X_1 + X_2$: we accept or reject the null hypothesis, or take another observation accordingly as $X_1 + X_2$ falls below B, above A, or between A and B inclusive. This is repeated at each step. If, after the nth observation, none of the partial sums $x_1 + \cdots + x_k$, $1 \le k \le n$ has exceeded A or fallen below B, then the test is terminated without a decision; however, as we shall see, this is very improbable if n is large.

The number of observations in the sequential test is not fixed beforehand but is itself a function on R^n. (It is similar to the statistic giving the size of the Sample in the two-Sample confidence interval (Sec. 8-3).) We let $N = N(\mathbf{x})$ stand for this function. In accordance with the above description of the sequential test, it is defined at each point $\mathbf{x}$ as the index j of the smallest coordinate for which the partial sum $x_1 + \cdots + x_j$ leaves the interval $[B, A]$, or as n if all the partial sums fall in $[B, A]$:

$$\begin{aligned} N(\mathbf{x}) &= 1, \quad \text{if} \quad x_1 > A \quad \text{or} \quad x_1 < B \\ &= k, \quad \text{if} \quad B \le x_1 + \cdots + x_j \le A, \quad j = 1, \ldots, k - 1 \\ &\qquad \text{and either} \quad x_1 + \cdots + x_k > A \quad \text{or} \quad < B, \\ &\qquad 2 \le k \le n - 1 \\ &= n, \quad \text{if} \quad B \le x_1 + \cdots + x_j \le A, \quad j = 1, \ldots, n - 1. \end{aligned}$$

Example 10-1. Take $n = 4$ and $A = -B = 3$; then

$$\begin{aligned} N(\mathbf{x}) &= 1, \quad \text{if} \quad x_1 > 3 \quad \text{or} \quad x_1 < -3 \\ &= 2, \quad \text{if} \quad -3 \le x_1 \le 3 \quad \text{and either} \quad x_1 + x_2 > 3 \\ &\qquad \text{or} < -3 \\ &= 3, \quad \text{if} \quad -3 \le x_1 \le 3, \ -3 \le x_1 + x_2 \le 3 \\ &\qquad \text{and either} \quad x_1 + x_2 + x_3 > 3 \quad \text{or} \quad < -3 \\ &= 4, \quad \text{if} \quad -3 \le x_1 \le 3, \ -3 \le x_1 + x_2 \le 3, \\ &\qquad -3 \le x_1 + x_2 + x_3 \le 3. \end{aligned}$$

If $\mathbf{x} = (1, -2, 5, 4)$, then $N(\mathbf{x}) = 3$ because $x_1 = 1$, $x_1 + x_2 = -1$, $x_1 + x_2 + x_3 = 4$. ∎

The set in R^n upon which N assumes the value k,

$$\{\mathbf{x}: N(\mathbf{x}) = k\},$$

is determined by the first k coordinates $x_1, \ldots, x_k$ of its member points; in other words, whether or not $\mathbf{x}$ belongs to this set depends only on the first k coordinates. This is evident in Example 10-1: any point whose first three coordinates are 1, -2, and 5 belongs to $\{\mathbf{x}: N(\mathbf{x}) = 3\}$. For every k, the set is also regular because it is obtained from the half-spaces

$$\{\mathbf{x}: x_1 + \cdots + x_j \le A\}, \quad \{\mathbf{x}: x_1 + \cdots + x_j \ge B\}, \quad j = 1, \ldots, k,$$

by a finite number of elementary set operations: the latter sets are regular and the class of regular sets is closed under the operations (Theorems 3-7, 3-8, and 3-9). In Example 10-1 the set $\{\mathbf{x}: N(\mathbf{x}) = 3\}$ is the intersection of

$$\{\mathbf{x}: -3 \leq x_1 \leq 3\}, \quad \{\mathbf{x}: -3 \leq x_1 + x_2 \leq 3\},$$

and

$$\{\mathbf{x}: x_1 + x_2 + x_3 < -3\} \cup \{\mathbf{x}: x_1 + x_2 + x_3 > 3\}.$$

The formal proof of the general case is similar to this and is left to the reader (Exercise 2).

In the definition of the sequential procedure it was noted that no decision about the hypothesis is made if none of the partial sums leave $[B, A]$. This is considered an undesirable outcome; however, we shall show that this is very "unlikely." The precise form of this statement is that the probability that $\mathbf{X}$ falls in the set

$$\{\mathbf{x}: B \leq x_1 + \cdots + x_k \leq A, \quad k = 1, \ldots, n\}$$

converges to 0 as $n \to \infty$. We call the latter probability the "probability that no decision is made after n observations."

THEOREM 10-1

The probability that no decision is made after n observations is at most equal to

$$1 - \Phi\left(\frac{B + n\mu}{\sqrt{n}}\right) \tag{10-2}$$

under the null hypothesis, and

$$\Phi\left(\frac{A - n\mu}{\sqrt{n}}\right) \tag{10-3}$$

under the alternative. Each of these bounds converges to 0 as $n \to \infty$.

Proof: The set $\{\mathbf{x}: B \leq x_1 + \cdots + x_k \leq A, \ k = 1, \ldots, n\}$ is the intersection of the n sets $\{\mathbf{x}: B \leq x_1 + \cdots + x_k \leq A\}$, $k = 1, \ldots, n$; thus, the first set is included in the last set, $\{\mathbf{x}: B < x_1 + \cdots + x_n \leq A\}$. It follows that the probability of the first—which is the probability that no decision is made—is less than or equal to the probability of the last set [property (3-32)]. Since $x_1 + \cdots + x_n$ has a normal distribution with mean $\mp n\mu$ and standard deviation

$\sqrt{n}$, the probability that $\mathbf{X}$ falls in $\{\mathbf{x}: B \leq x_1 + \cdots + x_n \leq A\}$ is equal to

$$\Phi\left(\frac{A + n\mu}{\sqrt{n}}\right) - \Phi\left(\frac{B + n\mu}{\sqrt{n}}\right)$$

under the null hypothesis $-\mu$, and

$$\Phi\left(\frac{A - n\mu}{\sqrt{n}}\right) - \Phi\left(\frac{B - n\mu}{\sqrt{n}}\right)$$

under the alternative. The former is less than (10-2) and the latter less than (10-3). ■

In advanced statistical theory the function N is defined on an infinite-dimensional normal Sample space, and it is shown that N assumes a finite value "with probability 1." In statistical terminology we say that the sequential test terminates in a decision after finitely many observations.

The sequential test on the standard deviation of a normal population with known mean μ is based on the partial sums $\sum_{i=1}^{k} (x_i - \mu)^2$, $k = 1, \ldots, n$. Suppose that $\mu = 0$. The null hypothesis is σ_0 and the alternative σ_1, with $\sigma_0 < \sigma_1$. By changing the scale (if necessary) we take $\sigma_0 = 1$, and put $\sigma = \sigma_1$; thus, the null hypothesis is 1 and the alternative $\sigma > 1$. The fixed interval $[B, A]$ in the test for the mean is replaced by a moving interval in the test for the standard deviation. Now let A and B be fixed numbers such that $B < 0 < A$, and form the sequence of intervals

$$I_k = \left[B + \frac{k\sigma^2 \log \sigma^2}{\sigma^2 - 1}, \quad A + \frac{k\sigma^2 \log \sigma^2}{\sigma^2 - 1}\right], \qquad k = 1, \ldots, n.$$

The test continues as long as $x_1^2 + \cdots + x_k^2$ falls in I_k. It terminates in acceptance of the null hypothesis as soon as $x_1^2 + \cdots + x_k^2$ falls to the left of I_k, in rejection to the right of I_k. No decision is made in case $x_1^2 + \cdots + x_k^2$ never leaves I_k; the probability that the Sample point falls in this set converges to 0 as $n \to \infty$ (Exercises 4–6).

EXERCISES

1. Using Formula 10-1 calculate the necessary number of observations for these tests:

(a) $\alpha = \beta = .01$, $\mu = 1$
(b) $\alpha = .02$, $\beta = .05$, $\mu = 2$
(c) $\alpha = \beta = .05$, $\mu = 1/2$.

2. Show that $\{\mathbf{x}: N(\mathbf{x}) = k\}$ is a regular set. (*Hint:* Express it as the intersection of k regular sets.)

3. *Prove:* For $x > 0$,

$$\int_x^\infty e^{-y^2/2}\,dy = \int_{-\infty}^{-x} e^{-y^2/2}\,dy \le (1/x)\,e^{-x^2/2}.$$

Use this to obtain asymptotic bounds on the expressions (10-2) and (10-3) for large n.

4. Prove the analogue of Theorem 10-1 (part I) for the sequential test on the standard deviation with (10-2) and (10-3) replaced by

$$\int_x^\infty \psi_n(y)\,dy, \quad x = A + \frac{n\sigma^2 \log \sigma^2}{\sigma^2 - 1}$$

and

$$\int_0^x \psi_n(y)\,dy, \quad x = \frac{B}{\sigma^2} + \frac{n \log \sigma^2}{\sigma^2 - 1},$$

respectively.

5. *Prove:*

$$\lim_{n\to\infty} \int_0^{tn} \psi_n(y)\,dy = 0, \qquad 0 < t < 1$$

$$= 1, \qquad t > 1.$$

[*Hint:* If $y < tn$, $0 < t < 1$, then $(y - n)^2 > n^2(1 - t)^2$, and so

$$\int_0^{tn} \psi_n(y)\,dy \le \frac{\int_0^\infty (y - n)^2 \psi_n(y)\,dy}{n^2(1 - t)^2}.$$

If $y > tn$, $0 < t < 1$, then $(y - n)^2 > n^2(1 - t)^2$, and so

$$\int_0^{tn} \psi_n(y)\,dy = 1 - \int_{tn}^\infty \psi_n(y)\,dy \ge 1 - \frac{1}{n^2(1 - t)^2} \cdot \int_0^\infty (y - n)^2 \psi_n(y)\,dy.]$$

6. By means of the inequalities

$$\log \sigma^2/(\sigma^2 - 1) < 1, \quad \sigma^2 \log \sigma^2/(\sigma^2 - 1) > 1, \qquad \text{for} \quad \sigma > 1,$$

and the result of Exercise 5, show that the bounds in Exercise 4 converge to 0 as $n \to \infty$.

10-2. SELECTION OF THE "BARRIERS" A AND B

The sequential test ends with a decision as soon as the first partial sum of the coordinates of the Sample point falls outside the interval $[B, A]$. The numbers A and B are pictured as "barriers" for the sequence of partial sums: the test terminates as soon as one of the partial sums oversteps either barrier. Under the null hypothesis, the partial sum $x_1 + \cdots + x_k$ has a normal distribution with mean $-k\mu$ and standard deviation $\sqrt{k}$. The density is centered at the negative value $-k\mu$; thus the sum is more likely to overstep the negative barrier B than the positive barrier A. By analogy, the sum is more likely to fall beyond the positive barrier under the alternative hypothesis.

The quantities A and B determine the probabilities of error: there is an error of the first kind if, when the mean is negative, the partial sum oversteps A before B; and, similarly, there is an error of the second kind if, when the mean is positive, the partial sum oversteps the negative barrier first. By intuitive considerations, we see that if A and B are too close to each other, there is a substantial probability of error because the partial sum can easily overstep the wrong barrier, leading to an incorrect decision; on the other hand, if A and B are far apart, it is unlikely that an error is made.

We do not infer that the barriers should be selected arbitrarily far from each other; for in that case it may take many observations in order for the partial sum to get large enough in magnitude to reach either barrier, and so the *cost of sampling* is large. In this section we show that for any fixed numbers α and β, $0 < \alpha, \beta < 1$, the barriers A and B can be so selected that the probabilities of error of the first and second kinds are less than or equal to α and β respectively; in the next section we shall determine the order of magnitude of the number of observations for the test with these barriers. The prescribed values of A and B are

$$A = \frac{-\log \alpha}{2\mu}, \quad B = \frac{\log \beta}{2\mu}, \tag{10-4}$$

where the logarithm is the natural one (base e). Note that A increases as α decreases: the positive barrier is chosen large to reduce the probability of error of the first kind; similarly, B is chosen large in the negative sense if β is small. The difference between A and B

is inversely proportional to the difference between the values of the mean under the two hypotheses. If $\alpha = \beta$, the barriers are placed symmetrically: $A = -B$.

THEOREM 10-2
In the sequential test for μ with the barriers A and B defined by (10-4), the probabilities of error of the first and second kinds are at most equal to α and β, respectively.

Proof: The proof is based on the following two remarks:

(i) The ratio of the likelihood function under the alternative hypothesis to that under the null hypothesis is an exponential function of the sum of the coordinates:

$$\frac{\prod_{i=1}^{k} \phi(x_i - \mu)}{\prod_{i=1}^{k} \phi(x_i + \mu)} = \exp\,[2\mu(x_1 + \cdots + x_k)],$$

$k = 1, \ldots, n$; this is verified by elementary algebra. An immediate consequence is:

$$x_1 + \cdots + x_k > A \quad or \quad < B \tag{10-5}$$

if and only if

$$\frac{\prod_{i=1}^{k} \phi(x_i - \mu)}{\prod_{i=1}^{k} \phi(x_i + \mu)} > 1/\alpha \quad or \quad < \beta,$$

respectively, where A and B are defined by (10-4).

(ii) The set of points in the Sample space for which the null (or alternative) hypothesis is rejected may be decomposed into the union of disjoint (regular) sets:

$\{\mathbf{x}:$ The null (or alternative) hypothesis is rejected, and $N(\mathbf{x}) = k\}$, $k = 1, \ldots, n$;

in other words, the set of points leading to rejection may be decomposed into the pieces upon which the number of observations is the same. Recall that a similar decomposition was used in the derivation of the two-Sample confidence interval (Theorem 8-3).

Let H be the set in the Sample space upon which the sequential test rejects the null hypothesis, and let H_k be its intersection with $\{\mathbf{x}: N(\mathbf{x}) = k\}$. In accordance with the remark (ii) above H is the union of the disjoint sets H_k, $k = 1, \ldots, n$; thus, by property (3-33), the probability under the alternative that the null hypothesis is rejected $\left(\text{the integral of } \prod_{i=1}^{n} \phi(x_i - \mu) \text{ over } H\right)$ is equal to the sum

$$\sum_{k=1}^{n} \int \cdots \int_{H_k} \prod_{i=1}^{n} \phi(x_i - \mu) \prod_{i=1}^{n} dx_i. \tag{10-6}$$

Write the integrand in the kth integral above as

$$\prod_{i=1}^{k} \phi(x_i - \mu) \cdot \prod_{i=k+1}^{n} \phi(x_i - \mu).$$

If $\mathbf{x}$ belongs to H_k, then $x_1 + \cdots + x_k > A$; thus by (10-5), the first subproduct above exceeds

$$(1/\alpha) \prod_{i=1}^{k} \phi(x_i + \mu);$$

therefore the sum (10-6) is at least equal to

$$(1/\alpha) \sum_{k=1}^{n} \int \cdots \int_{H_k} \prod_{i=1}^{k} \phi(x_i + \mu) \prod_{i=k+1}^{n} \phi(x_i - \mu) \prod_{i=1}^{n} dx_i. \tag{10-7}$$

As stated in Sec. 10-1 (following Example 10-1) the set $\{\mathbf{x}: N(\mathbf{x}) = k\}$ is determined by the first k coordinates of its member points; furthermore, the same is true of the set H_k because

$$H_k = \{\mathbf{x}: N(\mathbf{x}) = k\} \cap \{\mathbf{x}: x_1 + \cdots + x_k > A\}.$$

Any changes in the last $n - k$ coordinates of a point do not affect its membership or nonmembership in H_k; in particular H_k is unaltered by the transformation of the space under which the signs of the last $n - k$ coordinates of each point are changed. This implies that the kth integral in (10-7) is unchanged if the variables of integration $x_{k+1}, \ldots, x_n$ are replaced by $-x_{k+1}, \ldots, -x_n$; thus, by the symmetry of the function ϕ, the sum (10-7) is equal to

$$(1/\alpha) \sum_{k=1}^{n} \int \cdots \int_{H_k} \prod_{i=1}^{n} \phi(x_i + \mu) \prod_{i=1}^{n} dx_i. \tag{10-8}$$

The sum above (excluding the factor $1/\alpha$) is, by the additivity property (3-33), equal to

$$\int_H \cdots \int \prod_{i=1}^{n} \phi(x_i + \mu) \prod_{i=1}^{n} dx_i,$$

which is the probability of rejecting the null hypothesis when it is in fact true. Comparing (10-6), (10-7), and (10-8), we see that

> Probability under the alternative that the null hypothesis is rejected $\geq (1/\alpha) \cdot$ probability under the null hypothesis that the null hypothesis is rejected.

Since the quantity on the left-hand side of this inequality is not more than 1 (because it is a probability), so is the right-hand side not more than 1: the probability of an error of the first kind is at most α.

The proof that the probability of error of the second kind does not exceed β is analogous, and is left as an exercise (Exercise 1). ■

In the sequential test for the null hypothesis that the standard deviation is 1 against the alternative that it is equal to $\sigma > 1$, the numbers A and B in the moving barriers are

$$A = \frac{-2 \log \alpha}{1 - 1/\sigma^2}, \quad B = \frac{2 \log \beta}{1 - 1/\sigma^2}. \tag{10-9}$$

The probabilities of error are at most equal to α and β, respectively. The proof is similar to that of Theorem 10-2: the remark (i) is replaced by the observation that the ratio of the likelihood function under the alternative hypothesis to that under the null hypothesis is an exponential function of the sum of the squares of the coordinates:

$$\frac{\prod_{i=1}^{k} (1/\sigma)\phi(x_i/\sigma)}{\prod_{i=1}^{k} \phi(x_i)} = (\sigma^2)^{-n/2} \exp\left[\frac{\sum_{i=1}^{k} x_i^2}{2}\left(1 - \frac{1}{\sigma^2}\right)\right] \tag{10-10}$$

$k = 1, \ldots, n$; its consequence is

$$\sum_{i=1}^{k} x_i^2 > A + \frac{k\sigma^2 \log \sigma^2}{\sigma^2 - 1} \tag{10-11}$$

or

$$< B + \frac{k\sigma^2 \log \sigma^2}{\sigma^2 - 1}$$

if and only if

$$\frac{\prod_{i=1}^{k} (1/\sigma)\phi(x_i/\sigma)}{\prod_{i=1}^{k} \phi(x_i)} > 1/\alpha$$

or

$$< \beta,$$

respectively. The remark (ii) of the proof of Theorem 10-2 is unaltered. The invariance of H_k under the change of signs of the last $n - k$ variables is replaced by invariance under the transformation of scale: $x_i \rightarrow x_i\sigma$, $i = k + 1, \ldots, n$. The details of the proof are left to the reader (Exercise 2).

EXERCISES

1. Following the proof of Theorem 10-2, show that

   Probability under the null hypothesis that the null hypothesis is accepted $\geq (1/\beta) \cdot$ probability under the alternative that the null hypothesis is accepted.

2. Using the outline of the proof of Theorem 10-2 and the alterations outlined in the last paragraph, prove the version of Theorem 10-2 for the sequential test for the standard deviation.

10-3. EXPECTED SAMPLE SIZE IN THE SEQUENTIAL TEST

As stated in Sec. 10-1, the reason for using the sequential test instead of the fixed Sample size test is that the former usually requires fewer observations while having the same error probabilities. Now we compare the number of observations in the fixed sample size test with error probabilities α and β to the number in the sequential test with the barriers (10-4). A direct comparison is not possible because the Sample size in the sequential test is not fixed but varies with the Sample point: the index of the terminal observation, $N(\mathbf{x})$, is a function on the Sample space, assuming values $1, \ldots, n$. For this reason, we compare the number of observations in the fixed Sample size test to the "expected value" of N: this is a weighted average of the integers $1, \ldots, n$, where the integer k is weighted by the probability of the set $\{\mathbf{x}: N(\mathbf{x}) = k\}$, $k = 1, \ldots, n$. The expected value of N is denoted $E(N)$. Put

$$p_k = \text{probability of } \{\mathbf{x}: N(\mathbf{x}) = k\}; \tag{10-12}$$

then

$$E(N) = \sum_{k=1}^{n} kp_k. \tag{10-13}$$

In accordance with Sec. 2-2, $E(N)$ represents the center of gravity of a mass distribution $p_1, \ldots, p_n$ on the integers $1, \ldots, n$.

The justification of the use of $E(N)$ as a representative or typical value of N is in the "law of large numbers" of the theory of probability: If the sequential test is performed many times then the long-run average Sample size actually observed over time will, with high probability, be very close to $E(N)$. The version of this "weak law of large numbers" is found in most of the standard elementary books on probability; in particular, we refer to the books in the bibliography of Chapter 1. It is not necessary that the reader *now* have a good understanding of this principle because the theorem we shall prove involves only the definition (10-13) of $E(N)$ but not its justification.

There is a different value for $E(N)$ under the null and alternative hypotheses, respectively, because p_k, defined by (10-12), is computed with two different likelihood functions. The following proposition furnishes a useful estimate of the magnitude of $E(N)$:

THEOREM 10-3

In the sequential test for the mean of the normal population with the barriers (10-4), $E(N)$ is bounded above by

$$1 + \sum_{k=1}^{n-1} \left[\Phi\left(\frac{A \pm k\mu}{\sqrt{k}}\right) - \Phi\left(\frac{B \pm k\mu}{\sqrt{k}}\right) \right], \tag{10-14}$$

with $(+)$ and $(-)$ under the null and alternative hypotheses, respectively.

Proof: The first step in the proof is the identity

$$\sum_{k=1}^{n} kp_k = \sum_{k=1}^{n} \sum_{j=k}^{n} p_j. \tag{10-15}$$

To prove this write the double sum above as

$$\begin{aligned} p_1 + p_2 + p_3 + \cdots + p_n \\ + p_2 + p_3 + \cdots + P_n \\ + p_3 + \cdots + p_n \\ \vdots \\ + p_n. \end{aligned}$$

Now sum these down each column, and add the column subtotals: this is the sum on the left-hand side of (10-15).

The definition (10-13) and the identity (10-15) imply:

$$E(N) = \sum_{k=1}^{n} \sum_{j=k}^{n} p_j. \tag{10-16}$$

We claim:

$$\begin{aligned} \sum_{j=k}^{n} p_j &= 1, \qquad \text{for} \quad k = 1 \\ &\leq \Phi\left(\frac{A \pm (k-1)\mu}{\sqrt{k-1}}\right) - \Phi\left(\frac{B \pm (k-1)\mu}{\sqrt{k-1}}\right), \\ &\qquad\qquad 2 \leq k \leq n. \end{aligned} \tag{10-17}$$

The sets $\{\mathbf{x}: N(\mathbf{x}) = k\}$, $k = 1, \ldots, n$ are disjoint. The set $\{\mathbf{x}: N(\mathbf{x}) \geq k\}$ is the union of the former sets over indices j, $k \leq j \leq n$; thus, by the additivity property (3-33), the probability of the latter is equal to the sum in (10-17):

$$P\{\mathbf{x}: N(\mathbf{x}) \geq k\} = \sum_{j=k}^{n} p_j.$$

For $k = 1$ the equality in (10-17) holds because $N(\mathbf{x}) \geq 1$ for every $\mathbf{x}$. For $k > 1$, if $N(\mathbf{x}) \geq k$ then no decision is made in the sequential test before the kth observation; thus, the sum of the first $k - 1$ coordinates must be in the interval $[B, A]$; hence, by the property (3-32):

$$P\{\mathbf{x}: N(\mathbf{x}) \geq k\} \leq P\{\mathbf{x}: B \leq x_1 + \cdots + x_{k-1} \leq A\}.$$

The right-hand side is equal to the right-hand side of the inequality in (10-17) because $x_1 + \cdots + x_{k-1}$ has a normal distribution with mean $\pm(k - 1)\mu$ and standard deviation $\sqrt{k-1}$, $k > 1$ (Theorem 4-5).

The estimate (10-14) follows from (10-16) and (10-17). ∎

In the following example we compute the numerical value of the bound (10-14) for $E(N)$ and compare it to the number of observations in the fixed Sample size test with corresponding probabilities of error.

Example 10-2. Suppose $\mu = 1$, and $\alpha = \beta = .01$. From Table I, Appendix, we find $z_{.01} = 2.33$. In accordance with (10-1) the number of observations necessary in the fixed Sample size test is the smallest integer at least equal to $(4.66)^2/4 = 5.45$, which is 6. In the sequential test with approximately the same error probabilities the barriers are $A = -B = -\frac{1}{2}\log(.01) = 2.3$. The sequential test is performed as follows. We successively observe X_1, $X_1 + X_2, \ldots$, stopping as soon as one of the partial sums falls above 2.3, in which case we reject the null hypothesis, or falls below -2.3, in which case we accept it. Since $A = -B$, the integrals in (10-14) are the same for both choices of sign (Exercise 1); thus, the bound is the same under both hypotheses. The kth term in the sum in (10-11) is

$$\Phi\left(\frac{2.3 - k}{\sqrt{k}}\right) - \Phi\left(\frac{-2.3 - k}{\sqrt{k}}\right).$$

The values for $k = 1, \ldots, 10$ are obtained from Table I, Appendix:

k	$(2.3 - k)/\sqrt{k}$	$(-2.3 - k)/\sqrt{k}$	Φ − difference
1	1.30	−3.30	.9027
2	.21	−3.30	.5827
3	−.40	−3.06	.3435
4	−.85	−3.15	.1969
5	−1.19	−3.21	.1163
6	−1.52	−3.38	.0639
7	−1.77	−3.50	.0382
8	−2.04	−3.64	.0206
9	−2.23	−3.76	.0128
10	−2.44	−3.90	.0073
		Sum:	2.2849

The remaining terms of the series, of index greater than 10, contribute a negligible amount to the sum, so that they may be neglected; indeed it can be shown that the infinite series

$$\sum_{k=1}^{\infty} \Phi\left(\frac{A - k\mu}{\sqrt{k}}\right) \tag{10-18}$$

converges rapidly (Exercise 2). The ratio of the bound (10-14) for $E(N)$ to 6 (the number in the fixed Sample size test) is $3.2849/6 = .5475$. This shows that the expected Sample size in the sequential test is not more than 54 percent of the number in the fixed Sample size test. ■

The bound (10-14) is crude: by more advanced methods (Wald's equation), a smaller bound can be obtained; it is

$$\frac{B(1 - \alpha) + A\alpha}{-\mu} \tag{10-19}$$

under the null hypothesis, and

$$\frac{B\beta + A(1 - \beta)}{\mu} \tag{10-20}$$

under the alternative. These are much simpler to evaluate than is (10-14); however, they can be justified only by approximations requiring arguments outside the scope of this book. (The popular proof of the validity of (10-19) and (10-20) involves the neglect of the "excess over the barrier" at the time the partial sum crosses the barrier. The estimation of the excess is usually avoided because it depends on advanced methods.) In Example 10-2 the bounds (10-19) and (10-20) are equal to 2.26; its ratio to 6 is about 37 percent.

The calculation of the bound (10-14) can be simplified in many cases by noticing that (10-14) is hardly increased if, under the null hypothesis, the second Φ-term is dropped, and, under the alternative, the first Φ-term is replaced by 1. The resulting bounds are

$$1 + \sum_{k=1}^{n-1} \left[1 - \Phi\left(\frac{B + k\mu}{\sqrt{k}}\right)\right] \tag{10-21}$$

under the null hypothesis, and

$$1 + \sum_{k=1}^{n-1} \Phi\left(\frac{A - k\mu}{\sqrt{k}}\right) \tag{10-22}$$

under the alternative. The approximate equality of the original bound (10-14) and the modified bounds (10-21) and (10-22) is clear in the case of Example 10-2.

In the sequential test for the standard deviation with the

barriers (10-9), $E(N)$ is bounded by

$$1 + \sum_{k=1}^{n-1} \int_{B+k\sigma^2 \log \sigma^2/(\sigma^2-1)}^{A+k\sigma^2 \log \sigma^2/(\sigma^2-1)} \psi_k(y)\, dy \tag{10-23}$$

and

$$1 + \sum_{k=1}^{n-1} \int_{B+k\sigma^2 \log \sigma^2/(\sigma^2-1)}^{A+k\sigma^2 \log \sigma^2/(\sigma^2-1)} \psi_k(y/\sigma^2)\, dy/\sigma^2 \tag{10-24}$$

under the null and alternative hypothesis, respectively; the reasoning is, with the modification of the barriers, the same as for (10-14) (Exercise 4). We omit a numerical illustration involving these bounds because it depends on the complete table of the chi-square distribution.

EXERCISES

1. *Prove:* $\Phi(x + \mu) - \Phi(-x + \mu) = \Phi(x - \mu) - \Phi(-x - \mu)$, for all x, μ.
2. Using the result of Exercise 3, Sec. 10-1, show that the series (10-18) converges.
3. Compare the expected Sample size in the sequential test with the number of observations in the fixed Sample size test in each of the cases below; use the crude bounds (10-21) and (10-22) as well as the sharper bounds (10-19) and (10-20):
   (a) $\alpha = \beta = .02,\ \mu = 1$
   (b) $\alpha = \beta = .02,\ \mu = 2$
   (c) $\alpha = \beta = .05,\ \mu = 1$
   (d) $\alpha = \beta = .01,\ \mu = 4.$
4. Verify the bounds (10-23) and (10-24).

10-4. SEQUENTIAL TESTING FOR A GENERAL POPULATION

The sequential tests on the mean and standard deviation of a normal population are particular cases of the *sequential probability ratio test* for a general population with density $h(x;\theta)$. Suppose we want to sequentially test the null hypothesis $\theta = \theta_0$ against the alternative $\theta = \theta_1$. The test is based on the observation of the successive values of the likelihood ratio

$$\lambda_m(\mathbf{x}) = \frac{\prod_{i=1}^{m} h(x_i;\theta_1)}{\prod_{i=1}^{m} h(x_i;\theta_0)},$$

for $m = 1, 2, \ldots$. Let α and β be the desired probabilities of error. Two barriers A and B, with $0 < B < A$, are selected in accordance with the equations

$$A = 1/\alpha, \quad B = \beta.$$

(The conventionally used barriers are $A' = (1 - \beta)/\alpha$ and $B' = \beta/(1 - \alpha)$. If α and β are small the two pairs are practically equal.) The first observation is taken, and $\lambda_1(\mathbf{x})$ is computed: If it exceeds A, we stop sampling and reject the null hypothesis; if it falls below B, we immediately accept the null hypothesis; finally, if it falls between B and A, inclusive, we take another observation, and repeat the same procedure with $\lambda_2(\mathbf{x})$, and so on. The intuitive justification for this test is similar to that of the likelihood ratio test of Neyman and Pearson (Theorem 9-4) for a fixed number of observations: one or the other hypothesis is accepted on the basis of the relative likelihood of the observed Sample.

By referring to the proofs of Theorems 10-1 and 10-2 (for the normal case) we shall indicate why appropriate versions hold in the general case.

Let us begin with Theorem 10-1, which asserts that the sequential test cannot "go on forever." The ideas of the proof in the normal case are:

If no decision is reached by the conclusion of the nth observation, then the nth partial sum, $x_1 + \cdots + x_n$, must certainly fall in the interval $[B, A]$; and

As n tends to infinity the probability that the sum $x_1 + \cdots + x_n$ falls in the fixed interval $[B, A]$ tends to 0.

We shall show that the sequential test based on the successive values of $\lambda_m(\mathbf{x})$ is always equivalent to a sequential test based on partial sums of observations from a common population—not necessarily normal. Since $\log \lambda$ is a monotonic function of λ, the test is equivalent to one based on the logarithms,

$$\log \lambda_m(\mathbf{x}) = \sum_{i=1}^{m} \log\left[\frac{h(x_i;\theta_1)}{h(x_i;\theta_0)}\right]. \tag{10-25}$$

The test continues as long as the partial sums of logarithms in (10-25) remain in the closed interval [$\log B$, $\log A$]. The term

$$z_i = \log\left[\frac{h(x_i; \theta_1)}{h(x_i; \theta_0)}\right], \qquad i = 1, \ldots, n \tag{10-26}$$

is a function of the ith coordinate of the point $\mathbf{x}$ in the Sample space, that is, it is a "relabeled" value of the ith observation x_i. It follows that $\mathbf{z} = (z_1, \ldots, z_n)$ is a point in a Sample space of n observations from some population. By analogy to the first idea mentioned above in connection with the proof of Theorem 10-1, *if no decision is reached by the conclusion of the nth observation, then the nth partial sum, $z_1 + \cdots + z_n$, must fall in the interval* [$\log B$, $\log A$].

In order to complete the extension of Theorem 10-1, we have to show that the second idea in the proof of the theorem extends to the partial sum $z_1 + \cdots + z_n$, that is, the probability that

$$\log B \leq z_1 + \cdots + z_n \leq \log A \tag{10-27}$$

converges to 0 as $n \to \infty$. It can be shown that the distribution of the sum of n observations from a common population with positive variance tends to "flatten" as $n \to \infty$; in other words, the probability attached to any fixed interval tends to 0 (Exercise 4); thus, the probability that (10-27) holds converges to 0.

Example 10-3. In the case $h(x; \theta) = \phi(x - \mu)$, with $\theta_0 = -\mu$ and $\theta_1 = \mu$, it is seen that

$$\log \lambda_m(\mathbf{x}) = 2\mu(x_1 + \cdots + x_m),$$

and so the test given above in the normal case is indeed a special case of the general sequential test. (Note that the barriers A and B in the discussion of the normal case are transformed into their logarithms in the general formulation.) If $h(x; \theta)$ is the exponential density with mean θ, then the logarithm (10-25) is a linear function of $x_1 + \cdots + x_n$, and so the sequential test is based on the sum of the observations, as in the normal case. ■

The proof of Theorem 10-2 easily generalizes to the density $h(x; \theta)$; as a matter of fact, it is based upon the relation (10-5), linking the partial sums of the observations to the likelihood ratio, and most of the proof involves operations with the normal likelihood function that are valid also for a general likelihood.

The bound for the expected Sample size given in Theorem 10-3 is valid only for the normal population; however, the approximations (10-19) and (10-20), which were stated without proof, have the following form for a density $h(x; \theta)$:

$$E(N) = \frac{(1 - \alpha) \log B + \alpha \log A}{\int_{-\infty}^{\infty} \log [h(x; \theta_1)/h(x; \theta_0)] \, h(x; \theta_0) \, dx} \tag{10-28}$$

under $\theta = \theta_0$, and

$$E(N) = \frac{\beta \log B + (1 - \beta) \log A}{\int_{-\infty}^{\infty} \log [h(x; \theta_1)/h(x; \theta_0)] \, h(x; \theta_1) \, dx} \tag{10-29}$$

under $\theta = \theta_1$. (Recall the remark in Example 10-3 about the change from A and B to their logarithms.)

Example 10-4. If $h(x; \theta)$ is the exponential density with mean θ, then the denominators in (10-28) and (10-29) are equal to

$$\log (\theta_0/\theta_1) + 1 - (\theta_0/\theta_1),$$

and

$$\log (\theta_1/\theta_0) + 1 - (\theta_1/\theta_0),$$

respectively (Exercise 2). ■

EXERCISES

1. Let $h(x; \theta)$ be the Rayleigh density with scale factor θ:

$$\begin{aligned} h(x; \theta) &= 0, & x &\leq 0 \\ &= (x/\theta) \exp(-x^2/2\theta^2), & x &> 0. \end{aligned}$$

Show that $\log \lambda_m$ is a linear function of $x_1^2 + \cdots + x_m^2$.

2. Verify the result in Example 10-4.

3. Use the central limit theorem (Sec. 4-6) to show that the distribution of the sum of n observations from a population with the density $h(x)$ "flattens" as $n \to \infty$, under the condition that the variance is finite.

4. By the result of Exercise 3, the probability of no decision by the end of the nth observation converges to 0 as $n \to \infty$ if the random variables (10-26) are from a population with finite variance. This extension of Theorem 10-1 is valid even when

$$\int_{-\infty}^{\infty} x^2 h(x)\, dx = \infty .$$

The steps in the proof are:

(a) Under the last condition, for every $d > 0$, the inequality

$$\int_{-d}^{d} h(x)\, dx < 1$$

holds.

(b) For any $d > 0$, if the first n partial sums $x_1, x_1 + x_2, \ldots, x_1 + \cdots + x_n$ all fall between $-d/2$ and $d/2$, then all the coordinates x_i satisfy $|x_i| < d, i = 1, \ldots, n$.

(c) The probability that the Sample point falls in the set $\{\mathbf{x}: |x_i| < d, i = 1, \ldots, n\}$ is equal to

$$\left(\int_{-d}^{d} h(x)\, dx\right)^n$$

which, by the result (a), converges to 0 as $n \to \infty$.

10-5. A NUMERICAL EXAMPLE

The test on the mean of the normal population will be demonstrated by the use of normal deviates from Table VI, Appendix. In this example we put $\mu = 1$, $\alpha = \beta = .01$. The point $z_{.01}$ on the axis of the normal distribution is 2.33; thus, by (10-1), the minimum number of observations in the most powerful test of size .01 necessary to achieve power .99 is $n = 6$. (cf. Example 10-2.)

The barriers for the sequential test are

$$\mp \frac{\log .01}{2} = \pm 2.302.$$

(The logarithm has the base e.) We shall let each column of 50 normal deviates on the third and fourth pages of Table VI, Appendix represent a Sample of 50 observations from a standard normal population; these are the "scheduled" observations. If 1 is added to each normal deviate, then each represents an observation from a population with mean 1; thus, for example, when 1 is added to the first column, the observations are 1.064, 2.206, etc. We calculate the successive partial sums for each column, and stop as soon as the first partial sum exceeds 2.302 in magnitude. There is a "correct decision" if +2.302 is the overstepped barrier.

In the first column, the first two partial sums are 1.064, $1.064 + 2.206 = 3.270$; thus, the positive barrier is crossed on the *second* observation, and a correct decision is made. Similar calculations with the observations in the first 25 columns yield correct decisions in every single case; furthermore, the following numbers of observations are required for the 25 Samples:

2 9 2 6 2 1 3 2 3 2 2 2 2
3 3 2 2 1 1 3 2 2 2 3 5.

The average number is $67/25 = 2.68$; its ratio to the value $n = 6$ for the nonsequential most powerful test is .45.

EXERCISES (Optional)

1. Repeat the numerical example above when $\mu = -1$ is the true value, so that 1 is subtracted from each normal deviate.

2. Repeat the numerical example in the text for $\mu = 1/2$. What is the value of n for the corresponding nonsequential test with $\alpha = \beta = .01$?

BIBLIOGRAPHY

The fundamental work on the topic is Abraham Wald, *Sequential Analysis* (New York: Wiley, 1947).

CHAPTER 11

Sampling from Several Normal Populations

11-1. SAMPLING FROM n BOXES OF BALLS AND n NORMAL POPULATIONS

In the previous chapters we have been concerned with statistical inference on a single normal population based on a Sample of n observations. There are many problems of interest involving two or more populations.

We begin with an example involving two populations. An experimenter wants to test the effect of a nutrient on the growth of animals. He takes two groups of animals of comparable biological characteristics; for example, two groups of mice of the same age and weight. One group, called the control group, is not given the nutrient, while the other group—the experimental group—gets it. The control group is considered to be a Sample from the population of animals deprived of the nutrient, and the experimental group is considered to be a Sample from the population receiving the nutrient. The experimenter is interested in the differences, if any, between the two populations; for example, he wants to know if there is any difference between the average weights of the animals in the two groups. This suggests a general inference problem concerning two populations: testing whether the means are equal. Another is estimating the difference between the means.

The above problem involving two populations can be generalized to more than two. Suppose the experimenter wants to establish a relation between the *quantity* of the nutrient absorbed by an animal and the growth of the animal. He takes n experimental animals of comparable biological characteristics, and gives them varying amounts of the nutrient; then he observes the record of the growth of each animal. He may find a linear relationship between the quantity of the nutrient in the diet and the growth of the animal; such a relationship is described by the linear regression model discussed in Chapter 12. Each animal is considered to be a Sample of one observation from the population of animals furnished with the corresponding amount of the nutrient.

Statistical inference involving several populations is also used in the analysis of periodic records or *time series*. Suppose a governmental agency wants to know whether prices are rising or not over a given time period—for example, over a year. At the beginning of each week the prices of certain commodities are recorded in various geographic areas. The set of prices in all areas at the beginning of a week is considered to be a fixed population. Every week the new prices form a new population, and a fresh sample is taken. At the end of the year there are samples from each of the 52 "populations." The problem of statistical interest is whether the means of these populations increase with time or not.

Now we turn to the theory of sampling from several populations. As in the case of a single population, we first consider drawing labeled balls from a box. Consider n numbered boxes, each containing a set of balls. Let there be N_i balls in the ith box, $i = 1, \ldots, n$. A *sample from the set of n boxes or populations* is a collection of n balls, one from each of the boxes endowed with this order relation: the ball from box 1 is first, the ball from box 2 is second, ..., and the ball from the nth box is last.

Here is an example. Let there be three boxes containing balls identified as

A_1, B_1, and C_1 in box 1
A_2, B_2, and C_2 in box 2
A_3, B_3, and C_3 in box 3.

The samples from the set of boxes are (A_1, A_2, A_3), (A_1, B_2, A_3), (A_1, C_2, A_3), $(A_1, A_2, B_3), \ldots, (C_1, C_2, C_3)$.

Corresponding to Lemma 1-1 for a sample from one population, there is the following result for a sample from the set of populations:

LEMMA 11-1

The number of samples from the set of n populations is equal to the product of the numbers of balls in each, namely, $N_l \cdots N_n$.

The proof is analogous to that of Lemma 1-1. ∎

Suppose that each population is decomposed into mutually exclusive and exhaustive subsets. Let E_{ij} represent the jth subset of the ith population, and let N_{ij} be the number of balls in it; then

$$N_i = \sum_j N_{ij}.$$

A random sample—a "Sample from the set of n populations" is defined as in Sec. 1-2: the probability attached to each sample is the reciprocal of the number of samples,

$$1/N_1 \cdots N_n.$$

Let p_{ij} be the proportion of balls in the ith population which belong to the jth subset:

$$p_{ij} = N_{ij}/N_i;$$

this is equal to the probability that a ball selected at random from the ith population belongs to the jth subset of the population. The probability that the balls in the Sample belong to the subsets $E_{1,i_1}, \ldots, E_{n,i_n}$ in that order is, in accordance with the definition (1-2), equal to the product of the corresponding proportions,

$$p_{1,i_1} \cdots p_{n,i_n}. \tag{11-1}$$

Suppose, as in Chapter 1, that the balls are labeled by numbers, and that x_{ij} is the jth label in the ith box. Let $X_1, \ldots, X_n$ be random variables representing the labels of the balls drawn from the set of n boxes. If E_{ij} is the subset of the ith population consisting of the balls with label x_{ij}, then the probability that the random variables $X_1, \ldots, X_n$ assume the values $x_{1,j_1}, \ldots, x_{n,j_n}$, respectively, is equal to the product (11-1). The n-component point $\mathbf{X} = (X_1, \ldots, X_n)$ is called a Sample of n observations from the n respective populations.

The Sample space has almost the same definition as in Sec. 1-3: the only change is that the product (11-1) with double indices replaces the product (1-8) with single indices.

Suppose that there are n normal populations with means $\mu_1, \ldots, \mu_n$ and standard deviations $\sigma_1, \ldots, \sigma_n$, respectively. Let X_i be a Sample of one observation from the ith population, as intuitively defined in Sec. 2-6. These are combined—for $i = 1, \ldots, n$—into a single "Sample $\mathbf{X} = (X_1, \ldots, X_n)$ of n observations from the n respective normal populations." The Sample space is defined as in Sec. 3-6: the only change is that the likelihood function corresponding to the normal population with mean μ and standard deviation σ is replaced by the product of the corresponding n normal densities:

$$\prod_{i=1}^{n} \frac{1}{\sigma_i} \phi\left(\frac{x_i - \mu_i}{\sigma_i}\right).$$

This is analogous to the replacement of the product (1-8) with single indices by the doubly indexed product (11-1).

Most of the statistical theory of several normal populations has been developed under the hypothesis that the populations may differ in their means but have a common standard deviation σ:

$$\sigma = \sigma_1 = \cdots = \sigma_n.$$

This is known as the assumption of "homoscedasticity," and is assumed throughout the rest of this book. (In applications this must be verified by supplementary information about the populations.) The likelihood function takes the form

$$L(\mathbf{x}; \mu_1, \ldots, \mu_n, \sigma) = (2\pi\sigma^2)^{-n/2} e^{-(1/2\sigma^2) \sum_{i=1}^{n} (x_i - \mu_i)^2} \tag{11-2}$$

Put $\boldsymbol{\mu} = (\mu_1, \ldots, \mu_n)$; then this function may also be expressed in the more compact form

$$L(\mathbf{x}; \boldsymbol{\mu}, \sigma^2) = (2\pi\sigma^2)^{-n/2} e^{-(1/2\sigma^2)\|\mathbf{x} - \boldsymbol{\mu}\|^2}. \tag{11-3}$$

11-2. A DECOMPOSITION OF A SUM OF SQUARES, AND ITS GEOMETRIC INTERPRETATION

The distribution theory of the Sample space of n observations from a single normal population (Chapters 4–6) depended on two special mathematical tools: the decomposition of the sum of squares (4-12), and the diagonal rotation (Sec. 4-4). In order to build a more general theory—for populations with different means but a common standard deviation—we shall use a more general decomposition and rotation.

The following identity is a generalization of (4-12). Let $x_1, \ldots, x_n$ and $t_1, \ldots, t_n$ be real numbers with at least one t_i not equal to 0; put

$$b = \frac{\sum_{i=1}^{n} x_i t_i}{\sum_{i=1}^{n} t_i^2}; \tag{11-4}$$

then

$$\sum_{i=1}^{n} x_i^2 = \sum_{i=1}^{n} (x_i - bt_i)^2 + b^2 \sum_{i=1}^{n} t_i^2. \tag{11-5}$$

(This reduces to (4-12) for $t_i = 1$, $i = 1, \ldots, n$.) To prove (11-5), expand the sum of squares on the right-hand side, sum over i, and apply the definition (11-4) of b (Exercise 1).

The geometric meaning of this identity becomes apparent when the quantities are expressed in the notation of points in R^n:

$$\|\mathbf{x}\|^2 = \|\mathbf{x} - b\mathbf{t}\|^2 + b^2\|\mathbf{t}\|^2. \tag{11-6}$$

(Recall that $b\mathbf{t}$ is the point whose ith coordinate is bt_i, $i = 1, \ldots, n$.) The equation signifies that the line passing through $\mathbf{0}$ and $b\mathbf{t}$ is perpendicular to the line through $b\mathbf{t}$ and $\mathbf{x}$ (Sec. 3-1).

We illustrate this in the case $n = 2$. Let (x_1, x_2) and (t_1, t_2) be points in the plane. When a perpendicular from (x_1, x_2) is dropped onto the line through (0,0) and (t_1, t_2), it intersects the latter line at the point (bt_1, bt_2) (see Fig. 11-1).

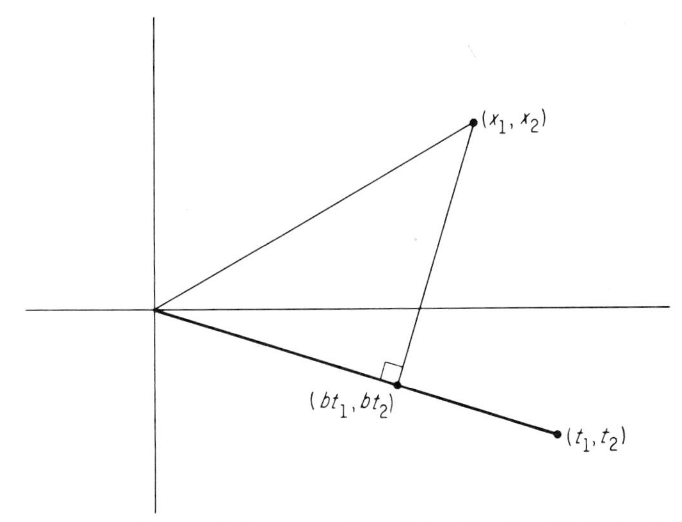

Fig. 11-1. Geometric interpretation of the identity (11-5) in the case of the plane.

Generalizing the definition of a hyperplane, given in formula (4-14), we call the set of points in R^n

$$\left\{\mathbf{x}: \sum_{i=1}^{n} x_i t_i = c\right\}, \tag{11-7}$$

where c is a constant, the "hyperplane with intercepts proportional to c/t_i, $i = 1, \ldots, n$." In the particular case $n = 2$ the hyperplane (11-7) is, for $c \neq 0$, a straight line passing through the first and second axes at the points $(c/t_1, 0)$ and $(0, c/t_2)$, respectively. For $c = 0$ the hyperplane is a line passing through the origin perpendicular to the line through the origin and (t_1, t_2) (Fig. 11-2).

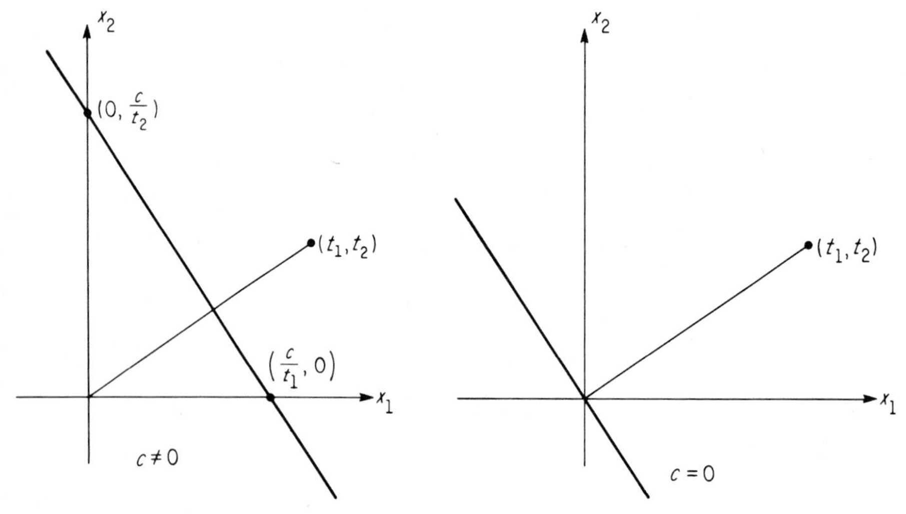

Fig. 11-2. The hyperplane $x_1 t_1 + x_2 t_2 = c$ for $c \neq 0$ and $c = 0$.

The hyperplane (11-7) is related to the decomposition (11-5) in the same way that the hyperplane (4-14) is related to the decomposition (4-12).

LEMMA 11-2

If $c = 0$, then a point $\mathbf{x}$ belongs to (11-7) if and only if

$$\|\mathbf{x} - \mathbf{t}\|^2 = \|\mathbf{x}\|^2 + \|\mathbf{t}\|^2; \tag{11-8}$$

in other words, if and only if the lines through the origin and $\mathbf{x}$ and $\mathbf{t}$, respectively, are perpendicular. If $c \neq 0$, then $\mathbf{x}$ belongs *to (11-7) if and only if*

$$\|\mathbf{x}\|^2 = \left\|\mathbf{x} - \frac{c}{\|\mathbf{t}\|^2}\mathbf{t}\right\|^2 + \frac{c^2}{\|\mathbf{t}\|^2}; \tag{11-9}$$

in other words, if and only if the line through $\mathbf{0}$ *and* $(c/\|\mathbf{t}\|^2)\mathbf{t}$ *is perpendicular to the one through* $\mathbf{x}$ *and* $(c/\|\mathbf{t}\|^2)\mathbf{t}$.

Proof: Expand the square on the left-hand side of (11-8):

$$\|\mathbf{x} - \mathbf{t}\|^2 = \|\mathbf{x}\|^2 - 2\sum_{i=1}^{n} x_i t_i + \|\mathbf{t}\|^2;$$

then (11-8) holds if and only if $\mathbf{x}$ belongs to the hyperplane (11-7) with $c = 0$.

Suppose $c \neq 0$, and that $\mathbf{x}$ belongs to the hyperplane (11-7); then, by the definition (11-4) of b, it follows that

$$c = \sum_{i=1}^{n} x_i t_i = b\|\mathbf{t}\|^2,$$

or $b = c/\|\mathbf{t}\|^2$; and so (11-9) follows from (11-6).

To prove the converse—that if $\mathbf{x}$ satisfies (11-9) with some $c \neq 0$, then it belongs to the hyperplane (11-7)—expand the first term on the right-hand side of (11-9):

$$\left\|\mathbf{x} - \frac{c}{\|\mathbf{t}\|^2}\mathbf{t}\right\|^2 = \|\mathbf{x}\|^2 - 2c\frac{\sum_{i=1}^{n} x_i t_i}{\|\mathbf{t}\|^2} + \frac{c^2}{\|\mathbf{t}\|^2},$$

and simplify Eq. 11-9:

$$c\sum_{i=1}^{n} x_i t_i = c^2.$$

Since $c \neq 0$, it follows that $\sum_{i=1}^{n} x_i t_i = c$, so that $\mathbf{x}$ belongs to (11-7). ■

In Sec. 4-3 we studied the effect of the diagonal rotation on the hyperplane $\left\{\mathbf{x}: \sum_{i=1}^{n} x_i = c\right\}$. Now we generalize this to the effect of another rotation on the hyperplane (11-7). For any $\mathbf{t} \neq 0$ in R^n there exists, by Theorem 3-1, a rotation carrying the point $\mathbf{t}$ onto the point $\|\mathbf{t}\|\mathbf{e}_1 = (\|\mathbf{t}\|, 0, \ldots, 0)$, which is on the first axis.

THEOREM 11-1
For $\mathbf{t} \neq \mathbf{0}$ *the image of the hyperplane (11-7) under a rotation carrying* $\mathbf{t}$ *onto* $\|\mathbf{t}\|\,\mathbf{e}_1$ *is the set of points whose first coordinates are equal to* $c/\|\mathbf{t}\|$.

Proof: Let $\mathbf{x}'$ and $\mathbf{t}'$ be the images of $\mathbf{x}$ and $\mathbf{t}$, respectively, under the rotation, and x_i' and t_i' the corresponding ith coordinates of the images.

Suppose $c = 0$. By Lemma 11-2 the point $\mathbf{x}$ belongs to the hyperplane (11-7) if and only if Eq. 11-8 holds; therefore, an analogous equation holds for the images of $\mathbf{x}$ and $\mathbf{t}$ because rotation preserves distance:

$$\|\mathbf{x}' - \mathbf{t}'\|^2 = \|\mathbf{x}'\|^2 + \|\mathbf{t}'\|^2$$

Apply the lemma to $\mathbf{x}'$ and $\mathbf{t}'$: the latter equation holds if and only if

$$\sum_{i=1}^{n} x_i' t_i' = 0.$$

The coordinates $t_2', \ldots, t_n'$ are all equal to 0 because $\|\mathbf{t}\|\,\mathbf{e}_1 = \mathbf{t}'$ is on the first axis; furthermore: $t_1' = \|\mathbf{t}\|$; therefore, the equation displayed above is equivalent to

$$x_1'\|\mathbf{t}\| = 0.$$

Since, by hypothesis, $\|\mathbf{t}\|$ is not equal to 0, we must have $x_1' = 0$.

Suppose $c \neq 0$. The image of $(c/\|\mathbf{t}\|^2)\mathbf{t}$ is

$$\frac{c}{\|\mathbf{t}\|^2}\mathbf{t}' = \frac{c}{\|\mathbf{t}\|}\mathbf{e}_1 = \left(\frac{c}{\|\mathbf{t}\|}, 0, \ldots, 0\right).$$

By Lemma 11-2, $\mathbf{x}$ belongs to (11-7) if and only if (11-9) holds; thus the equation

$$\|\mathbf{x}'\|^2 = \left\|\mathbf{x}' - \frac{c}{\|\mathbf{t}'\|^2}\mathbf{t}'\right\|^2 + \frac{c^2}{\|\mathbf{t}'\|^2}$$

characterizes the image of a point in the hyperplane. By virtue of the form of $(c/\|\mathbf{t}\|^2)\mathbf{t}'$ given above, the last equation is equivalent to

$$\sum_{i=1}^{n} x_i'^2 = \left(x_1' - \frac{c}{\|\mathbf{t}\|}\right)^2 + \sum_{i=2}^{n} x_i'^2 + \frac{c^2}{\|\mathbf{t}\|^2},$$

or to

$$x_1'^2 = \left(x_1' - \frac{c}{\|\mathbf{t}\|}\right)^2 + \frac{c^2}{\|\mathbf{t}\|^2}.$$

This holds for $c \neq 0$ if and only if $x_1' = c/\|\mathbf{t}\|$ (Exercise 2). ■

We illustrate the result of Theorem 11-1 in the special case $n = 2$. Let (t_1, t_2) be a point not equal to $(0, 0)$, and consider the line in the x_1x_2-plane: $x_1t_1 + x_2t_2 = c$. Apply the rotation carrying (t_1, t_2) onto $(\sqrt{t_1^2 + t_2^2}, 0)$. In accordance with Fig. 11-2, the line perpendicular to $x_1t_1 + x_2t_2 = c$ is carried onto the x_2-axis if $c = 0$, and onto the line $x_1 = c/\sqrt{t_1^2 + t_2^2}$ if $c \neq 0$.

EXERCISES

1. Furnish the details of the verification of the identity (11-5).

2. Verify the last step in the proof of Theorem 11-1.

3. *Prove:* For any real number d, and for b defined by (11-4):

$$\|\mathbf{x} - d\mathbf{t}\|^2 \geq \|\mathbf{x} - b\mathbf{t}\|^2.$$

Give the geometric interpretation of this inequality.

4. *Prove:*

$$\sum_{i=1}^{n} x_i t_i = (1/4)[\|\mathbf{x} + \mathbf{t}\|^2 - \|\mathbf{x} - \mathbf{t}\|^2].$$

5. Let $\mathbf{x}'$ and $\mathbf{t}'$ be the images of $\mathbf{x}$ and $\mathbf{t}$ under some rotation. Using the result of Exercise 4, show that

$$\sum_{i=1}^{n} x_i t_i = \sum_{i=1}^{n} x_i' t_i'.$$

11-3. ESTIMATION OF A LINEAR FUNCTION OF THE MEANS OF SEVERAL NORMAL POPULATIONS

As indicated in Sec. 11-1, an experimenter may be interested in testing for the existence of relations among the means of several normal populations. The simplest kind of relation is the linear relation; its theory is well developed and is applicable to the linear regression model (Chapter 12) and analysis of variance (Chapter

13). We postpone a discussion of the practical applications to the relevant chapters. In this section we shall prove that the "efficient Estimator" of a linear combination of the means of several normal populations is the corresponding linear combination of the observations from those populations. This will be applied in Chapter 12 to proving the optimality of the classical prediction procedure in the linear regression model.

Consider n normal populations with means $\mu_1, \ldots, \mu_n$ and common standard deviation σ; let $\mathbf{X} = (X_1, \ldots, X_n)$ be a Sample of n observations from the respective populations. Suppose that the means (and standard deviation) are all unknown, but that we want to estimate the first one, μ_1. It is intuitively apparent that the observations $X_2, \ldots, X_n$ from the last $n - 1$ populations provide the statistician with no "information" about the first population if he knows of no relations between it and the other populations; thus, inference about the first population should be based only on X_1, the first observation; therefore, the last $n - 1$ observations are irrelevant to the analysis, and so an Estimator $f(\mathbf{x})$ of μ_1 should be a function only of the first variable x_1. The Sample mean is the efficient Estimator of the population mean in sampling from a single population; thus, in the case of one observation X_1 from the first population, the efficient Estimator of μ_1 is X_1. We shall now furnish the mathematical proof of these heuristic remarks.

The following theory of unbiased estimation is a modification of the corresponding theory for the mean of a single normal population (Chapter 7). An Estimator of μ_1 based on a Sample of n observations is a square-integrable function $f(\mathbf{x})$ on R^n; it satisfies

$$\int_{R^n} \cdots \int (f(\mathbf{x}))^2 L(\mathbf{x}; \boldsymbol{\mu}, \sigma)\, d\mathbf{x} < \infty$$

for all $\boldsymbol{\mu}$ and $\sigma > 0$. For fixed $\sigma > 0$, an Estimator f of μ_1 (f may depend on σ) is unbiased if

$$\int_{R^n} \cdots \int f(\mathbf{x})\, L(\mathbf{x}; \boldsymbol{\mu}, \sigma)\, d\mathbf{x} = \mu_1, \qquad \text{for all } \boldsymbol{\mu}. \tag{11-10}$$

The variance of f is the integral

$$\int_{R^n} \cdots \int (f(\mathbf{x}) - \mu_1)^2 L(\mathbf{x}; \boldsymbol{\mu}, \sigma)\, d\mathbf{x}. \tag{11-11}$$

An unbiased Estimator is efficient if among all square-integrable functions satisfying (11-10) it minimizes (11-11). By Theorem 7-1, an efficient Estimator is unique.

An example of a square-integrable function f which is an unbiased Estimator of μ_1 is $f(\mathbf{x}) = x_1$; in this case the variance is equal to σ^2 (Exercise 1). In the following theorem it is shown that this Estimator is, *for each* σ, the efficient Estimator of μ_1.

THEOREM 11-2

The function $f(\mathbf{x}) = x_1$ is the efficient Estimator of μ_1 for any $\sigma > 0$.

Proof: Fix $\sigma > 0$, and let $f(\mathbf{x})$ be an arbitrary unbiased Estimator of μ_1. Let $g(x_1) = g(x_1; \mu_2, \ldots, \mu_n, \sigma)$ be the function obtained by integrating $f \cdot L$ over the last $n - 1$ variables:

$$g(x_1) = \int_{-\infty}^{\infty} \cdots \int_{-\infty}^{\infty} f(\mathbf{x})(1/\sigma)^{n-1} \prod_{i=2}^{n} \phi\left(\frac{x_i - \mu_i}{\sigma}\right) \prod_{i=2}^{n} dx_i \tag{11-12}$$

where the first factor has been omitted from the likelihood function. By the Extension of Theorem 3-3, g is a continuous function of x_1; indeed, even though the latter theorem was stated for integration over a rectangle, its conclusion can be extended to the integral of a square-integrable function over R^n. By the Cauchy-Schwarz inequality g is also square-integrable:

$$\int \cdots \int_{R^n} (g(x_1))^2 L(\mathbf{x}; \boldsymbol{\mu}, \sigma)\, d\mathbf{x} \leq \int \cdots \int_{R^n} (f(\mathbf{x}))^2 L(\mathbf{x}; \boldsymbol{\mu}, \sigma)\, d\mathbf{x} \tag{11-13}$$

(Exercise 2). When g is substituted for f in Eq. 11-10, the multiple integral reduces, by Theorem 4-1, to a single integral with respect to g, and the equation becomes

$$\int_{-\infty}^{\infty} g(x_1) \frac{1}{\sigma} \phi\left(\frac{x_1 - \mu_1}{\sigma}\right) dx_1 = \mu_1. \tag{11-14}$$

A similar operation on (11-13) changes it to

$$\int_{-\infty}^{\infty} (g(x_1))^2 \frac{1}{\sigma} \phi\left(\frac{x_1 - \mu_1}{\sigma}\right) dx_1$$

$$\leq \int \cdots \int_{R^n} (f(\mathbf{x}))^2 L(\mathbf{x}; \boldsymbol{\mu}, \sigma)\, d\mathbf{x}.$$

This and Eqs. 11-10 and 11-14 imply

$$\int_{-\infty}^{\infty} (g(x_1) - \mu_1)^2 \frac{1}{\sigma} \phi\left(\frac{x_1 - \mu_1}{\sigma}\right) dx_1$$

$$\leq \int \cdots \int_{R^n} (f(\mathbf{x}) - \mu_1)^2 L(\mathbf{x}; \boldsymbol{\mu}, \sigma)\, d\mathbf{x} \qquad (11\text{-}15)$$

(Exercise 3). By Eq. 11-14, g is an unbiased Estimator of μ_1 based on a Sample of one observation; thus, by the previous result on estimation (Theorem 7-2), the variance of g—the integral appearing on the left-hand side of (11-15)—is at least equal to σ^2 (here $n = 1$). It follows from (11-15) that the variance of $f(\mathbf{x})$ is also at least equal to σ^2. As noted before the statement of the theorem, this lower bound is attained by the variance of the unbiased Estimator $f(\mathbf{x}) = x_1$; thus the latter is efficient. ■

The result of Theorem 11-2 can be slightly extended in this manner.

THEOREM 11-3

For any real number $c \neq 0$, the efficient Estimator of $c\mu_l$, the multiple of the unknown parameter μ_l, is the corresponding multiple cx_l of the efficient Estimator of μ_l:

Efficient Estimator of $c\mu_l$ = $c \cdot$ Efficient Estimator of μ_l.

Proof: In sampling from a normal population with mean μ_1 and standard deviation σ, the statistic cx_1 has a normal distribution with mean $c\mu_1$ and standard deviation $c\sigma$ (Example 4-1); thus, cx_1 is an unbiased Estimator of $c\mu_1$, and has variance $c^2\sigma^2$.

Now let $f(\mathbf{x})$ be an arbitrary unbiased Estimator of $c\mu_1$:

$$\int \cdots \int_{R^n} f(\mathbf{x})\, L(\mathbf{x}; \boldsymbol{\mu}, \sigma)\, d\mathbf{x} = c\mu_1;$$

then, $f(\mathbf{x})/c$ is an unbiased Estimator of μ_1. By Theorem 11-2, its variance is at least equal to that of x_1, which is σ^2:

$$\int \cdots \int_{R^n} \left(\frac{f(\mathbf{x})}{c} - \mu_1\right)^2 L(\mathbf{x}; \boldsymbol{\mu}, \sigma)\, dx \geq \sigma^2.$$

Multiply by c^2 on both sides of this inequality:

$$\int \cdots \int_{R^n} (f(\mathbf{x}) - c\mu_1)^2 L(\mathbf{x}; \boldsymbol{\mu}, \sigma)\, d\mathbf{x} \geq c^2\sigma^2.$$

This signifies that the variance of f as an unbiased Estimator of $c\mu_1$ is at least $c^2\sigma^2$. Since f is an arbitrary unbiased Estimator of $c\mu_1$, it follows that $c^2\sigma^2$ is a lower bound for the variance of any unbiased Estimator. Since cx_1 has this variance, it is efficient. ■

The main object of our interest is the estimation of linear functions of the means of several normal populations. For any set $t_1, \ldots, t_n$ of n real numbers not all equal to 0, consider the linear combination $\sum_{i=1}^{n} t_i\mu_i$. Our main result is that in sampling from the n respective normal populations, the efficient Estimator of this linear combination is the corresponding linear function on R^n, $\sum_{i=1}^{n} t_i x_i$. The special case $t_1 = c$, $t_2 = \cdots = t_n = 0$ has already been proved in Theorem 11-3.

First we derive the distribution of this linear function:

THEOREM 11-4

In a Sample of n observations from n respective normal populations with means $\mu_1, \ldots, \mu_n$, and common standard deviation σ, the statistic $\sum_{i=1}^{n} t_i x_i$ has a normal distribution with mean $\sum_{i=1}^{n} t_i \mu_i$ and standard deviation $\sigma \|\mathbf{t}\|$.

Proof: The distribution of this statistic at the point y is, by definition, the integral of $L(\mathbf{x}; \boldsymbol{\mu}, \sigma)$ over the set $\left\{\mathbf{x}: \sum_{i=1}^{n} t_i x_i \leq y\right\}$. The regularity of this set is a consequence of the fact that by a suitable rotation, given below, it can be transformed into a set which is known to be regular (Theorem 3-5).

Consider the rotation carrying $\mathbf{t}$ onto $\|\mathbf{t}\|\mathbf{e}_1$. It carries the set

$$\left\{\mathbf{x}: \sum_{i=1}^{n} t_i x_i \leq y\right\}; \tag{11-16}$$

onto

$$\{\mathbf{x}: x_1 \leq y / \|\mathbf{t}\|\}; \tag{11-17}$$

indeed, (11-16) is the union of the hyperplanes $\left\{x: \sum_{i=1}^{n} t_i x_i = w\right\}$, $w \leq y$, and each of these, by Theorem 11-1, has the image $\{\mathbf{x}: x_1 = w/\|\mathbf{t}\|\}$.

To integrate L over the set (11-16), change the variables of integration from x_i to $x_i - \mu_i$, $i = 1, \ldots, n$: the integral becomes

$$\int_{\left\{\mathbf{x}: \sum_{i=1}^{n} t_i x_i \leq y - \sum_{i=1}^{n} t_i \mu_i\right\}} \cdots \int L(\mathbf{x}; 0, \sigma)\, d\mathbf{x}.$$

By Theorem 4-4 this integral is invariant under the rotation of the domain of integration. Now apply the rotation carrying $\mathbf{t}$ onto $\|\mathbf{t}\|\mathbf{e}_1$: since the image of (11-16) is (11-17), it follows that the image of

$$\left\{\mathbf{x}: \sum_{i=1}^{n} t_i x_i \leq y - \sum_{i=1}^{n} t_i \mu_i\right\}$$

is

$$\left\{\mathbf{x}: x_1 \leq \left(\frac{y - \sum_{i=1}^{n} t_i \mu_i}{\|\mathbf{t}\|}\right)\right\};$$

therefore, the integral above is equal to

$$\int_{\left\{\mathbf{x}: x_1 \leq \left(y - \sum_{i=1}^{n} t_i \mu_i\right)/\|\mathbf{t}\|\right\}} \cdots \int L(\mathbf{x}; 0, \sigma)\, d\mathbf{x}.$$

This integral represents the distribution of the statistic $f(\mathbf{x}) = x_1$ in sampling from a single normal population with mean 0 and standard deviation σ, where the distribution is evaluated at the point $\left(y - \sum_{i=1}^{n} t_i \mu_i\right) \Big/ \|t\|$; thus it is equal to

$$\Phi\left(\frac{y - \sum_{i=1}^{n} t_i \mu_i}{\sigma \|\mathbf{t}\|}\right).$$

By construction, this is also the distribution of $\sum_{i=1}^{n} t_i x_i$ at the point y. ■

Now we generalize the previous discussion about the estimation of μ_1 to the estimation of the linear combination $\sum_{i=1}^{n} t_i \mu_i$. Equation 11-10 for unbiasedness becomes

$$\int_{R^n} \cdots \int f(\mathbf{x})\, L(\mathbf{x}; \boldsymbol{\mu}, \sigma)\, d\mathbf{x} = \sum_{i=1}^{n} t_i \mu_i, \quad \text{for all } \boldsymbol{\mu}, \qquad (11\text{-}18)$$

and the integral representing the variance is now

$$\int_{R^n} \cdots \int \left(f(\mathbf{x}) - \sum_{i=1}^{n} t_i \mu_i \right)^2 L(\mathbf{x}; \boldsymbol{\mu}, \sigma)\, d\mathbf{x}. \qquad (11\text{-}19)$$

THEOREM 11-5

Consider a Sample of n observations from n normal populations with means $\mu_1, \ldots, \mu_n$, respectively, and common standard deviation σ. For any $t_1, \ldots, t_n$ the efficient Estimator of $\sum_{i=1}^{n} t_i \mu_i$ is $\sum_{i=1}^{n} t_i x_i$.

Proof: The idea of the proof is to perform a rotation of the variables of integration in (11-18) and (11-19) in order to reduce this estimation problem to the one considered in Theorems 11-2 and 11-3.

Let $\mathbf{x}'$ and $\boldsymbol{\mu}'$ be the images of $\mathbf{x}$ and $\boldsymbol{\mu}$, respectively, under a rotation carrying $\mathbf{t}$ onto $\|\mathbf{t}\| \mathbf{e}_1$. By Theorem 11-1 the first coordinates of $\mathbf{x}'$ and $\boldsymbol{\mu}'$ are

$$x_1' = \frac{\sum_{i=1}^{n} t_i x_i}{\|\mathbf{t}\|} \quad \text{and} \quad \mu_1' = \frac{\sum_{i=1}^{n} t_i \mu_i}{\|\mathbf{t}\|},$$

respectively. From Eq. 11-3 we see that $L(\mathbf{x}; \boldsymbol{\mu}, \sigma)$ is a function of $\|\mathbf{x} - \boldsymbol{\mu}\|$ and σ; thus it is unchanged when $\mathbf{x}'$ and $\boldsymbol{\mu}'$ are substituted for $\mathbf{x}$ and $\boldsymbol{\mu}$, respectively. From this and from the form of μ_1'

it follows that the condition (11-18) for unbiasedness is equivalent to

$$\int_{R^n} \cdots \int f(\mathbf{x})\, L(\mathbf{x}'; \boldsymbol{\mu}', \sigma)\, d\mathbf{x} = \| \mathbf{t} \| \mu_1'.$$

Define the function f^* on R^n by means of the equation $f^*(\mathbf{x}') = f(\mathbf{x})$ (cf. statement of Theorem 5-3); then f^* is square-integrable (Exercise 4), and the latter equation for unbiasedness becomes

$$\int_{R^n} \cdots \int f^*(\mathbf{x}')\, L(\mathbf{x}', \boldsymbol{\mu}', \sigma)\, d\mathbf{x} = \| \mathbf{t} \| \mu_1'.$$

By the invariance of integration under rotation (Theorem 3-4), the integral above is unchanged when the prime is removed from the variable $\mathbf{x}$ in the integrand:

$$\int_{R^n} \cdots \int f^*(\mathbf{x})\, L(\mathbf{x}; \boldsymbol{\mu}', \sigma)\, d\mathbf{x} = \| \mathbf{t} \| \mu_1'.$$

Since, by the hypothesis (11-18), this equation is valid for all μ, it is also valid for all μ when the prime is removed:

$$\int_{R^n} \cdots \int f^*(\mathbf{x})\, L(\mathbf{x}; \boldsymbol{\mu}, \sigma)\, d\mathbf{x} = \| \mathbf{t} \| \mu_1. \tag{11-20}$$

This equation signifies that f^* is an unbiased Estimator of $\| \mathbf{t} \| \mu_1$; thus, by Theorem 11-3, its variance is at least equal to $\| \mathbf{t} \|^2 \sigma^2$:

$$\int_{R^n} \cdots \int (f^*(\mathbf{x}) - \| \mathbf{t} \| \mu_1)^2\, L(\mathbf{x}; \boldsymbol{\mu}, \sigma)\, d\mathbf{x} \geq \| \mathbf{t} \|^2 \sigma^2. \tag{11-21}$$

The integral (11-19) representing the variance of f is transformed in a similar way into

$$\int_{R^n} \cdots \int (f^*(\mathbf{x}) - \| \mathbf{t} \| \mu_1')^2\, L(\mathbf{x}; \boldsymbol{\mu}', \sigma)\, d\mathbf{x};$$

since (11-21) holds for *all* μ, it certainly holds for all $\boldsymbol{\mu}'$, and so the latter integral is at least equal to $\| \mathbf{t} \|^2 \sigma^2$. This signifies that the variance of any unbiased Estimator of $\sum_{i=1}^{n} t_i \mu_i$ is at least equal to $\| \mathbf{t} \|^2 \sigma^2$. By Theorem 11-4, the Estimator $\sum_{i=1}^{n} t_i x_i$ is unbiased and its variance is equal to $\| \mathbf{t} \|^2 \sigma^2$; thus it is efficient. ■

There is a question of the uniqueness of the efficient Estimator, but this is easily put aside. It is conceivable that there is *another* point **s** such that

$$\sum_{i=1}^{n} s_i \mu_i = \sum_{i=1}^{n} t_i \mu_i$$

for all $\mu_1, \ldots, \mu_n$, so that $s_1 x_1 + \cdots + s_n x_n$ is the efficient Estimator; however, this cannot happen if the equation above holds for *all* $\mu_1, \ldots, \mu_n$.

EXERCISES

1. Why is $f(x) = x_1$ an unbiased Estimator of μ, with variance σ^2?
2. Verify the inequality (11-13).
3. Verify (11-15).
4. *Prove:* If f is square-integrable, then so is the function f^* defined in the proof of Theorem 11-5.
5. Suppose that the means $\mu_1, \ldots, \mu_n$ are known but σ is not. Show that $(1/n) \sum_{i=1}^{n} (x_i - \mu_i)^2$ is the efficient Estimator of σ^2, and that its variance is $2\sigma^4/n$.
6. Use Theorem 11-4 to construct a confidence interval for $\sum_{i=1}^{n} t_i \mu_i$ when σ is known.
7. Suppose that μ_1 is known. What is the efficient Estimator of $\sum_{i=1}^{n} t_i \mu_i$, and how does its variance compare with that of the efficient Estimator when μ_1 is unknown?

11-4. THE NONCENTRAL CHI-SQUARE DISTRIBUTION

Theorem 4-2 states that the statistic $x_1^2 + \cdots + x_n^2$ has the chi-square distribution with n degrees of freedom in sampling from a standard normal population. The proof depended in part on the fact that the likelihood function is constant over spheres about the origin. Now we derive the distribution of this statistic in a Sample from populations with different means and a common standard deviation. This distribution is called the "noncentral chi-square distribution with n degrees of freedom." In addition to the degrees of freedom the density contains a "noncentrality parameter" τ, depending on the means and standard deviation of the underlying populations. Although we do not need a table of values of this distribution, we shall make use of its explicit analytic form.

THEOREM 11-6

In sampling from n normal populations with means $\mu_1, \ldots, \mu_n$, respectively, and common standard deviation σ, the statistic $(x_1^2 + \cdots + x_n^2)/\sigma^2$ has the density function

$$\psi_n(y; \tau) = 0, \qquad y \leq 0 \tag{11-22}$$

$$\frac{1}{2}\int_0^y (y-u)^{-1/2}\psi_{n-1}(u)[\phi(\sqrt{y-u}-\tau)+\phi(\sqrt{y-u}+\tau)]\,du, \qquad y > 0$$

where the parameter $\tau \geq 0$ is defined by

$$\tau = \|\boldsymbol{\mu}\|/\sigma. \tag{11-23}$$

(We remark that if all the means are equal to 0, then $\tau = 0$, and the density (11-22) is equal to the ordinary chi-square density with n degrees of freedom (Exercise 1).)

Proof: Since the statistic is nonnegative, and assumes the value 0 with probability 0 (Exercise 2), the density is 0 for $y \leq 0$.

The distribution of the statistic at a point $y > 0$ is the integral of the likelihood function over the ball centered at $\mathbf{0}$ and of radius $\sqrt{y}\sigma$:

$$\int\cdots\int_{\{\mathbf{x}:\,\|\mathbf{x}\|^2 \leq y\sigma^2\}} L(\mathbf{x}; \boldsymbol{\mu}, \sigma)\,d\mathbf{x}.$$

Let $\mathbf{x}'$ and $\boldsymbol{\mu}'$ be the images of $\mathbf{x}$ and $\boldsymbol{\mu}$ under the rotation carrying $\boldsymbol{\mu}$ onto $\|\boldsymbol{\mu}\|\mathbf{e}_1$, so that $\boldsymbol{\mu}' = \|\boldsymbol{\mu}\|\mathbf{e}_1$. By the form (11-3) of L, it depends on $\mathbf{x}$ and $\boldsymbol{\mu}$ through $\|\mathbf{x}-\boldsymbol{\mu}\|$; thus $L(\mathbf{x}; \boldsymbol{\mu}, \sigma) = L(\mathbf{x}'; \boldsymbol{\mu}', \sigma)$, and the integral above is equal to

$$\int\cdots\int_{\{\mathbf{x}:\,\|\mathbf{x}\|^2 \leq y\sigma^2\}} L(\mathbf{x}'; \boldsymbol{\mu}', \sigma)\,d\mathbf{x}.$$

By the invariance of integration under rotation (Theorem 3-4) the integral is unchanged when $\mathbf{x}$ is substituted for $\mathbf{x}'$ in the integrand:

$$\int_{\{\mathbf{x}:\, \|\mathbf{x}\|^2 \le y\sigma^2\}} \cdots \int L(\mathbf{x}; \boldsymbol{\mu}', \sigma)\, d\mathbf{x}.$$

Now $\mu_1' = \|\boldsymbol{\mu}\|$ and $\mu_j' = 0$, $j = 2, \ldots, n$ by the definition of the rotation; thus

$$L(\mathbf{x}; \boldsymbol{\mu}', \sigma) = \frac{1}{\sigma}\, \phi\left(\frac{x_1 - \|\boldsymbol{\mu}\|}{\sigma}\right) \cdot \frac{1}{\sigma^{n-1}} \prod_{i=2}^{n} \phi\left(\frac{x_i}{\sigma}\right).$$

Substitute this in the integrand and evaluate by iterated integration: fix x_1, $0 \le x_1^2 \le y\sigma^2$, integrate with respect to $x_2, \ldots, x_n$ over

$$\left\{(x_2, \ldots, x_n)\colon 0 \le \sum_{i=2}^{n} x_i^2 \le y\sigma^2 - x_1^2\right\},$$

and finally integrate x_1 over $[0, y\sigma^2]$; the integral is equal to

$$\int_{-\sqrt{y}\sigma}^{\sqrt{y}\sigma} \frac{1}{\sigma}\, \phi\left(\frac{x_1 - \|\boldsymbol{\mu}\|}{\sigma}\right) \left\{\int_{\left\{(x_2,\ldots,x_n)\colon \frac{x_2^2 + \cdots + x_n^2}{\sigma^2} \le y - \frac{x_1^2}{\sigma^2}\right\}} \cdots \int \prod_{i=2}^{n} \frac{1}{\sigma}\, \phi\left(\frac{x_i}{\sigma}\right) \prod_{i=2}^{n} dx_i\right\} dx_1 \qquad (11\text{-}24)$$

The inner $(n - 1)$-fold integral is recognizable as the integral defining the distribution function (at $y - x_1^2/\sigma^2$) of the statistic $(x_2^2 + \cdots + x_n^2)/\sigma^2$ in sampling from a normal population with mean 0 and standard deviation σ; thus by Theorem 4-2 (with $n - 1$ in place of n) it is the chi-square distribution, and (11-24) takes the form

$$\int_{-\sqrt{y}\sigma}^{\sqrt{y}\sigma} \frac{1}{\sigma}\, \phi\left(\frac{x_1 - \|\boldsymbol{\mu}\|}{\sigma}\right) \int_0^{y - x_1^2/\sigma^2} \psi_{n-1}(u)\, du \; dx_1.$$

Change the variable of integration: $x = x_1/\sigma$, and employ the definition (11-23) of τ; the integral above is equal to

$$\int_{-\sqrt{y}}^{\sqrt{y}} \phi(x - \tau)\left[\int_0^{y - x^2} \psi_{n-1}(u)\, du\right] dx.$$

Split the domain of integration with respect to x into its positive and negative parts, and add the integrals over each:

$$\int_0^{\sqrt{y}} [\phi(x + \tau) + \phi(x - \tau)]\left(\int_0^{y-x^2} \psi_{n-1}(u)\, du\right) dx.$$

Change the variable of integration in the outer integral from x to $t = y - x^2$:

$$\frac{1}{2}\int_0^y [\phi(\sqrt{y-t} + \tau) + \phi(\sqrt{y-t} - \tau)]$$

$$\cdot\left(\int_0^t \psi_{n-1}(u)\, du\right)(y - t)^{-1/2}\, dt. \qquad (11\text{-}25)$$

To show that (11-22) is the derivative of (11-25) for $y > 0$, integrate the former from 0 to y and compare the resulting double integral to the latter (Exercises 3 ,4 ,5). ■

EXERCISES

1. Show that the noncentral chi-square density reduces to the ordinary chi-square density when $\tau = 0$. (*Hint:* Exercise 8, Sec. 4-3.)
2. Show that the probability that **X** falls in the set containing only the origin is 0.
3. Show that (11-25) may be written in the form

$$-\int_0^y \left\{\frac{d}{dt}\int_0^{\sqrt{y-t}} [\phi(x + \tau) + \phi(x - \tau)]\, dx\right\}\left\{\int_0^t \psi_{n-1}(u)\, du\right\} dt.$$

4. By integration by parts show that the integral in Exercise 3 is equal to

$$\int_0^y \left\{\int_0^{\sqrt{y-t}} [\phi(x + \tau) + \phi(x - \tau)]\, dx\right\} \psi_{n-1}(t)\, dt.$$

5. By either appealing to Leibniz's rule for differentiation of the integral in Exercise 4, or else by a direct computation of the limit of the difference quotient, prove that the derivative of the latter integral with respect to y is equal to the expression (11-22).

11-5. TESTING HOMOGENEITY AND INVARIANT TESTS (OPTIONAL)

Consider n normal populations with means $\mu_1, \ldots, \mu_n$ and common known standard deviation σ. For simplicity assume

$\sigma = 1$. The means are unknown and we wish to test the simple null hypothesis that they are all equal to 0 against the composite alternative hypothesis that they are not all equal to 0, that is, at least one of these is different from 0. This is called a *test of homogeneity*: under the null hypothesis all the populations are standard normal. Let $\boldsymbol{\mu}$ be the point in R^n with coordinates $\mu_1, \ldots, \mu_n$; then the null hypothesis is $\|\boldsymbol{\mu}\| = 0$ and the alternative is $\|\boldsymbol{\mu}\| > 0$. We are given a Sample $\mathbf{X}$ of n observations from the respective populations. The likelihood function is

$$L(\mathbf{x}; \mu, 1) = (2\pi)^{-n/2} e^{-(1/2)\|\mathbf{x}-\boldsymbol{\mu}\|^2}.$$

In order to test the null hypothesis, we select a level of significance α, and then a rejection region of corresponding size. The Sample point is observed, and the null hypothesis is rejected or accepted accordingly as the point falls within or without the region. For any rejection region A, the level of significance is

$$\int_A \cdots \int (2\pi)^{-n/2} e^{-\|\mathbf{x}\|^2/2}\, d\mathbf{x} = \alpha. \tag{11-26}$$

The power of the test against the alternative $\|\boldsymbol{\mu}\|$ is

$$\int_A \cdots \int (2\pi)^{-n/2} e^{-(1/2)\|\mathbf{x}-\boldsymbol{\mu}\|^2/2}\, d\mathbf{x} = P(\boldsymbol{\mu}). \tag{11-27}$$

For this testing problem there does not exist a uniformly most powerful test; indeed, in the case $n = 1$, this is the problem of a two-sided test on the mean of a normal population, and a uniformly most powerful test does not exist (Sec. 9-3). In the latter case we restricted the class of tests under consideration to unbiased tests, and found a best one among these. In the present homogeneity testing problem, we shall demonstrate that a certain test also has a qualified optimality property.

We present a heuristic justification of the proposed test. If $\|\boldsymbol{\mu}\| = 0$, then the likelihood function has its maximum at $\mathbf{x} = \mathbf{0}$ and is constant over spheres about $\mathbf{0}$, so that the origin is the "center" of the distribution of the Sample point. Under the alternative $\|\boldsymbol{\mu}\| > 0$, the likelihood function has its maximum at $\boldsymbol{\mu}$. It follows that the distance of $\mathbf{X}$ from the origin is more likely to be smaller under the null hypothesis than under the alternative. This suggests the procedure: reject or accept the null hypothesis accordingly as the distance of the Sample point from the origin is larger than some given constant or not; in other words, accordingly as the Sample point falls outside some ball about $\mathbf{0}$ or not. The radius of the ball is determined by the level of significance. Let $\chi^2_{n,\alpha}$ be the upper α-percentile of the chi-square distribution with n degrees of freedom:

$$\int_{\chi^2_{n,\alpha}}^{\infty} \psi_n(u)\,du = \alpha.$$

The rejection region is taken to be the complement of the ball of radius $\sqrt{\chi^2_{n,\alpha}}$. This test is of size α: under the null hypothesis $\boldsymbol{\mu} = \mathbf{0}$ the probability that $\mathbf{X}$ falls outside the ball is equal to the probability that the statistic $\|\mathbf{x}\|^2$ exceeds $\chi^2_{n,\alpha}$; the latter probability, by Theorem 4-2 on the distribution of $\|\mathbf{x}\|^2$, and by the definition of $\chi^2_{n,\alpha}$, is equal to α.

The optimal property of this test is that it has "largest average power over spheres in the parameter space." The precise meaning is: For each α, and $r > 0$, the average of the power function of any test of size α over the sphere of radius r is less than or equal to the spherical average of the power function of the test above (the one whose rejection region is the complement of the ball). Now the spherical average was defined in Sec. 6-1 for functions continuous over spheres; thus in order to justify the above description of the optimal property, we must show that the power function (11-27) of any test is continuous as a function of $\boldsymbol{\mu}$. The proof is very similar to that of Lemma 9-2; the details are left to the reader (Exercise 1).

The problem of determining the test of size α of largest average power may be put in this precise form: Among all regular sets A for which the condition (11-26) holds, find the one—if it exists—for which the average of (11-27) is maximized over every sphere about $\mathbf{0}$. We proceed as in Chapter 9, first deriving an associated inequality involving integrals of continuous functions and then applying the result to integrals over regular sets.

In the integrals in (11-26) and (11-27) we replace the regular set A by a continuous function in the integrand. Let $f(\mathbf{x})$ be a continuous function on R^n such that

$$0 \le f(\mathbf{x}) \le 1. \tag{11-28}$$

For each such f satisfying the condition

$$\int \cdots \int_{R^n} f(\mathbf{x})\, L(\mathbf{x}; \mathbf{0}, 1)\, d\mathbf{x} = \alpha \tag{11-29}$$

form the integral

$$Q(\boldsymbol{\mu}) = \int \cdots \int_{R^n} f(\mathbf{x})\, L(\mathbf{x}; \boldsymbol{\mu}, 1)\, dx; \qquad (11\text{-}30)$$

and let $\bar{Q}(r^2)$ be the average of Q over the sphere $\{\boldsymbol{\mu}: \|\boldsymbol{\mu}\| = r\}$.

THEOREM 11-7

Let $f(\mathbf{x})$ be a continuous function satisfying (11-28) and (11-29); then the spherical average $\bar{Q}(r^2)$ satisfies

$$\bar{Q}(r^2) \le \int_{\chi^2_{n,\alpha}}^{\infty} \psi_n(x; r)\, dx, \qquad (11\text{-}31)$$

where the function in the integrand is the noncentral chi-square density with $\tau = r$, defined by (11-22).

The proof of this theorem is long and is broken up into several lemmas. The main point of the proof is that in forming the spherical average of $Q(\boldsymbol{\mu})$ in (11-30) we may replace f by its spherical average.

LEMMA 11-3

Let $f(\mathbf{x})$ be a continuous function on R^n, and $L(y)$ a continuous function of a real variable y. Let B and D be balls in R^n, centered at $\mathbf{0}$, and $\bar{f}$ the spherical average of f. Then the integral

$$\int \cdots \int_D \bar{f}(\|\mathbf{x}\|^2)\, L(\|\mathbf{x} - \boldsymbol{\mu}\|)\, d\mathbf{x} \qquad (11\text{-}32)$$

depends on $\boldsymbol{\mu}$ only through $\|\boldsymbol{\mu}\|$, and

$$\int \cdots \int_B \int \cdots \int_D f(\mathbf{x}) L(\|\mathbf{x} - \boldsymbol{\mu}\|)\, d\mathbf{x}\, d\boldsymbol{\mu}$$

$$= \int \cdots \int_B \int \cdots \int_D \bar{f}(\|\mathbf{x}\|^2) L(\|\mathbf{x} - \boldsymbol{\mu}\|)\, d\mathbf{x}\, d\boldsymbol{\mu}. \qquad (11\text{-}33)$$

Proof: For each r the spherical average of $L(\|\mathbf{x} - \boldsymbol{\mu}\|)$ over $\{\boldsymbol{\mu}: \|\boldsymbol{\mu}\| = r\}$ depends on $\mathbf{x}$ only through $\|\mathbf{x}\|$. To see this, let $\mathbf{x}$ and $\mathbf{x}'$ be two points at a common distance from the origin. Let $\boldsymbol{\mu}'$ be the image of $\boldsymbol{\mu}$ under a rotation carrying $\mathbf{x}$ onto $\mathbf{x}'$. (Such a rotation exists by virtue of Theorem 3-1.) Since $\boldsymbol{\mu}'$ is on the same sphere as $\boldsymbol{\mu}$, the spherical averages of $L(\|\mathbf{x} - \boldsymbol{\mu}\|)$ and $L(\|\mathbf{x} - \boldsymbol{\mu}'\|)$ over $\{\boldsymbol{\mu}: \|\boldsymbol{\mu}\| = r\}$ coincide; furthermore: $L(\|\mathbf{x} - \boldsymbol{\mu}\|) = L(\|\mathbf{x}' - \boldsymbol{\mu}'\|)$; thus the spherical average of $L(\|\mathbf{x} - \boldsymbol{\mu}\|)$ is the same as that of $L(\|\mathbf{x}' - \boldsymbol{\mu}\|)$. By interchanging the roles of $\mathbf{x}$ and $\boldsymbol{\mu}$ we see that the spherical average of $L(\|\mathbf{x} - \boldsymbol{\mu}\|)$ over a sphere $\{\mathbf{x}: \|\mathbf{x}\| = t\}$ is a function of $\|\boldsymbol{\mu}\|$ alone.

We prove that (11-32) depends only on $\|\boldsymbol{\mu}\|$. Since $\bar{f}$ is constant over spheres, the spherical average of the integrand in (11-32) is equal to $\bar{f}$ times the spherical average of L (with respect to $\mathbf{x}$) (by the result of Exercise 5, Sec. 6-1). The latter spherical average depends on $\boldsymbol{\mu}$ only through $\|\boldsymbol{\mu}\|$ (a conclusion of the previous paragraph). By Eq. 6-4 the integral (11-32) is unchanged when the integrand is replaced by its spherical average; therefore the integral depends only on $\|\boldsymbol{\mu}\|$.

In proving Eq. 11-33 we shall interchange the order of integration with respect to $\mathbf{x}$ and $\boldsymbol{\mu}$. This is permited by the Extension of Theorem 3-3; indeed, even though the latter is stated for rectangles it is also valid for balls.

Let us integrate on the left-hand side of (11-33) first with respect to $\boldsymbol{\mu}$:

$$\int_D \cdots \int \left(\int_B \cdots \int L(\|\mathbf{x} - \boldsymbol{\mu}\|)\, d\boldsymbol{\mu} \right) f(\mathbf{x})\, d\mathbf{x}. \qquad (11\text{-}34)$$

By Eq. 6-4 the integrand in the inner multiple integral may be replaced by its spherical average over $\boldsymbol{\mu}$ (with fixed $\mathbf{x}$). By the conclusion of the first paragraph of this proof, this spherical average depends on $\mathbf{x}$ through $\|\mathbf{x}\|$ alone; therefore, the inner integral over B is also a function of $\|\mathbf{x}\|$ only. It follows as before that the spherical average of

$$\left(\int_B \cdots \int L(\|\mathbf{x} - \boldsymbol{\mu}\|)\, d\boldsymbol{\mu} \right) f(\mathbf{x})$$

is equal to

$$\left(\int_B \cdots \int L(\|\mathbf{x} - \boldsymbol{\mu}\|)\, d\boldsymbol{\mu} \right) \bar{f}(\|\mathbf{x}\|^2)$$

(Exercise 5, Sec. 6-1); therefore, by Eq. 6-4, the integral (11-34) is equal to

$$\int_D \cdots \int \left(\int_B \cdots \int L(\| \mathbf{x} - \boldsymbol{\mu} \|) \, d\boldsymbol{\mu} \right) \bar{f}(\| \mathbf{x} \|^2) \, d\mathbf{x}. \quad (11\text{-}35)$$

Interchange the order of integration; then (11-35) is equal to the expression on the right-hand side of (11-33). ■

LEMMA 11-4

The spherical average $\bar{Q}(r^2)$ is equal to

$$\int_{R^n} \cdots \int \bar{f}(\| \mathbf{x} \|^2) \, L(\mathbf{x}; \boldsymbol{\mu}, 1) \, d\mathbf{x} \qquad (11\text{-}36)$$

on the sphere $\| \boldsymbol{\mu} \| = r$, where (11-36) depends on $\boldsymbol{\mu}$ only through $\| \boldsymbol{\mu} \|$.

Proof: By Formula (11-3), $L(\mathbf{x}; \boldsymbol{\mu}, 1)$ is a function of $\| \mathbf{x} - \boldsymbol{\mu} \|$ alone.

We recall from Sec. 6-1 that $\bar{Q}(r^2)$ is obtained from $Q(\boldsymbol{\mu})$ by integrating it over a ball of radius r, differentiating the integral with respect to r, and multiplying the derivative by an appropriate function of r; therefore, in order to prove the two assertions of the theorem, it suffices to prove that

$$\int_B \cdots \int Q(\boldsymbol{\mu}) \, d\boldsymbol{\mu} = \int_B \cdots \int \int_{R^n} \cdots \int \bar{f}(\| \mathbf{x} \|^2) \, L(\mathbf{x}; \boldsymbol{\mu}, 1) \, d\mathbf{x} \, d\boldsymbol{\mu}, \qquad (11\text{-}37)$$

for any ball B, and that the integral (11-36) depends only on $\| \boldsymbol{\mu} \|$.

Let $L(\mathbf{x}; \boldsymbol{\mu}, 1)$ be the function $L(\| \mathbf{x} - \boldsymbol{\mu} \|)$ in Eq. 11-33, and let the radius of D become infinite. This implies the validity of (11-37); indeed, the condition (11-28) on the boundedness and non-negativity of f permits the passage to the limit inside the sign of integration (Exercise 2).

For any ball D, the integral

$$\int_D \cdots \int \bar{f}(\| \mathbf{x} \|^2) \, L(\mathbf{x}; \boldsymbol{\mu}, 1) \, d\mathbf{x}$$

depends only on $\| \boldsymbol{\mu} \|$ (Lemma 11-3); thus, by letting the radius of D become infinite, we conclude that (11-36) has the same property. ■

LEMMA 11-5
The integral (11-36) is equal to

$$\int_0^{\infty} \bar{f}(u)\,\psi_n(u;r)\,du, \tag{11-38}$$

for $\|\boldsymbol{\mu}\| = r$.

Proof: By Theorem 11-6, the statistic $\|\mathbf{x}\|^2$ has the noncentral chi-square distribution; thus the multiple integral (11-36) is reduced to the single integral (11-38) (Theorem 4-1). ■

LEMMA 11-6
For $\tau > 0$, *the ratio of the noncentral chi-square density with noncentrality parameter* τ *to the ordinary chi-square density,*

$$\psi_n(y;\tau)/\psi_n(y),$$

is a nondecreasing function of y *for* $y > 0$.

Proof: Refer to the formula (11-22) for $\psi_n(y;\tau)$. We write

$$\phi(\sqrt{y-u}-\tau) + \phi(\sqrt{y-u}+\tau) = \phi(\sqrt{y-u})\,e^{-\tau^2/2}(e^{\tau\sqrt{y-u}} + e^{-\tau\sqrt{y-u}}),$$

and

$$\psi_{n-1}(u)\,\phi(\sqrt{y-u}) = \text{constant}\cdot u^{(n-3)/2}e^{-y/2};$$

thus the expression for $\psi_n(y;\tau)$ for $y > 0$ is a constant multiple of

$$\int_0^y (y-u)^{-1/2}u^{(n-3)/2}\cosh(\tau(y-u)^{1/2})\,du,$$

where $\cosh x = (e^x + e^{-x})/2$. Change the variable of integration from u to $v = u/y$; then the ratio $\psi_n(y;\tau)/\psi_n(y)$ is a constant multiple of

$$\int_0^1 (1-v)^{-1/2}v^{(n-3)/2}\cosh(\tau\sqrt{y(1-v)})\,dv.$$

This increases with y for $\tau > 0$ because $\cosh x$ increases for $x > 0$. ■

The rest of the proof of Theorem 11-7 is now undertaken. The function $\bar{f}(\|\mathbf{x}\|^2)$ satisfies (11-28) because $f(\mathbf{x})$ does. By Eq. 6-8, the integral in (11-29) may be simplified by replacing f by its spherical average, and then reducing the multiple integral to a single integral: the condition (11-29) becomes

$$\int_0^\infty \bar{f}(u)\,\psi_n(u)\,du = \alpha. \tag{11-39}$$

By Lemmas 11-4 and 11-5, $\bar{Q}(r^2)$ is equal to (11-38). We seek a bound on $\bar{Q}(r^2)$. This problem is analogous to the one considered in Sec. 9-1 in testing a simple hypothesis: for all continuous functions $\bar{f}(u)$, $u \geq 0$, assuming values between 0 and 1, and satisfying (11-39), we seek a bound on the integral (11-38). We apply the Neyman-Pearson lemma (Lemma 9-1): by virtue of Lemma 11-6, the hypotheses of the Neyman-Pearson lemma is satisfied with

$$p_0(x) = \psi_n(x), \quad p_1(x) = \psi_n(x;\tau), \quad x > 0.$$

(Even though Lemma 9-1 is stated for densities which are positive everywhere, its statement and proof are valid for densities which are positive on the positive axis.) We infer

$$\bar{Q}(r^2) = \int_0^\infty \bar{f}(u)\,\psi_n(u;r)\,du \leq \int_{\chi^2_{n,\alpha}}^\infty \psi_n(u;r)\,du. \tag{11-40}$$

This completes the proof of Theorem 11-7.

We now finish the proof that the proposed test has the largest average power over spheres:

THEOREM 11-8

A Sample of n observations is taken from normal populations with means $\mu_1, \ldots, \mu_n$, respectively, and common unit standard deviation. We wish to test the simple null hypothesis $\|\boldsymbol{\mu}\| = 0$ against the composite alternative $\|\boldsymbol{\mu}\| > 0$. Let $\chi^2_{n,\alpha}$ be the upper α-percentile of the chi-square distribution with n degrees of freedom. Among all tests of size α, the one with the rejection region

$$M_0 = \{\mathbf{x}: \|\mathbf{x}\|^2 > \chi^2_{n,\alpha}\}$$

has the power function with the largest average over spheres $\{\boldsymbol{\mu}: \|\boldsymbol{\mu}\| = r\}$, $r > 0$.

Proof: Let A be a regular set in R^n satisfying (11-26), and let $f(\mathbf{x})$ be an arbitrary function in $\mathfrak{U}(A)$ (f is continuous, assumes values between 0 and 1 inclusive, and is equal to 1 on A). From the property (3-35) of the integral it follows that

$$\int_{R^n}\cdots\int f(\mathbf{x})\,L(\mathbf{x};\boldsymbol{\mu},1)\,d\mathbf{x} \geq \int_{A}\cdots\int L(\mathbf{x};\boldsymbol{\mu},1)\,d\mathbf{x} \qquad (11\text{-}41)$$

for all $\boldsymbol{\mu}$; in particular, for $\boldsymbol{\mu} = \mathbf{0}$:

$$\alpha' = \int_{R^n}\cdots\int f(\mathbf{x})\,L(\mathbf{x};\mathbf{0},1)\,d\mathbf{x} \geq \int_{A}\cdots\int L(\mathbf{x};\mathbf{0},1)\,d\mathbf{x} = \alpha. \qquad (11\text{-}42)$$

The inequality (11-41) is also valid for the spherical averages: averaging each member of the inequality, and applying the equality in (11-40), we get

$$\text{Average of} \int_{A}\cdots\int L(\mathbf{x};\boldsymbol{\mu},1)\,d\mathbf{x} \quad \text{over} \quad \|\boldsymbol{\mu}\| = \tau$$

$$\leq \int_0^\infty \bar{f}(y)\,\psi_n(y;\tau)\,dy. \qquad (11\text{-}43)$$

By analogy to (11-39), the first equality in (11-42) is equivalent to

$$\int_0^\infty \bar{f}(y)\,\psi_n(y)\,dy = \alpha'.$$

Apply Theorem 11-7 with α' in place of α: the right-hand side of (11-43)—and so the left-hand side—is not more than

$$\int_{\chi^2_{n,\alpha'}}^\infty \psi_n(y;\tau)\,dy.$$

In this bound, the value α' can be replaced by the original value α. By the property (3-35) of the integral, the greatest lower bound of the integral on the left-hand side of (11-42) is equal to the integral on the right-hand side; therefore:

$$\alpha = \text{glb}\,\alpha';$$

thus, by the continuity of the chi-square distribution:

$$\int_{\chi^2_{n,\alpha}}^\infty \psi_n(y;\tau)\,dy = \text{glb}\int_{\chi^2_{n,\alpha'}}^\infty \psi_n(y;\tau)\,dy,$$

and so the average of $\int \cdots \int_A L(\mathbf{x}; \boldsymbol{\mu}, 1)\, d\mathbf{x}$ over the sphere of radius τ is not more than

$$\int_{\chi^2_{n,\alpha}}^{\infty} \psi_n(y; \tau)\, dy. \tag{11-44}$$

This represents an upper bound on the spherical average of the power function of any test of size α.

In order to see that the test with the rejection region M_0 is of size α, and that its power is constant and equal to (11-44) over the sphere of radius τ, note that $\| \mathbf{x} \|^2$ has the density ψ_n under the null hypothesis and $\psi_n(\cdot; \tau)$ under the alternative. ■

EXERCISES

1. *Prove:* For any regular set M the integral

$$\int \cdots \int_M L(\mathbf{x}; \boldsymbol{\mu}, 1)\, d\mathbf{x}$$

is a continuous function of $\boldsymbol{\mu}$.

2. Show that the limits of the integrals in (11-33) as the radius of D becomes infinite are equal to the corresponding integrals with $D = R^n$; in other words, it is permissible to pass to the limit under the sign of integration over B. (*Hint:* Note that $0 \leq f \leq 1$, so that

$$\left| \int \cdots \int_B \int \cdots \int_{R^n} f L\, d\mathbf{x}\, d\boldsymbol{\mu} - \int \cdots \int_B \int \cdots \int_D f L\, d\mathbf{x}\, d\boldsymbol{\mu} \right|$$
$$\leq \int \cdots \int_B \int \cdots \int_{\text{exterior of } D} L(\mathbf{x}; \boldsymbol{\mu}, 1)\, d\mathbf{x}\, d\boldsymbol{\mu}.$$

3. Verify the algebraic identities in the proof of Lemma 11-6.

11-6. SAMPLING FROM SEVERAL GENERAL POPULATIONS AND LINEAR ESTIMATION OF THE MEANS (OPTIONAL)

The Sample space of n observations from n normal populations is now generalized to arbitrary populations. Let $h_1(x), \ldots, h_n(x)$ be n density functions. The Sample space of n observations from the respective populations is defined as for normal populations with the modification that the likelihood function is now

$$L(\mathbf{x}) = \prod_{i=1}^{n} h_i(x_i), \tag{11-45}$$

in place of the normal density product. As noted in Sec. 5-5 the distribution theory does not extend to general populations because of the unique relation between rotation and the normal distribution; however, the result on estimation in Sec. 11-3 has a useful version in the general case.

Suppose that μ_i and σ_i^2 are the mean and variance, respectively, of the population with density $h_i(x)$. We shall assume as before that the variances have a common value: $\sigma_i^2 = \sigma^2$, $i = 1, \ldots, n$; and that $\mu_1, \ldots, \mu_n$ and σ^2 are all unknown. Let $t_1, \ldots, t_n$ be a fixed set of numbers, and suppose that we want to estimate the linear combination $t_1\mu_1 + \cdots + t_n\mu_n$. Theorem 11-5 states that in the normal case the efficient Estimator is the corresponding linear combination of the coordinates $t_1x_1 + \cdots + t_nx_n$.

This result for the normal case suggests the definition of the class of linear Estimators even in the general (not necessarily normal) case. An Estimator $f(\mathbf{x})$ is called *linear* if it is a linear function of the coordinates of $\mathbf{x}$:

$$f(\mathbf{x}) = s_1x_1 + \cdots + s_nx_n,$$

where $s_1, \ldots, s_n$ are real numbers. By computations very much like those in Sec. 7-5, Exercise 10, it can be shown that the mean of the distribution of this linear Estimator is the corresponding linear combination of the means:

$$\int_{R^n} \cdots \int \left(\sum_{i=1}^{n} s_i x_i\right) L(\mathbf{x})\, d\mathbf{x} = \sum_{i=1}^{n} s_i \mu_i \qquad \text{(Exercise 1).} \tag{11-46}$$

It follows that a linear Estimator of $t_1\mu_1 + \cdots + t_n\mu_n$ is unbiased if and only if

$$\sum_{i=1}^{n} s_i\mu_i = \sum_{i=1}^{n} t_i\mu_i, \qquad \textit{for all } \mu_1, \ldots, \mu_n. \tag{11-47}$$

By substitution of appropriate values of n-tuples $(\mu_1, \ldots, \mu_n)$—first $(1, 0, \ldots, 0)$, then $(0, 1, 0, \ldots, 0), \ldots$—it follows that

$$s_i = t_i, \qquad i = 1, \ldots, n;$$

therefore, in the general case, *$t_1x_1 + \cdots + t_nx_n$ is the only linear unbiased Estimator of $t_1\mu_1 + \cdots + t_n\mu_n$.*

The variance of the linear combination is given by

$$\text{Var}\left(\sum_{i=1}^{n} s_i x_i\right) = \sigma^2 \sum_{i=1}^{n} s_i^2; \qquad (11\text{-}48)$$

this follows directly from the multiple integral

$$\int \cdots \int \left(\sum_{i=1}^{n} s_i(x_i - \mu_i)\right)^2 L(\mathbf{x})\, d\mathbf{x} \qquad (11\text{-}49)$$

defining the variance (Exercises 2, 3).

The result that $s_1x_1 + \cdots + s_nx_n$ has the mean (11-46) and the variance (11-48) is different from Theorem 11-4: The latter proposition states that if the Sample space is *normal* then the linear combination has a *normal* distribution with the indicated mean and variance.

EXERCISES

1. Verify (11-46) by termwise integration.

2. Show that (11-49) is indeed the variance of $\Sigma s_i x_i$.

3. Verify that (11-49) is equal to the right-hand side of Eq. 11-48. (Expand the square and integrate termwise.)

BIBLIOGRAPHY

The noncentral chi-square distribution is derived in T. W. Anderson, *Introduction to Multivariate Statistical Analysis* (New York: Wiley, 1958).

Invariant tests of homogeneity are discussed in generality in E. L. Lehmann, *Testing Statistical Hypotheses* (New York: Wiley, 1959).

CHAPTER 12

Linear Regression

12-1. EMPIRICAL BACKGROUND OF THE LINEAR REGRESSION MODEL

In the statistical model of sampling from several normal populations (Chapter 11) the point $\boldsymbol{\mu}$ representing the means was arbitrary. In many problems of practical importance it is assumed that there are functional relations among the means. One of the simple assumptions about the means is that they are linearly related in the following way: There are real numbers α, β and $t_1, \ldots, t_n$ such that

$$\mu_i = \alpha + \beta t_i, \qquad i = 1, \ldots, n. \tag{12-1}$$

The numbers $t_1, \ldots, t_n$ are assumed to be known constants, and are called *independent variables*. The constants α and β are unknown parameters. The common standard deviation σ is also assumed to be unknown. These hypotheses define the *linear regression model*.

It is helpful to visualize the relation (12-1) in terms of coordinate geometry. Consider the line $\mu = \alpha + \beta t$ in the (t, μ)-plane: the means μ_i represent the ordinates corresponding to the abscissa values t_i (Fig. 12-1).

Here are some applications. Short term economic time series are considered to be Samples from normal populations with means given by the linear regression model. Consider the price of a particular commodity, recorded on the Monday of each week. Suppose that the price is not subject to seasonal fluctuations. In accordance with the remarks in Sec. 11-1, the prices observed over the various Mondays are like a Sample from various populations. Let $t_1, \ldots, t_n$ be the indices of the time points when the various prices were recorded; for example, since the time points are equally spaced, we may put $t_i = i, i = 1, \ldots, n$. We assume that the means of the popu-

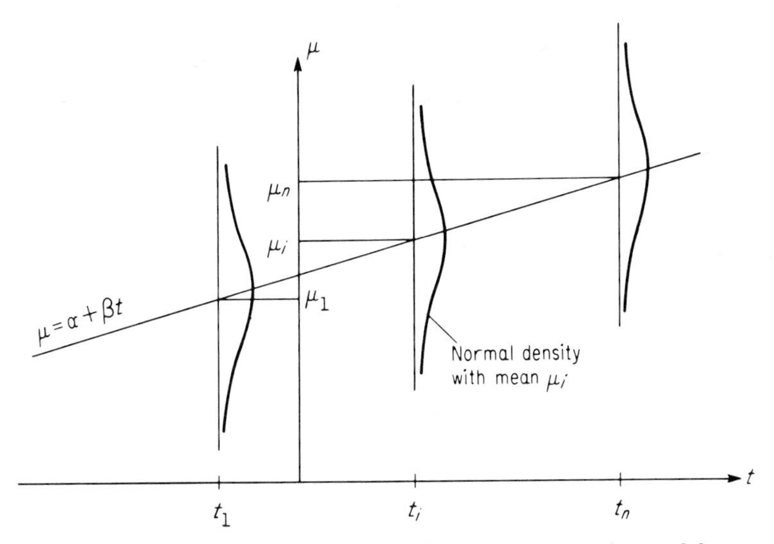

Fig. 12-1. Geometric interpretation of the linear regression model.

lationsvary in accordance with (12-1):

$$\mu_i = \alpha + \beta i, \qquad i = 1, \ldots, n;$$

in other words, the "mean price" varies linearly with time.

The assumption of a linear relation between the mean price and the time is realistic in the case of short term economic time series but not *necessarily* long-term series. As an example of a short-term series, consider the price of a commodity on an exchange during a period of several hours of trading. It is reasonable to suppose that under stable economic conditions the *mean* price varies "smoothly" with time; however, it must be added that the actual price does not necessarily behave in this way. Let $m(t)$ be the mean price at time t; then, it can be supposed that the graph of $m(t)$ is "smooth," that is, the function has one or more continuous derivatives. In accordance with Taylor's theorem, the function $m(t)$ is "locally linear": near each point t_0, it is expressible as a linear function plus a small remainder:

$$m(t) = m(t_0) + m'(t_0)(t - t_0) + R.$$

We neglect R and put $\alpha = m(t_0) - t_0 m'(t_0)$, $\beta = m'(t_0)$; thus, $m(t) = \alpha + \beta t$ for t near t_0.

The line $\alpha + \beta t$ is called the regression line, or *trend* line of the series. The coefficient β is the slope of the trend line: If $\beta > 0$, there is a positive "trend" in the series. Analogous statements hold for $\beta = 0$ and $\beta < 0$. The economist is interested in problems of the following kind:

(a) Estimation of the unknown parameters α, β, and σ, and confidence intervals for them.

(b) Testing hypotheses—for example, testing the hypothesis that the trend is 0 ($\beta = 0$).

(c) Prediction: Given the values of the Sample at the time points $t_1, \ldots, t_n$, predict the value of the next observation, that is, the coördinate of the Sample point arising from the population with mean $\alpha + \beta t_{n+1}$.

The linear regression model is also applicable to problems in agriculture and industry where there is a process with a controlled "input" and a randomly varying "output." In the growth of crops, the treatment of the soil with fertilizer is considered a controlled input and the actual crop yield is considered a random output. Let $t_1, \ldots, t_n$ represent various levels of fertilizer input (e.g., grams per unit of soil), and let $X_1, \ldots, X_n$ be the corresponding crop yields. If the inputs are not extremely different, then the yields may be regarded as a Sample of observations from n normal populations with means (12-1). (The justification of the linear model is the same as for economic time series.) The problems of interest to the agricultural experimenter are similar to the corresponding problems of the economist, mentioned above.

We give an example of linear regression in zoology. In the article "The Effect of Temperature upon the Heterozygotes in the Bar Series of *Drosophila*," by A. H. Hersh, *Journal of Experimental Zoology*, Vol. 39 (1924), pp. 55–71, it is shown that there is an inverse relation between the temperature in which certain flies live and the numbers of their "facets." In an experiment the flies were put at various temperatures from 15°C through 31°C, and the number of facets on each fly was counted. Table 12-1 shows the distribution of the number of facets in the flies at each temperature level; note that the distributions are roughly bell-shaped. The means and standard deviations for the distribution at each temperature level are given in Table 12-2. The means appear to vary approximately linearly with the temperature, and the standard deviations are fairly constant; thus the assumptions for the linear regression model are satisfied.

TABLE 12-1
The Data of Table 12-2 on Ultra-bar Females.
***The Class Which Contains the Mean Value.**

Classes in Factorial Units	Classes in Facets	Distribution of Flies at Each Temperature								
		15°	17°	19°	21°	23°	25°	27°	29°	31°
−18.93	8									1
−17.93	9							1		0
−16.93	10							1	2	1
−15.93	11							0	3	2
−14.93	12							4	5	2
−13.93	13+					1	1	1.5	1	2.5
−12.93	15+					1	1	5.5	9	9.5*
−11.93	16–17					0	4	9	7	4
−10.93	18–19				1	0	12	9	12*	8
−9.93	20–21				5	7	11	17*	9	4
−8.93	22–23			1	5	6	21	10	4	7
−7.93	24–26			3	11	12	42*	13	4	1
−6.93	27–29			7	11	11*	20	6	1	1
−5.93	30–32		1	13	12	17	7	6		
−4.93	33–35		0	9	20*	7	2	0		
−3.93	36–39		0	23	20	6	2	1		
−2.93	40–43		4	10*	11	3	0			
−1.93	44–48	7	9	17	8	2	1			
−0.93	49–53	2	13	5	2					
+0.07	54–59	8	10*	5	1					
+1.07	60–65	12*	8	2						
+2.07	66–72	5	4	2						
+3.07	73–80	6	2							
+4.07	81–88	3	2							
+5.07	89–97	2	1							
+6.07	98–107	1								
+7.07	108–118	1								
Totals..........		47	54	97	107	73	124	84	57	43

TABLE 12-2
Summary of Data of Ultra-bar Females. These Data Are Seen in Table 12-1. The Last Two Columns Are in Terms of Factorial Units.

Temperature	Number of Flies	Arithmetical Mean	Mean	Standard Deviation
15	47	65.3	+1.33 ± 0.22	2.19 ± 0.16
17	50	56.0	−0.19 ± 0.18	2.00 ± 0.13
19	97	40.0	−3.64 ± 0.16	2.30 ± 0.11
21	107	33.6	−5.29 ± 0.15	2.32 ± 0.11
23	73	29.1	−6.75 ± 0.18	2.26 ± 0.13
25	124	24.4	−8.35 ± 0.11	1.82 ± 0.08
27	84	21.1	−9.92 ± 0.19	2.63 ± 0.13
29	57	17.6	−11.67 ± 0.21	2.39 ± 0.15
31	43	17.4	−11.73 ± 0.36	2.49 ± 0.18

12-2. LEAST SQUARES AND ITS GEOMETRIC INTERPRETATION

Consider the following problem in the numerical analysis of data. Let $t_1, \ldots, t_n$ be given values of an independent variable, and $x_1, \ldots, x_n$ the corresponding outputs; for example, the t's are time points, and the x's are the quoted prices of a commodity. The points (t_i, x_i), $i = 1, \ldots, n$ are plotted in the (t, x)-plane, as in Fig. 12-2. We do not necessarily assume any statistical relation among the t's and x's, or even that the latter are a random Sample. Suppose that we want to fit a straight line to the given data—that is, determine a straight line $x = a + bt$ which "most closely" resembles the configuration of points (Fig. 12-2).

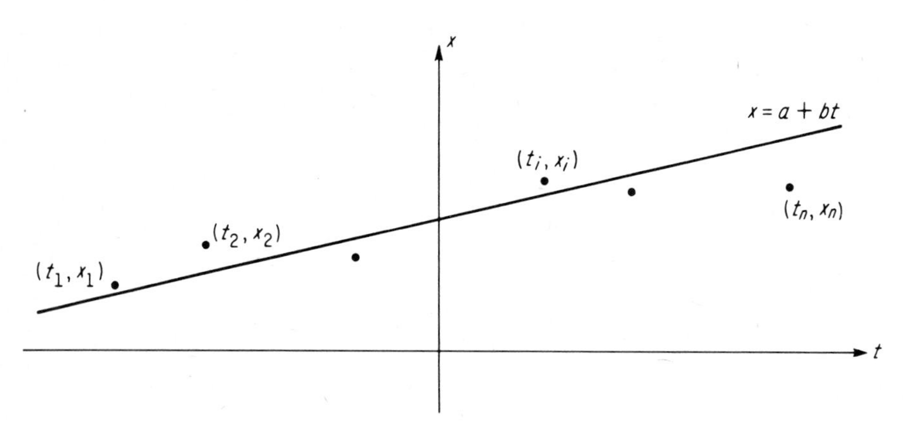

Fig. 12-2. A straight line fitted to n points in the plane.

For any given set of n points in the plane there are many lines that can be drawn which seem to "fit" the data. The best-fitting line is defined in terms of the principle of "least squares." For each line $x = \alpha + \beta t$ in the plane, consider the sum of squares of the vertical distances from the points (x_i, t_i) to the corresponding point $\alpha + \beta t_i$ on the line: it is equal to

$$\sum_{i=1}^{n} (x_i - \alpha - \beta t_i)^2. \tag{12-2}$$

The least-squares line is the one for which this sum of squares is a minimum; in other words, the constants α and β are chosen to minimize the sum of squares.

In the following theorem the minimizing values are obtained under the condition that the sum of the t's is equal to 0.

THEOREM 12-1
Under the condition

$$t_1 + \cdots + t_n = 0 \tag{12-3}$$

the sum of squares (12-2) is a minimum with respect to α *and* β *for the values*

$$\alpha = \bar{x}, \qquad \beta = \frac{\sum_{i=1}^{n} x_i t_i}{\sum_{i=1}^{n} t_i^2} = b; \tag{12-4}$$

furthermore, the following decomposition of sum of squares holds:

$$\sum_{i=1}^{n} x_i^2 = \sum_{i=1}^{n} (x_i - \bar{x} - bt_i)^2 + n\bar{x}^2 + b^2 \sum_{i=1}^{n} t_i^2, \tag{12-5}$$

where b is given by (12-4).

Proof: Apply the original decomposition (4-12) to the sum of squares (12-2) with $x_i - \alpha - \beta t_i$ in place of x_i: under the condition (12-3), we obtain

$$\sum_{i=1}^{n} (x_i - \alpha - \beta t_i)^2 = \sum_{i=1}^{n} (x_i - \bar{x} - \beta t_i)^2 + n(x - \alpha)^2 \tag{12-6}$$

(Exercise 3). Now decompose the sum of squares on the right hand side of Eq. 12-6. Define

$$b' = \frac{\sum_{i=1}^{n} (x_i - \bar{x} - \beta t_i)\, t_i}{\sum_{i=1}^{n} t_i^2},$$

and apply the decomposition (11-5) with $x_i - \bar{x} - \beta t_i$, t_i, and b' in the places of x_i, t_i, and b, respectively:

$$\sum_{i=1}^{n} (x_i - \bar{x} - \beta t_i)^2 = \sum_{i=1}^{n} (x_i - \bar{x} - \beta t_i - b' t_i)^2 + b'^2 \sum_{i=1}^{n} t_i^2. \tag{12-7}$$

By condition (12-3) and the definition (12-4) of b, we have:

$$b' = \frac{\sum_{i=1}^{n} x_i t_i}{\sum_{i=1}^{n} t_i^2} - \frac{\bar{x} \sum_{i=1}^{n} t_i}{\sum_{i=1}^{n} t_i^2} - \beta = b - \beta;$$

therefore (12-7) may be written as

$$\sum_{i=1}^{n} (x_i - \bar{x} - \beta t_i)^2 = \sum_{i=1}^{n} (x_i - \bar{x} - bt_i)^2 + (b - \beta)^2 \sum_{i=1}^{n} t_i^2. \qquad (12\text{-}8)$$

Substitute the right-hand side of (12-8) for the first term on the right-hand side of (12-6):

$$\sum_{i=1}^{n} (x_i - \alpha - \beta t_i)^2 = \sum_{i=1}^{n} (x_i - \bar{x} - bt_i)^2 + (b - \beta)^2 \sum_{i=1}^{n} t_i^2 + n(\bar{x} - \alpha)^2. \qquad (12\text{-}9)$$

It is clear that the expression on the right-hand side of this equation is minimized with respect to α and β for the values given by (12-4); thus the first assertion of the theorem is proved. The identity (12-5) is a special case of (12-9) for $\alpha = \beta = 0$. ■

The geometric interpretation of the identity (12-5) is seen by writing its members in terms of lengths of lines in R^n:

$$\|\mathbf{x}\|^2 = \|\mathbf{x} - \bar{x}\mathbf{1} - b\mathbf{t}\|^2 + \|\bar{x}\mathbf{1}\|^2 + \|b\mathbf{t}\|^2. \qquad (12\text{-}10)$$

This states that the square of the distance from $\mathbf{0}$ to $\mathbf{x}$ is equal to the sum of the squares of the lengths of three lines: the lines from $\bar{\mathbf{x}}$ to $x\mathbf{1} + b\mathbf{t}$, from $\mathbf{0}$ to $\bar{x}\mathbf{1}$, and from $\mathbf{0}$ to $b\mathbf{t}$. By virtue of (12-3), the last two lines are perpendicular:

$$\|x\mathbf{1} - b\mathbf{t}\|^2 = \|\bar{x}\mathbf{1}\|^2 + \|b\mathbf{t}\|^2; \qquad (12\text{-}11)$$

in other words, the line segments from $\mathbf{0}$ to $\bar{x}\mathbf{1}$ and from $\mathbf{0}$ to $\mathbf{t}$ are perpendicular if $\Sigma t_i = 0$. The relations among these points are illustrated in Fig. 12-3 in the case $n = 3$. The point $\bar{x}\mathbf{1} + b\mathbf{t}$ is the one in the plane spanned by the diagonal and by the line through $\mathbf{0}$ and $\mathbf{t}$ and whose distance from $\mathbf{x}$ is smallest; this is the meaning of the minimization of (12-2). By substituting $\|\bar{x}\mathbf{1} + b\mathbf{t}\|^2$ for the sum of the last two terms in (12-10) we see that

$$\|\mathbf{x}\|^2 = \|\mathbf{x} - \bar{x}\mathbf{1} - b\mathbf{t}\|^2 + \|x\mathbf{1} + b\mathbf{t}\|^2,$$

so that the line from $\mathbf{x}$ to $\bar{x}\mathbf{1} + b\mathbf{t}$ is perpendicular to the line from $\mathbf{0}$ to $\bar{x}\mathbf{1} + b\mathbf{t}$.

The decompositions (4-12) and (11-5) were used in particular rotations which were applied to the transformation of certain integrals. The first rotation carried the diagonal line onto the first coordinate axis; and the second one carried a point $\mathbf{t}$ onto $\|\mathbf{t}\|\mathbf{e}_1$.

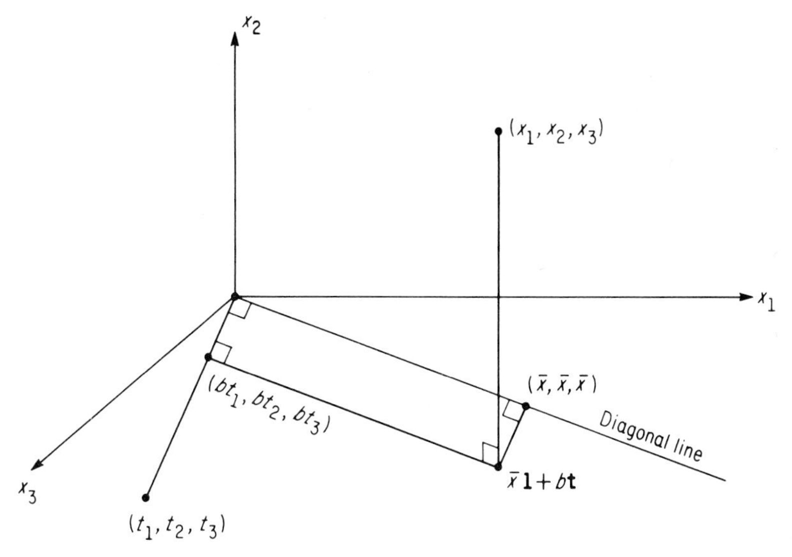

Fig. 12-3. Geometric interpretation of Eq. 12-10 for $n = 3$.

Now we are going to use the decomposition (12-5) in a rotation which combines the effects of both of the previous ones: it carries a *pair* of perpendicular lines onto a pair of coordinate axes. Such a rotation exists by virtue of Theorem 3-2.

THEOREM 12-2

Let $\mathbf{t} \neq \mathbf{0}$ be a point in R^n, $n \geq 3$, whose coordinates satisfy (12-3). Consider a rotation which carries $\mathbf{1}$ and $\mathbf{t}$ onto $\sqrt{n}\mathbf{e}_1$ and $\|\mathbf{t}\|\ \mathbf{e}_2$, respectively. [Such a rotation exists because the lines from the origin to $\mathbf{1}$ and $\mathbf{t}$, respectively, are perpendicular by virtue of (12-11).] The images of the hyperplanes

$$\left\{\mathbf{x}: \sum_{i=1}^{n} x_i = c\right\} \quad \text{and} \quad \left\{\mathbf{x}: \sum_{i=1}^{n} t_i x_i = d\right\}$$

under this rotation are

$$\{\mathbf{x}: x_1 = c/\sqrt{n}\} \quad \text{and} \quad \{\mathbf{x}: x_2 = d/\|\mathbf{t}\|\},$$

respectively, for all c and d.

Proof: The statement about the first hyperplane follows from the general result of Theorem 4-3 on a diagonal rotation. The result about the second hyperplane follows from Theorem 11-1: while it is stated for a rotation carrying **t** onto $\|\mathbf{t}\|\mathbf{e}_1$ its statement and proof are valid for a rotation carrying **t** onto $\|\mathbf{t}\|\mathbf{e}_2$. ∎

This theorem furnishes the relations between the coordinates of a point **x** and those of its image **x**′ under the rotation:

$$x_1' = \sqrt{n}\bar{x}, \qquad x_2' = b\|\mathbf{t}\|, \tag{12-12}$$

$$\sum_{i=3}^{n} x_i'^2 = \sum_{i=1}^{n} (x_i - \bar{x} - bt_i)^2.$$

The first two of these are immediate consequences of Theorem 12-2 and of the definition (12-4) of b (Exercise 4). The third relation follows from the identity (12-5), the preservation of distance under rotation, and the first two relations in (12-12) (Exercise 5).

EXERCISES

1. Compute the least-squares lines to fit the following data:

(a)

t	−2	−1	0	1	2
x	−1	−1	+5	.5	.5

(b)

t	−3	−2	−1	0	1	2	3
x	5	3.5	3	2	1.5	1	−1

2. Compute the sum of the squares of the vertical deviations from the least squares lines in Exercise 1. In order to avoid incorporating a large rounding error, use this identity, based on Eq. 12-5:

$$\sum_{i=1}^{n} (x_i - \bar{x} - bt_i)^2 = \sum_{i=1}^{n} \bar{x}_i^2 - n\bar{x}^2 - b^2 \sum_{i=1}^{n} t_i^2.$$

3. Verify Eq. 12-6.

4. Verify the first two relations in (12-12).

5. Verify the third relation in (12-12).

6. Minimize (12-2) by partial differentiation.

12-3. STATISTICAL INFERENCE AND DISTRIBUTION THEORY IN THE LINEAR REGRESSION MODEL

The statistical theory of the linear regression model is based on the Sample space associated with sampling from n normal populations (Sec. 11-1). Let **X** be a Sample of observations from the normal populations with means $\mu_i = \alpha + \beta t_i$, $i = 1, \ldots, n$ and common standard deviation σ. In accordance with (11-2), the likelihood function is

$$L = (2\pi\sigma^2)^{-n/2} e^{-(1/2\sigma^2) \sum_{i=1}^{n} (x_i - \alpha - \beta t_i)^2}$$

There is no loss of generality in assuming that $t_1 + \cdots + t_n = 0$ [condition (12-3)]. If the t's do not originally satisfy this condition, then the parameter α and the t's can be transformed in such a way that the latter do satisfy it. Let $\bar{t}$ be the average of the t's; then we have the identity

$$\alpha + \beta t_i = (\alpha + \beta\bar{t}) + \beta(t_i - \bar{t}).$$

Put $\alpha' = \alpha + \beta\bar{t}$ and $t_i' = t_i - \bar{t}$; then

$$\alpha + \beta t_i = \alpha' + \beta t_i',$$

where $t_1' + \cdots + t_n' = 0$. In this way the system of means (12-1) can always be reparametrized so that condition (12-3) is satisfied.

Suppose that α, β, and σ^2 are unknown, and we wish to estimate these parameters on the basis of a Sample of n observations. Although, as we noted, there is no sampling theory in the least-squares principle, the solution to the fitting problem suggests the best statistical procedure in the linear regression model: the coefficients $\bar{x}$ and b are the efficient Estimators of the unknown parameters α and β, respectively, and the average sum of squares about the fitted regression line, $\Sigma(x_i - \bar{x} - bt_i)^2/(n-2)$, is an unbiased Estimator of σ^2:

THEOREM 12-3

In sampling from n normal populations with means (12-1) and common standard deviation σ, the statistics $\bar{x}$ and b (defined by (12-4)) are the efficient Estimators of α and β, respectively, and

$$s^2 = \frac{\sum_{i=1}^{n} (x_i - \bar{x} - bt_i)^2}{n-2} \qquad (12\text{-}13)$$

is an unbiased Estimator of σ^2 with variance $2\sigma^4/(n-2)$.

Proof: The parameter α is the average of $\mu_1, \ldots, \mu_n$:

$$\frac{1}{n}\sum_{i=1}^{n} \mu_i = \frac{1}{n}\sum_{i=1}^{n} (\alpha + \beta t_i) = \alpha,$$

by condition (12-3). The parameter β is the linear combination of the means

$$\beta = \frac{\sum_{i=1}^{n} \mu_i t_i}{\sum_{i=1}^{n} t_i^2}.$$

One is tempted to invoke Theorem 11-5 which states that the efficient Estimator of a linear combination of the means is the corresponding linear combination of the observations, and conclude that $\bar{x}$ and b are efficient Estimators of α and β. This cannot be done because Theorem 11-5 is not applicable: no relations among the means were there assumed, but here all the means are expressible as a linear combination of any two; indeed, the first two equations (12-1) can be solved for α and β in terms of μ_1 and μ_2, and then μ_j can be expressed in terms of these for $j > 3$.

In order to prove that $\bar{x}$ and b are really efficient, we use the technique of Theorem 11-5. An Estimator $f(\mathbf{x})$, $\mathbf{x} \in R^n$, of α is unbiased if

$$\int_{R^n}\cdots\int f(\mathbf{x})(2\pi\sigma^2)^{-n/2} \exp\left\{-\frac{1}{2\sigma^2}\sum_{i=1}^{n} (x_i - \alpha - \beta t_i)^2\right\} d\mathbf{x} = \alpha$$

for all parameter values. Apply the decomposition (12-9) to the sum of squares in the exponent:

$$\int_{R^n}\cdots\int f(\mathbf{x})(2\pi\sigma^2)^{-n/2} \exp\left\{-\frac{1}{2\sigma^2}[(n-2)s^2 + (b-\beta)\|\mathbf{t}\|^2 + n(\bar{x}-\alpha)^2]\right\} d\mathbf{x} = \alpha.$$

Change the variables of integration from $\mathbf{x}$ to $\mathbf{x}'$, where the latter is the image of the former under the rotation defined in Theorem 12-2. Substitute in the exponent in accordance with (12-12), and then remove the primes from the variables in the integrand:

$$\int_{R^n}\cdots\int f^*(\mathbf{x})(2\pi\sigma^2)^{-n/2}\exp\left\{-\frac{1}{2\sigma^2}\left[\sum_{i=3}^{n} x_i^2 + (x_2 - \beta\|\mathbf{t}\|)^2 + (x_1 - \sqrt{n}\alpha)^2\right]\right\}d\mathbf{x} = \alpha.$$

Integrate over the variables $x_2, \ldots, x_n$, obtaining:

$$\int_{-\infty}^{\infty} g(x)(2\pi\sigma^2)^{-1/2}\exp\left\{-\frac{1}{2\sigma^2}(x - \sqrt{n}\alpha)^2\right\}dx = \alpha.$$

Change the variable of integration from x to $\sqrt{n}x$:

$$\int_{-\infty}^{\infty} g(\sqrt{n}x)(2\pi\sigma^2/n)^{-1/2}\exp\left\{-\frac{1}{2\sigma^2/n}(x - \alpha)^2\ dx\right\} = \alpha.$$

As in the proof of Theorem 11-2, we conclude from the last equation (by analogy to Eq. 11-14) that the variance of any unbiased Estimator of α is at least σ^2/n. Now $\bar{x}$ is a *particular* unbiased Estimator of α, and its variance *is* σ^2/n; this follows from Theorem 11-4 with $t_i = 1/n$. This completes the proof that $\bar{x}$ is efficient.

The proof of the efficiency of b as an Estimator of β is almost the same: just interchange the roles of α and β and of x_1 and x_2, respectively, and replace the variance σ^2/n by $\sigma^2\|\mathbf{t}\|^2$.

The sum of squares $\Sigma(x_i - \bar{x} - bt_i)^2$ has a distribution which is independent of the particular values α and β; this can be seen by changing the variables of integration in the multiple integral defining the distribution function (Exercise 1). For this reason we can put $\alpha = \beta = 0$ in calculating the distribution. By the rotation in Theorem 12-2, and by the third equation in (12-12), the sum of squares $\Sigma(x_i - \bar{x} - bt_i)^2$ is transformed into $\sum_{i=3}^{n} x_i^2$; thus, by Theorem 4-4, the two have the same distribution. We find $\sum_{i=3}^{n} x_i^2/(n-2)$ to be unbiased, and its variance to be $2\sigma^4/(n-2)$ by the fact (Theorem 4-2) that $\sum_{i=3}^{n} x_i^2/\sigma^2$ has the chi-square distribution with $n - 2$ degrees of freedom (Exercise 2). ■

The fundamental theorem upon which is based much of the theory for sampling from a common population is Theorem 5-4 on the independence of $\bar{x}$ and s^2, and their respective distributions.

Now we prove a more general result covering the joint distribution of the Estimators of α, β, and σ^2.

THEOREM 12-4

In sampling from n normal populations with means (12-1) and common standard deviation σ, and where the t's satisfy (12-3), the three statistics

$$\sqrt{n}(\bar{x} - \alpha)/\sigma, \quad (b - \beta)\|\mathbf{t}\|/\sigma \quad \textit{and} \quad (n - 2)s^2/\sigma^2$$

are independent, the first two having standard normal distributions, and the third having a chi-square distribution with $n - 2$ degrees of freedom. Here s^2 is given by (12-13).

Proof: It is sufficient to prove this for the particular case $\alpha = \beta = 0$, $\sigma = 1$; the proof of the more general case follows by extension (Exercise 3). For this reason we now suppose that the Sample is from a *common standard normal population.* Put

$$f_1(\mathbf{x}) = \sqrt{n}\bar{x}, \quad f_2(\mathbf{x}) = b\|\mathbf{t}\|, \quad f_3(\mathbf{x}) = (n - 2)s^2,$$

where b and s^2 are the functions of $\mathbf{x}$ defined in (12-4) and (12-13), respectively. Let $\mathbf{x}'$ be the image of $\mathbf{x}$ under a rotation carrying $\mathbf{1}$ and $\mathbf{t}$ onto $\sqrt{n}\mathbf{e}_1$ and $\|\mathbf{t}\|\mathbf{e}_2$, respectively (Theorem 12-2). Define:

$$f_i^*(\mathbf{x}') = f_i(\mathbf{x}), \qquad i = 1, 2, 3.$$

By Theorem 5-3 (more exactly, by its statement for *three* functions), the statistics f_i, $i = 1, 2, 3$ have the same joint distribution as f_i^*, $i = 1, 2, 3$. The particular forms of the latter functions are obtained from (12-12):

$$f_1^*(\mathbf{x}') = x_1', \quad f_2^*(\mathbf{x}') = x_2', \quad f_3^*(\mathbf{x}') = \sum_{i=3}^{n} x_i'^2$$

These three functions depend on nonoverlapping sets of coordinates; thus, by Theorem 5-2, they are independent. The functions f_1^* and f_2^* each have the standard normal distribution; and, by Theorem 4-2, f_3^* has the chi-square distribution with n $-$ 2 degrees of freedom. ■

From this we obtain:

THEOREM 12-5
Each of the ratio statistics

$$\frac{\sqrt{n}(\bar{x} - \alpha)}{s} \tag{12-14}$$

$$\frac{\|\mathbf{t}\|(b - \beta)}{s} \tag{12-15}$$

$$\frac{(\bar{x} + bt - \alpha - \beta t)\left(\frac{1}{n} + \frac{t^2}{\|\mathbf{t}\|^2}\right)^{-1/2}}{s}, \quad -\infty < t < \infty, \tag{12-16}$$

has the t-distribution with $n - 2$ degrees of freedom. (The real number t appearing in (12-16) is arbitrary.)

Proof: The proof is based on the fact that the numerator and denominator in each ratio are independent, the numerator (divided by σ) having a standard normal distribution, and the denominator (divided by σ) being the square root of a statistic with a chi-square distribution divided by the number of degrees of freedom. The fact that the ratios have the t-distribution follows as in the proof of Theorem 8-2; we shall give the details only in the most complicated case, which is the ratio (12-16).

As in the proof of Theorem 12-4 we assume that $\alpha = \beta = 0$, and that $\sigma = 1$. The ratio (12-16) becomes

$$\frac{(\bar{x} + bt)\left(\frac{1}{n} + \frac{t^2}{\|\mathbf{t}\|^2}\right)^{-1/2}}{s} = \frac{\left(\frac{f_1(\mathbf{x})}{\sqrt{n}} + \frac{tf_2(\mathbf{x})}{\|\mathbf{t}\|}\right)\left(\frac{1}{n} + \frac{t^2}{\|\mathbf{t}\|^2}\right)^{-1/2}}{\left(\frac{f_3(\mathbf{x})}{n - 2}\right)^{1/2}},$$

where f_i, $i = 1, 2, 3$ are defined in the proof of Theorem 12-4. As in that proof, the distribution is unchanged when the rotated functions f_i^* are substituted for the original functions f_i: the distribution of the ratio is the same as that of

$$\frac{\left(\frac{x_1}{\sqrt{n}} + \frac{tx_2}{\|\mathbf{t}\|}\right)\left(\frac{1}{n} + \frac{t^2}{\|\mathbf{t}\|^2}\right)^{-1/2}}{\left(\frac{\sum_{i=3}^{n} x_i^2}{n - 2}\right)^{1/2}}.$$

The numerator and denominator are independent because they depend on nonoverlapping sets of coordinates (Theorem 5-2). The numerator is the linear combination

$$x_1 \frac{1}{\sqrt{n}\left(\frac{1}{n} + \frac{t^2}{\|\mathbf{t}\|^2}\right)^{1/2}} + x_2 \frac{\frac{t}{\|\mathbf{t}\|}}{\left(\frac{1}{n} + \frac{t^2}{\|\mathbf{t}\|^2}\right)^{1/2}};$$

thus, by Theorem 11-5, it has a normal distribution with mean 0 and standard deviation

$$\left[\frac{1}{n\left(\frac{1}{n} + \frac{t^2}{\|\mathbf{t}\|^2}\right)} + \frac{\frac{t^2}{\|\mathbf{t}\|^2}}{\frac{1}{n} + \frac{t^2}{\|\mathbf{t}\|^2}}\right]^{1/2} = 1$$

The denominator—with $n - 2$ in place of $n - 1$—has the same distribution as the denominator in Example 5-4; thus, as in the proof of Theorem 8-2, the ratio has the t-distribution with $n - 2$ degrees of freedom. ■

Here are some applications of Theorem 12-5 to confidence intervals.

(i) With probability $1 - \epsilon$, the interval

$$\bar{x} \pm \frac{t_{n-2,\epsilon/2}s}{\sqrt{n}} \tag{12-17}$$

contains the value of the unknown parameter α (Exercise 5); here $0 < \epsilon < 1$ and $t_{n-2,\epsilon/2}$ is the upper $\epsilon/2$ percentile of the t-distribution with $n - 2$ degrees of freedom.

(ii) With probability $1 - \epsilon$, the interval

$$b \pm t_{n-2,\epsilon/2} \frac{s}{\|\mathbf{t}\|} \tag{12-18}$$

contains the value of the unknown parameter β (Exercise 6).

(iii) For each t and with probability $1 - \epsilon$, the interval

$$\bar{x} + bt \pm t_{n-2,\epsilon/2}s\left(\frac{1}{n} + \frac{t^2}{\|\mathbf{t}\|^2}\right)^{1/2} \tag{12-19}$$

contains the ordinate $\alpha + \beta t$ on the regression line with unknown coefficients α and β (Exercise 7).

Example 12-1. Let us apply this analysis to the following data. A Sample of seven observations is taken from normal populations with means (12-1) for $n = 7$. Suppose that the coordinates of the Sample point corresponding to the various values of t are

t	-3	-2	-1	0	1	2	3
x	4	2	3	2	2	1	0

These are plotted in Fig. 12-4.

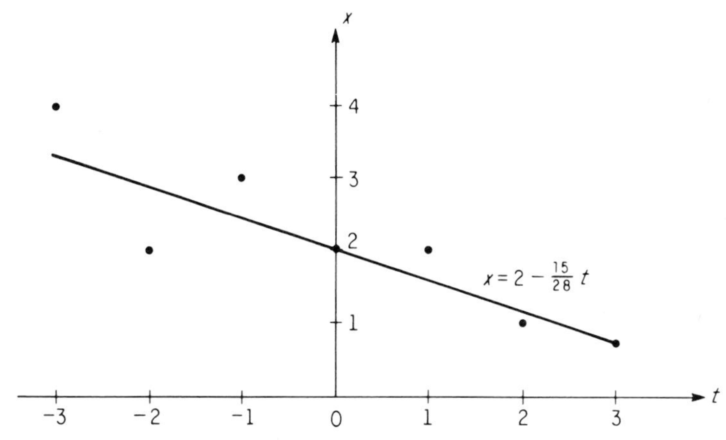

Fig. 12-4. Plotted points and least squares line.

We compute $\bar{x}$, b, and $\Sigma(x_i - \bar{x} - bt_i)^2$:

x	t	xt	t^2	x^2
4	-3	-12	9	16
2	-2	-4	4	4
3	-1	-3	1	9
2	0	0	0	4
2	1	2	1	4
1	2	2	4	1
0	3	0	9	0
$\Sigma x_i = 14$		$\Sigma xt = -15$	$\Sigma t^2 = 28$	$\Sigma x^2 = 38$
$\bar{x} = 2$		$b = -15/28.$		

The least-squares line has slope $-15/28$ and x-intercept 2; it is drawn in Fig. 12-4. The sum of the squares of the vertical deviations is calculated by means of the decomposition (12-5):

$$\begin{aligned}\Sigma(x_i - \bar{x} - bt_i)^2 &= \Sigma x_i^2 - n\bar{x}^2 - b^2\Sigma t_i^2 \\ &= 38 - 7(4) - (15/28)^2(28) \\ &= 38 - 28 - 225/28 = 2;\end{aligned}$$

thus $s^2 = .4$. For $\epsilon = .05$, the confidence interval for α of coefficient $1 - \epsilon = .95$ is

$$\bar{x} \pm 2.571\sqrt{.4/7} = 2 \pm .435$$

because $t_{5,.025} = 2.571$. The confidence interval for β of coefficient .95 is

$$b \pm 2.571\sqrt{.4/28} = -15/28 \pm .215.$$

The confidence interval for the point $2 - 15t/28$ on the regression line is

$$2 - 15t/28 \pm (2.571)\sqrt{.4}(1/7 + t^2/28)^{1/2}.$$

Note that the length of the confidence interval is smallest at $t = 0$ and increases with t^2. The two curves defining the confidence interval for the various values of t have the shapes indicated in Fig. 12-5. ∎

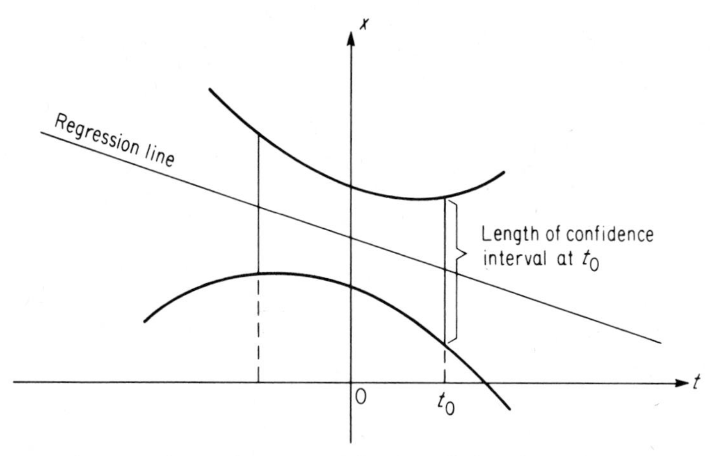

Fig. 12-5. Confidence intervals of fixed coefficient for various points on regression line.

EXERCISES

1. Show that the distribution of $\Sigma(x_i - \bar{x} - bt_i)^2$ is independent of the particular values α and β.

2. Show that $\sum_{i=3}^{n} x_i^2/(n - 2)$ is an unbiased Estimator of σ^2 when $\alpha = \beta = 0$, and that its variance is $2\sigma^4/(n - 2)$. (*Hint:* Exercises 1 and 2, Sec. 7-3.)

3. Show that the proof of Theorem 12-4 in the case $\alpha = \beta = 0$, $\sigma = 1$ implies the general case.

4. Following the proof of Theorem 12-5 for the ratio (12-16), derive the distributions of the simpler ratios (12-14) and (12-15).

5. Using Theorem 12-5, and the method of proof of Theorem 8-2, show that (12-17) is a confidence interval for α of confidence coefficient $1 - \epsilon$.

6. Repeat Exercise 5 for the confidence interval (12-18) for β.

7. Repeat Exercise 5 for the confidence interval (12-19) for $\alpha + \beta t$.

8. Assume that the x-values in the data of Exercise 1a, Sec. 12-2, are observations from normal populations of the linear regression model. Find confidence intervals of coefficient .95 for α, β, and $\alpha + \beta t$.

9. Repeat Exercise 8 for the data in Exercise 1b, Sec. 12-2.

10. Consider testing the null hypothesis $\alpha = \alpha_0$ against the two-sided composite alternative $\alpha \neq \alpha_0$ when σ is unknown. Show that the test with the rejection region $\{\mathbf{x}: \sqrt{n}\,|\,\bar{x} - \alpha_0\,| > st_{n-2,\epsilon/2}\}$ is of level of significance ϵ for all $\sigma > 0$.

11. For testing the null hypothesis $\beta = \beta_0$ against the two-sided composite alternative $\beta \neq \beta_0$, show that the test with the rejection region $\{\mathbf{x}: \|\mathbf{t}\| \cdot |\,b - \beta_0\,| > st_{n-2,\epsilon/2}\}$ is of size ϵ for all $\sigma > 0$.

12. (Multiple linear regression) Consider a Sample of n observations from the populations with means $\mu_1, \ldots, \mu_n$, respectively, and common standard deviation σ. Let $\mathbf{t}_1, \ldots, \mathbf{t}_k$ be k points in R^n, such that t_{ij} represents the jth coordinate of $\mathbf{t}_i$, $j = 1, \ldots, n$, $i = 1, \ldots, k$. Suppose that the lines from $\mathbf{0}$ to $\mathbf{t}_i$, $i = 1, \ldots, k$ are all perpendicular, and that we want to estimate the k linear functions

$$\sum_{j=1}^{n} t_{ij}\mu_j, \qquad i = 1, \ldots, k.$$

Why is $\sum_j t_{ij}x_j$ the efficient Estimator of $\sum_j t_{ij}\mu_j$, $i = 1, \ldots, k$? Show that these Estimators are independent and have normal distributions.

13. Assume $1 \leq k < n$. Define

$$b_i = \frac{\sum_{j=1}^{n} t_{ij}x_j}{\sum_{j=1}^{n} t_{ij}^2}, \qquad i = 1, \ldots, k.$$

Prove: $$\sum_{j=1}^{n}\left(x_j - \sum_{i=1}^{k} b_i t_{ij}\right)^2 = \sum_{j=1}^{n} x_j^2 - \sum_{i=1}^{k} b_i^2\left(\sum_{j=1}^{n} t_{ij}\right)^2.$$

(*Hint:* Expand the square on the left-hand side, sum over j, and apply the definition of b_i and the perpendicularity of the lines from $\mathbf{0}$ to $\mathbf{t}_i$, $i = 1, \ldots, k$.)

14. Define

$$s^2 = \frac{\sum_{j=1}^{n}\left(x_j - \sum_{i=1}^{k} b_i t_{ij}\right)^2}{n - k}.$$

Show that the $k + 1$ statistics

$$\frac{\sum_{j=1}^{n} t_{ij}(x_j - \mu_j)}{\sigma \|\mathbf{t}\|}, \qquad i = 1, \ldots, k, \quad (n - k)s^2/\sigma^2$$

are mutually independent, and that the last has the chi-square distribution with $n - k$ degrees of freedom. (*Hint:* Apply the rotation carrying $\mathbf{t}_1, \ldots, \mathbf{t}_k$ onto the first k coordinate axes, respectively, and extend the method of proof of Theorem 12-4 using the sum of squares decomposition in Exercise 13.)

15. State and prove the appropriate generalization of Theorem 12-5.

12-4. PREDICTION

One of the main reasons for recording and analyzing time series is to use the data for predicting future observations. In economic time series such as price records we are given the values of the observations at times $t_1, \ldots, t_n$, and wish to predict the value of an observation at time $t > t_1, \ldots, t_n$. In the records of agricultural experiments with fertilizers (Sec. 12-1) we are given the outputs at levels $t_1, \ldots, t_n$, and want to predict the output for any level $t \neq t_1, \ldots, t_n$, where t is not necessarily larger than the other values t_j. The main result of this section is the determination of the "best" prediction procedure. It happens that the best procedure is to extrapolate the estimated regression line, and predict the value at t to be $\bar{x} + bt$. A "prediction interval," analogous to the confidence interval, is also constructed.

The problem of prediction is analyzed in the setting of an $(n + 1)$-dimensional normal Sample space. We have $n + 1$ normal populations with means

$$\alpha + \beta t_1, \ldots, \alpha + \beta t_n, \quad \alpha + \beta t$$

and common standard deviation σ; as before, we assume the condition (12-3) that the sum of the first n t's is equal to 0. A Sample of $n + 1$ observations is taken from the respective populations. The first n coordinates (from the first n populations) are actually observed but not the last one, X_{n+1}. A Predictor is a function on the Sample space R^{n+1} which depends on a point $\mathbf{x}$ only through its first n coordinates—not the $n + 1$st—and which estimates or predicts the value of the last coordinate; in other words, it is a function $f(\mathbf{x})$, $\mathbf{x} \in R^{n+1}$, which actually depends only on $x_1, \ldots, x_n$, and which, for each such set of n coordinates, furnishes an estimate of x_{n+1}.

The framework of prediction theory is very similar to that of estimation (Chapter 7). The Predictor $f(\mathbf{x})$ is assumed to be square-integrable—that is, it is continuous and satisfies

$$\int_{R^{n+1}} \cdots \int |f(x)|^2 L \cdot \phi \, d\mathbf{x} < \infty$$

for all α, β, and σ, where L is the likelihood function (12-13) corresponding to the first n observations, and ϕ is the density function of the normal distribution with mean $\alpha + \beta t$ and standard deviation σ:

$$\phi = \frac{1}{\sigma} \phi \left(\frac{x_{n+1} - \alpha - \beta t}{\sigma} \right).$$

The Predictor is said to be unbiased if

$$\int_{R^{n+1}} \cdots \int f(\mathbf{x}) L\phi \, d\mathbf{x} = \alpha + \beta t \tag{12-20}$$

for all α and β (σ fixed). As in the definition of unbiasedness of Estimators this condition means (when f is a statistic) that the center of gravity of the distribution of the Predictor is equal to the mean of the predicted observation. The analogue of the variance of an Estimator is the "expected squared error of prediction":

$$\int_{R^{n+1}} \cdots \int (f(\mathbf{x}) - x_{n+1})^2 L\phi \, d\mathbf{x}, \tag{12-21}$$

the weighted average of the squared difference between the Predictor and the value of the predicted observation. One unbiased Predictor is considered better than another if its expected squared error is smaller than that of the latter. An unbiased Predictor is called a "best" Predictor if, among all unbiased Predictors, it minimizes the expected squared error (12-21).

The statistical problem of finding the best Predictor is equivalent to the following mathematical problem: Among all square-integrable functions $f(\mathbf{x})$ which depend on $\mathbf{x}$ only through the first n coordinates, and which satisfy (12-20) for all α and β, find the one, if it exists, minimizing the integral (12-21); f may depend on σ and t. It happens, as in the case of estimation of μ (Chapter 7), that the minimizing function does not depend on σ, and so minimizes (12-21) for all $\sigma > 0$.

The first step in the solution of the prediction problem is the decomposition of the "squared error" (12-21) into a component "due to the variance" and a component "due to prediction."

LEMMA 12-1

An unbiased Predictor of the observation with mean $\alpha + \beta t$ is an unbiased Estimator of that mean; and the expected squared error of prediction is equal to σ^2 plus the variance of the Predictor as an Estimator of $\alpha + \beta t$:

$$\int_{R^{n+1}} \cdots \int (f(\mathbf{x}) - x_{n+1})^2 L\phi \, d\mathbf{x}$$

$$= \sigma^2 + \int_{R^n} \cdots \int (f(\mathbf{x}) - \alpha - \beta t)^2 L \, d\mathbf{x} \qquad (12\text{-}22)$$

($f(\mathbf{x})$ is formally a function of $n + 1$ variables on the left-hand side, and a function of n variables on the right-hand side.)

Proof: Since $f(\mathbf{x})$ does not depend on x_{n+1}, the latter variable may be integrated out of the multiple integral on the left-hand side of (12-20), and so it becomes

$$\int_{R^n} \cdots \int f(\mathbf{x}) L \, d\mathbf{x} = \alpha + \beta t; \qquad (12\text{-}23)$$

thus f is an unbiased Estimator of $\alpha + \beta t$.

Now we prove (12-22). Write

$$f(\mathbf{x}) - x_{n+1} = [f(x) - \alpha - \beta t] - [x_{n+1} - \alpha - \beta t],$$

insert this in the integrand in (12-21), expand the square and integrate each of the resulting three terms. The first term is

$$\int_{R^{n+1}} \cdots \int (f(\mathbf{x}) - \alpha - \beta t)^2 L\phi \, d\mathbf{x}.$$

Since f does not depend on x_{n+1}, the argument leading to (12-23) implies that the integral may be reduced to one over R^n:

$$\int_{R^n} \cdots \int (f(\mathbf{x}) - \alpha - \beta t)^2 L \, d\mathbf{x}.$$

This accounts for the second term on the right-hand side of (12-22).

The second term arising from the decomposition of the left-hand side of (12-21) is

$$-2 \int_{R^{n+1}} \cdots \int (f(\mathbf{x}) - \alpha - \beta t)(x_{n+1} - \alpha - \beta t) \, L\phi \, d\mathbf{x}.$$

Integrate first over x_{n+1}: since f does not depend on this variable, the inner integral is

$$\int_{-\infty}^{\infty} (x - \alpha - \beta t) \frac{1}{\sigma} \phi \left(\frac{x - \alpha - \beta t}{\sigma} \right) dx = \sigma \int_{-\infty}^{\infty} x\phi(x) \, dx = 0,$$

thus the second term vanishes.

The third term arising on the left-hand side of (12-21) is

$$\int_{R^{n+1}} \cdots \int (x_{n+1} - \alpha - \beta t)^2 L \frac{1}{\sigma} \phi \left(\frac{x_{n+1} - \alpha - \beta t}{\sigma} \right) d\mathbf{x}.$$

Since L does not depend on x_{n+1}, we integrate first over that variable:

$$\int_{-\infty}^{\infty} (x - \alpha - \beta t)^2 \frac{1}{\sigma} \phi \left(\frac{x - \alpha - \beta t}{\sigma} \right) dx = \sigma^2 \int_{-\infty}^{\infty} x^2 \phi(x) \, dx = \sigma^2$$

[see (2-13)]. The integral of L over the first n coordinates is equal to 1; therefore, the third term is equal to σ^2. This completes the proof of (12-22). ■

THEOREM 12-6

The best Predictor of X_{n+1}, the observation from the population with mean $\alpha + \beta t$, is $\bar{x} + bt$, which is the ordinate on the estimated regression line corresponding to the abscissa t.

Proof: By Lemma 12-1 an unbiased Predictor of x_{n+1} is an unbiased Estimator of $\alpha + \beta t$; the converse is also true. A function f minimizing the variance of unbiased estimation [the integral on the right-hand side of (12-22)] necessarily minimizes the expected squared error of prediction (on the left-hand side). Such a function is $\bar{x} + bt$; indeed, $\alpha + \beta t$ may be expressed as a linear combination of the means (12-1):

$$\alpha + \beta t = \sum_{j=1}^{n} \left(\frac{1}{n} + \frac{tt_j}{\sum_{j=1}^{n} t_j^2} \right) \mu_j,$$

It follows, as in the proof of Theorem 12-3, that the conclusion of Theorem 11-5 holds even though the means are linearly related, so that $\bar{x} + bt$ is the efficient Estimator of $\alpha + \beta t$. (The details of the extension of the proof of Theorem 12-3 are left to the reader (Exercises 6, 7).)

The uniqueness of the best Predictor follows from the uniqueness of the efficient Estimator. ■

As in the theory of estimation it is useful to have a measure of "how good" the best Predictor is. Now we construct a prediction interval, analogous to the confidence interval for the value of an unknown parameter.

THEOREM 12-7

The interval

$$(\bar{x} + bt) \pm st_{n-2,\epsilon/2} \left(\frac{1}{n} + \frac{t^2}{\|\mathbf{t}\|^2} + 1 \right)^{1/2} \tag{12-24}$$

contains, with probability $1 - \epsilon$, the value of the predicted observation X_{n+1}; in other words, the probability that **X** *falls in the set*

$$\left\{ \mathbf{x} : \mathbf{x} \in R^{n+1},\ |x_{n+1} - \bar{x} - bt| \leq st_{n-2,\epsilon/2} \left(\frac{1}{n} + \frac{t^2}{\|\mathbf{t}\|^2} + 1 \right)^{1/2} \right\}$$

is equal to $1 - \epsilon$. *The interval (12-24) is called a prediction interval.*

Proof: We shall prove that the statistic on the $(n + 1)$-dimensional Sample space

$$\left(\frac{\bar{x} + bt - x_{n+1}}{s}\right)\left(\frac{1}{n} + \frac{t^2}{\|\mathbf{t}\|^2} + 1\right)^{-1/2} \tag{12-25}$$

has the t-distribution with $n - 2$ degrees of freedom. As in previous proofs we assume with no loss of generality that $\alpha = \beta = 0$, $\sigma = 1$, so that we are considering $n + 1$ observations from a standard normal population. The joint distribution of the four statistics

$$\sqrt{n}\bar{x}, \quad b\|\mathbf{t}\|, \quad (n-2)s^2, \quad x_{n+1} \tag{12-26}$$

is identical with that of the corresponding four statistics

$$x_1, \quad x_2, \quad \sum_{i=3}^{n} x_i^2, \quad x_{n+1}; \tag{12-27}$$

the proof is almost the same as of Theorem 12-4 (Exercise 3). It follows that the statistic (12-25) has the same distribution as

$$\frac{\left(\dfrac{x_1}{\sqrt{n}} + \dfrac{x_2 t}{\|\mathbf{t}\|} - x_{n+1}\right)\left(\dfrac{1}{n} + \dfrac{t^2}{\|\mathbf{t}\|^2} + 1\right)^{-1/2}}{\left(\dfrac{\sum_{i=3}^{n} x_i^2}{n-2}\right)^{1/2}} \tag{12-28}$$

The proof that this has the t-distribution with $n - 2$ degrees of freedom is similar to that of Theorem 12-5 (Exercise 4).

The probability that the value X_{n+1} of the predicted observation falls within

$$st_{n-2,\epsilon/2}\left(\frac{1}{n} + \frac{t^2}{\|\mathbf{t}\|^2} + 1\right)^{1/2}$$

units of the Predictor $\bar{x} + bt$ is the same as the probability that the statistic (12-25) falls in the interval from $-t_{n-2,\epsilon/2}$ to $t_{n-2,\epsilon/2}$. Since the statistic has the t-distribution the probability is equal to $1 - \epsilon$. ■

Note that the length of the prediction interval is

$$2st_{n-2,\epsilon/2}\left(\frac{1}{n} + \frac{t^2}{\|\mathbf{t}\|^2} + 1\right)^{1/2}.$$

It is minimized at $t = 0$ and increases with t^2; this means that the Predictor is most reliable at $t = 0$, the average value of the t's, and that its realiability decreases as t moves away from 0. The prediction interval is valid only for a single predicted observation but not for a set of more than one: in order to predict x_{n+2} after x_{n+1} has been observed, one should recompute $\bar{x}$, b, and s for $n + 1$ observations.

Example 12-2. For a numerical illustration we refer to the data in Example 12-1. A prediction interval containing the predicted value with probability .95 is

$$2 - 15t/28 \pm (2.571)\sqrt{.4}(8/7 + t^2/28)^{1/2};$$

for example, for $t = 4$, the prediction interval is $-1/7 \pm 2.2$. ■

EXERCISES

1. Find the prediction interval containing, with probability .95, the value of X_8 for the data in Exercise 1a, Sec. 12-2, at the points $t = .5, 3$. (Use the computations from Exercise 8, Sec. 12-3.)
2. Repeat Exercise 1 for the data in Exercise 1b of Sec. 12-2 at the points $t = -.5, -4$.
3. Show that the four statistics (12-26) have the same joint distribution as (12-27).
4. Show that the ratio (12-28) has the t-distribution with $n - 2$ degrees of freedom.
5. (Prediction in multiple regression) Let $X_1, \ldots, X_n, X_{n+1}$ be a Sample of $n + 1$ observations from the populations with means

$$\mu_j = \sum_{i=1}^{k} \beta_i t_{ij}, \quad j = 1, \ldots, n, \qquad \mu_{n+1} = \sum_{i=1}^{k} \beta_i \tau_i,$$

respectively, where the numbers (t_{ij}) satisfy the conditions in Exercise 12, Sec. 12-3, and where $\beta_1, \ldots, \beta_k$ are unknown parameters. Note that

$$\beta_i = \frac{\sum_{j=1}^{n} t_{ij}\mu_j}{\sum_{j=1}^{n} t_{ij}^2}, \qquad i = 1, \ldots, k.$$

Generalize Theorems 12-6 and 12-7: Show that the best Predictor of X_{n+1} based on the first n observations is

$$\sum_{i=1}^{k} b_i \tau_i,$$

where b_i is the Estimator of β_i:

$$b_i = \frac{\sum_{j=1}^{n} t_{ij}x_j}{\sum_{j=1}^{n} t_{ij}^2}, \qquad i = 1, \ldots, k.$$

Construct a prediction interval for X_{n+1} based on the t-distribution with $n - k$ degrees of freedom.

6. Prove that $\bar{x} + bt$ is the efficient Estimator of $\alpha + \beta t$. The proof is as follows: Apply the rotation in Theorem 12-2 (as in the proof of Theorem 12-3), and use the special case of Theorem 11-5 with $n = 2$ and $t_1 = 1$, $t_2 = t$.

12-5. LINEAR REGRESSION WITHOUT THE ASSUMPTION OF NORMALITY

Suppose that the Sample of n observations is from n populations with densities $h_1(x), \ldots, h_n(x)$, respectively. The only assumptions are that the means are of the form (12-1), that (12-3) is satisfied, and that the populations have a common variance σ^2. As before, α, β, and σ^2 are unknown. Much of the theory of estimation developed above in the normal case can be reformulated and applied to the more general setting.

Theorem 12-3 states that the efficient Estimators of α and β in the normal case are particular linear functions of the observations, namely, those furnished by the least squares solution. This suggests consideration of the class of *linear* Estimators in the general case, and the search for a "best" linear unbiased Estimator. (Cf. Sec. 11-6.) We will say that f is a best linear unbiased Estimator if, among all linear, unbiased Estimators, it has minimum variance for all parameter values. A famous result known as the Gauss-Markov theorem asserts that the efficient Estimators of α and β in the normal case are identical with the best linear unbiased Estimators in the general case; this will be proved below. It is also true that s^2, defined by Eq. 12-13, is an unbiased Estimator of σ^2 for any densities $h_1, \ldots, h_n$.

THEOREM 12-8 (Gauss-Markov Theorem)
The functions $\bar{x}$ and b are best linear unbiased Estimators of α and β, respectively; and s^2 is an unbiased Estimator of σ^2.

Proof: Let $\mathbf{c}$ be a point in R^n whose coordinates define a linear Estimator $c_1x_1 + \cdots + c_nx_n$. The latter is an unbiased Estimator of α if

$$\sum_{i=1}^{n} c_i(\alpha + \beta t_i) = \alpha$$

for all α and β (cf. Eq. 11-47). This condition holds if and only if

$$\sum_{i=1}^{n} c_i = 1, \quad \sum_{i=1}^{n} c_i t_i = 0. \tag{12-29}$$

The variance of this linear Estimator is

$$\sigma^2(c_1^2 + \cdots + c_n^2) = \sigma^2 \| \mathbf{c} \|^2 \tag{12-30}$$

(Sec. 11-6, Exercise 2). A best linear Estimator is one which, among all those satisfying (12-29), minimizes (12-30).

We shall prove:

If $\mathbf{c}$ *satisfies* (12-29), *then* $\| \mathbf{c} \|^2 \geq 1/n$.

As a matter of fact, if just the first condition in (12-29) holds, then the conclusion follows. This is a consequence of the fundamental sum of squares decomposition (4-12) applied to $\mathbf{c}$ in place of $\mathbf{x}$:

$$\sum_{i=1}^{n} c_i^2 = \sum_{i=1}^{n} (c_i - \bar{c})^2 + n\bar{c}^2.$$

If $\mathbf{c}$ satisfies the first part of (12-29), then $\bar{c} = 1/n$, and the identity above becomes

$$\sum_{i=1}^{n} c_i^2 = \sum_{i=1}^{n} \left(c_i - \frac{1}{n}\right)^2 + \frac{1}{n},$$

and so $\| \mathbf{c} \|^2 > 1/n$.

It is simple to see that if the point **c** has equal coordinates $1/n$, then condition (12-29) is satisfied, and $\|\mathbf{c}\|^2$ assumes the minimum value $1/n$. It follows that the linear function $\bar{x}$ of the coordinates of **x** is a best linear unbiased Estimator of α.

Next we shall prove that

$$b = \frac{\sum_{i=1}^{n} x_i t_i}{\sum_{i=1}^{n} t_1^2}$$

is a best linear unbiased Estimator of β. Let $d_1 x_1 + \cdots + d_n x_n$ be an arbitrary linear Estimator of β. The condition for unbiasedness is

$$\sum_{i=1}^{n} d_i(\alpha + \beta t_i) = \beta$$

for all α and β; this holds if and only if

$$\sum_{i=1}^{n} d_i = 0, \quad \sum_{i=1}^{n} d_i t_i = 1. \tag{12-31}$$

The variance of this Estimator is

$$\sigma^2 \|\mathbf{d}\|^2. \tag{12-32}$$

A best linear Estimator is one minimizing (12-32) among those satisfying (12-31).

We shall prove:

If **d** *satisfies* (12-31), *then* $\|\mathbf{d}\|^2 > 1/\|\mathbf{t}\|^2$.

To verify this, apply the decomposition (12-5) with **d** in place of **x** and

$$\frac{\sum_{i=1}^{n} d_i t_i}{\sum_{i=1}^{n} t_i^2}$$

in place of b. Under the condition (12-31) we have

$$\bar{d} = 0, \quad b = 1/\|\mathbf{t}\|^2.$$

It follows from (12-5) that

$$\|\mathbf{d}\|^2 = \sum_{i=1}^{n} \left(d_i - \frac{t_i}{\|\mathbf{t}\|^2}\right)^2 + \frac{1}{\|\mathbf{t}\|^2} \geq \frac{1}{\|\mathbf{t}\|^2}.$$

It is seen that if the point $\mathbf{d}$ is chosen with coordinates $t_i/\|\mathbf{t}\|^2$, then conditions (12-31) are satisfied, and the variance (12-32) is, in fact, equal to the minimum variance $1/\|\mathbf{t}\|^2$. It follows that b is the best linear unbiased Estimator of β.

In order to prove that s^2 is an unbiased Estimator of σ^2 we recall the identity (12-9):

$$\sum_{i=1}^{n} (x_i - \alpha - \beta t_i)^2 = (n-2)s^2 + (b-\beta)^2\|\mathbf{t}\|^2 + n(\bar{x}-\alpha)^2.$$

Multiply every member by the likelihood function and integrate over R^n. Recalling the definition of the variance, and the values of the variances of b and $\bar{x}$ obtained above, we get

$$n\sigma^2 = (n-2)\int_{R^n}\cdots\int s^2 L(\mathbf{x})\,d\mathbf{x} + 2\sigma^2 \qquad (12\text{-}33)$$

(Exercise 1). The equation for unbiasedness follows by solving (12-33) for σ^2. The proof of the theorem is complete. ■

The distribution theory of the Estimators is not valid for non-normal populations, so that the test procedures for α and β based on the t-distribution may not be used. However, if the number of observations is large, then, for any population satisfying the conditions of the central limit theorem, the statistics $\bar{x}$ and b have approximately normal distributions. It even can be shown that their joint distribution is approximately a product of normal distributions so that they are "approximately independent." As in the case of a Sample from a single population (Sec. 8-4) tests on α and β involving the t-distribution are "robust" if n is large.

Prediction theory can also be extended to the case of general populations. Attention is restricted to the class of *linear* unbiased Predictors. By analogy to estimation, it can be shown that the best unbiased Predictor in the normal case is, in fact, the best linear unbiased Predictor in the general case (Exercises 2–5).

EXERCISES

1. Derive (12-33) by integration of the equation preceding it.

2. Let $c_1x_1 + \cdots + c_nx_n$ be a linear Predictor of X_{n+1}, where the last coordinate has mean $\alpha + \beta t$. Prove that the Predictor is unbiased if

$$\sum_{i=1}^{n} c_i = 1, \quad \sum_{i=1}^{n} c_i t_i = t.$$

3. Show that the expected squared error of the linear Predictor is $\sigma^2 \| \mathbf{c} \|^2 + \sigma^2$.

4. Show that the expected squared error of a linear unbiased Predictor is at least equal to

$$\sigma^2 \left(\frac{1}{n} + \frac{t^2}{\| \mathbf{t} \|^2} + 1 \right).$$

The proof is contained in these steps:
(a) If $\Sigma c_i = 1$ and $\Sigma c_i t_i = t$, then

$$\| \mathbf{c} \|^2 \geq \frac{1}{n} + \frac{t^2}{\| \mathbf{t} \|^2}.$$

(*Hint:* Apply (12-5) with **c** in place of **x**.)
(b) Apply the results of Exercises 2 and 3.

5. Show that the linear Predictor $\bar{x} + bt$ is unbiased, and has the expected squared error given in Exercise 4.

12-6. NUMERICAL EXAMPLE OF NORMAL LINEAR REGRESSION

Here is a numerical example of the estimation of the regression coefficients α and β in sampling from normal populations. Twenty-five Samples of seven normal deviates are taken from Table VI, Appendix (the last seven numbers in each column of the first three pages); this corresponds to the very special normal linear regression model where α and β are known to be 0 and $\sigma = 1$. This will be later extended to general α and β. We put

$$\mu_i = \alpha + \beta i, \qquad i = -3, -2, -1, 0, 1, 2, 3.$$

The Estimators of α and β are

$$\bar{x} = \frac{1}{7}(x_{-3} + \cdots + x_3),$$

$$b = \sum_{i=-3}^{3} i x_i \Big/ \sum_{i=-3}^{3} i^2 = \sum_{i=-3}^{3} i x_i / 28,$$

respectively. The twenty-five values of $\bar{x}$ and b, the Estimators of $\alpha = 0$ and $\beta = 0$, are given in Table 12-3.

TABLE 12-3

Values of $\bar{x}$ and b for 25 Samples of Seven Observations When $\alpha = \beta = 0$.

Sample	$\bar{x}$	b	Sample	$\bar{x}$	b
1	.059	.088	14	−.084	.097
2	−.092	−.080	15	−.432	.160
3	−.797	.026	16	−.591	.132
4	.556	.011	17	−.072	−.489
5	−.065	−.113	18	.557	.251
6	.022	−.002	19	.340	−.017
7	−.101	.111	20	.305	−.029
8	−.510	.423	21	−.753	−.134
9	−.510	.027	22	.114	.114
10	.136	.289	23	−.087	−.191
11	.358	.177	24	.159	−.310
12	−.124	−.065	25	−.435	.064
13	−.877	.111			

Suppose that α and β are arbitrary. If x_i is a normal deviate in Table VI, then the linear function of x_i, $x_i + \alpha + \beta i$, represents a random variable from a normal population with mean $\alpha + \beta i$ and standard deviation 1, for $i = -3, \ldots, +3$ (cf. Example 4-1). The efficient Estimators of α and β are

$$\frac{1}{7} \sum_{i=-3}^{3} (x_i + \alpha + \beta i) = \bar{x} + \alpha$$

and

$$\frac{1}{28} \sum_{i=-3}^{3} i(x_i + \alpha + \beta i) = b + \beta,$$

respectively; thus, *the values of the Estimators differ from the corresponding parameters by the values of $\bar{x}$ and b, the Estimators in the case $\alpha = \beta = 0$.* In this way the observed values of $\bar{x}$ and b in Table 12-3 can be interpreted not only as examples of Estimators in the special case but also, by addition of the constants α and β, as Estimators in the general case; for example, in Sample 1,

$$.059 + \alpha \quad \textit{and} \quad .088 + \beta$$

are the observed values of the Estimators of α and β, respectively.

Note that, in the example in Table 12-3, b seems to be more reliable than $\bar{x}$ as an Estimator of 0. This is explained by the fact that the variance of $\bar{x}$ is four times that of b.

EXERCISE (Optional)

1. Construct a table like Table 12-3 for the seven observations at the *top* of the columns on the first three pages of Table VI, Appendix. Estimate σ^2 using the same data and the Estimator (12-13) (cf. Exercise 2, Sec. 12-2).

BIBLIOGRAPHY

An introductory text on statistics with an account of linear regression is Alexander M. Mood and Franklin A. Graybill, *Introduction to the Theory of Statistics*, 2nd ed. (New York: McGraw-Hill, 1963).

Many examples and techniques are given in G. Udny Yule and M. G. Kendall, *An Introduction to the Theory of Statistics*, 14th ed. (New York: Hafner, 1950); George W. Snedecor and William G. Cochran, *Statistical Methods*, 6th ed. (Ames, Iowa: Iowa State U. P., 1967).

CHAPTER 13

One-Way Analysis of Variance

13-1. EMPIRICAL BACKGROUND, REPLICATION, SUM-OF-SQUARES DECOMPOSITION

The analysis of variance is a method of comparing the means of several normal populations when the standard deviation is unknown. Consider k normal populations with means $\mu_1, \ldots, \mu_k$ and common standard deviation σ; and let us test the null hypothesis that the means have a common though unknown value μ against the alternative hypothesis that they do not necessarily all have the same value. The null and alternative hypotheses are composite. This is different from the homogeneity testing problem in Sec. 11-5 because there the null hypothesis was simple: the means were assumed equal to 0 and the standard deviation was known.

We give several examples which illustrate the practical importance of such a problem.

We first refer to the biological experiment mentioned at the beginning of Sec. 11-1. The experimenter wants to know whether a nutrient has an effect on the growth of animals. He takes a homogeneous group of nk animals (n and k are positive integers) and divides it into k subgroups or classes of n animals. Those in a particular class are given a particular amount of the nutrient: those in the class i are given the amount t_i, $i = 1, \ldots, k$. The weight gain of each animal at the end of the period of the experiment is observed. The scientific hypothesis being tested is that the nutrient affects growth. This is translated into a statistical hypothesis testing problem. Let X_{ij} be the weight gain of the jth animal in the ith class, $i = 1, \ldots, k$, $j = 1, \ldots, n$. These are conceived as a Sample of nk observations, n from each of k identical populations with a common unknown standard deviation. The null hypothesis is that the means of the populations are equal, that is, the "mean growths" of the animals in the various classes are the same. The null hypothesis states that this nutrient does not affect growth. The alternative hypothesis is that the means of the populations are not all the same; this signifies that the nutrient does have an effect.

For a second example we refer to an industrial problem. An engineer wants to test the effect of a chemical additive on the strength of a building material. He takes a homogeneous group of nk blocks of material, divides it into k classes of n blocks, and puts an amount t_i of the chemical in the blocks in the ith class, $i = 1, \ldots, k$. The strength of each block is measured. This is usually done by measuring the amount of force that the block can take without breaking. Let X_{ij} be the strength of the jth block in the ith class. The null hypothesis is that the observed strengths X_{ij} are a Sample from a common population; this signifies that the additive does not affect the strength. The alternative hypothesis is that the observed strengths are a Sample from k populations with different means; this signifies that the chemical additive has some effect on the strength of the material.

In both of these examples we assume not only that the populations are normal, but also that they have a common standard deviation σ. The latter assumption is called *homoscedasticity*. It may be difficult to verify in applications.

The sampling theory differs from that in Sec. 11-1 because of the introduction of *replication*; this means that a Sample of *several* observations is taken from *each* of the various populations. Suppose we have k normal populations with means $\mu_1, \ldots, \mu_k$, respectively, and common standard deviation σ, and we take a Sample of n observations from each. The Sample space is obtained by a specialization of that in Sec. 11-1: the observations are considered to come from nk populations which fall into k groups of n identical populations with means μ_i, $i = 1, \ldots, k$. The Sample space in nk-dimensional. The coordinates of each point $\mathbf{x}$ in R^{nk} are labeled by a double set of indices: $\mathbf{x} = (x_{ij})$, where x_{ij} is the coordinate corresponding to the jth observation from the ith population, $i = 1, \ldots, k$, $j = 1, \ldots, n$. The Sample point is denoted $(X_{ij}, i = 1, \ldots, k, j = 1, \ldots, n)$. The likelihood function is the product of the likelihood functions for each of the k populations:

$$L = (2\pi\sigma^2)^{-nk/2} \exp\left[(-1/2\sigma^2) \sum_{i=1}^{k} \sum_{j=1}^{n} (x_{ij} - \mu_i)^2\right]. \qquad (13\text{-}1)$$

The problem of interest is testing the null hypothesis that the means are all equal,

$$\mu_1 = \cdots = \mu_k,$$

against the alternative that at least one of these is not equal to any other. It is convenient to state the null and alternative hypothesis in a slightly different way. Let μ be the average of $\mu_1, \ldots, \mu_k$; then

$$\mu_i = \mu + (\mu_i - \mu) = \mu + \tau_i,$$

where $\tau_i = \mu_i - \mu$. The null hypothesis may be put in the form

$$\tau_i = 0, \qquad \text{for all } i = 1, \ldots, k, \tag{13-2}$$

and the alternative in the form

$$\tau_i \neq 0, \qquad \text{for some } i = 1, \ldots, k. \tag{13-3}$$

The advantage of stating the hypotheses in these forms is that the τ's satisfy

$$\tau_1 + \cdots + \tau_k = 0, \tag{13-4}$$

so that some of the results of the linear regression model are applicable.

For the purpose of deriving the distributions of certain statistics we prove an extension of the decomposition of the sum of squares (4-12). Put

$$\bar{x}_i = \frac{1}{n} \sum_{j=1}^{n} x_{ij}, \qquad i = 1, \ldots, k;$$

$$\bar{x} = \frac{1}{k} \sum_{i=1}^{k} \bar{x}_i = \frac{1}{nk} \sum_{i,j} x_{ij};$$

we prove:

$$\sum_{i=1}^{k} \sum_{j=1}^{n} x_{ij}^2 = \sum_{i=1}^{k} \sum_{j=1}^{n} (x_{ij} - \bar{x}_i)^2 + n \sum_{i=1}^{k} (\bar{x}_i - \bar{x})^2 + nk\bar{x}^2. \tag{13-5}$$

For the proof apply (4-12) to $x_{ij}, j = 1, \ldots, n$ for each fixed i:

$$\sum_{j=1}^{n} x_{ij}^2 = \sum_{j=1}^{n} (x_{ij} - \bar{x}_i)^2 + n\bar{x}_i^2.$$

Sum over i:

$$\sum_{i=1}^{k} \sum_{j=1}^{n} x_{ij}^2 = \sum_{i=1}^{k} \sum_{j=1}^{n} (x_{ij} - \bar{x}_i)^2 + n \sum_{i=1}^{k} \bar{x}_i^2. \tag{13-6}$$

Apply (4-12) to the numbers $\bar{x}_i$, $i = 1, \ldots, k$, noting that $\bar{x}$ is their average:

$$\sum_{i=1}^{k} \bar{x}_i^2 = \sum_{i=1}^{k} (\bar{x}_i - \bar{x})^2 + k\bar{x}^2.$$

Substitute the right-hand side of this expression in the corresponding term of (13-6); this yields (13-5).

The first sum on the right-hand side of (13-5) is the sum of the squares of the deviations of the coordinates about the respective class averages; it is called the "sum of squares due to error" and is denoted by

$$\text{SSE} = \sum_{i=1}^{k} \sum_{j=1}^{n} (x_{ij} - \bar{x}_i)^2. \tag{13-7}$$

The second term on the right-hand side is n times the sum of the squares of the deviations of the class averages from the overall average; it is called the "sum of squares between classes" and is denoted

$$\text{SSB} = n \sum_{i=1}^{k} (\bar{x}_i - \bar{x})^2. \tag{13-8}$$

The third term on the right-hand side is simply a multiple of the square of the mean; it is called the "square due to the mean" and denoted

$$\text{SSM} = nk\bar{x}^2.$$

The identity (13-5) states that the sum of the squares of the coordinates is equal to the sums of squares due to error, between classes, and due to the mean, respectively.

EXERCISES

1. Prove that $\bar{x}_i$ is the efficient Estimator of μ_i, $i = 1, \ldots, k$.

2. Put $(n - 1)s_i^2 = \sum_{j=1}^{n} (x_{ij} - \bar{x}_i)^2$. Define the conditional expectation of a continuous function $f(\mathbf{x})$, $\mathbf{x} \in R^{nk}$ given $\sqrt{n}\bar{x}_i$, $(n-1)s_i^2$, $i = 1, \ldots, k$, as a spherical average of f over each of the k sets of n coordinates; more specifically, f is averaged over the first n coordinates, then over the second n coordinates, and so on. State and prove an appropriate extension of Theorem 6-1.

3. Let $\bar{f} = \bar{f}(\sqrt{n}\bar{x}_1, \ldots, \sqrt{n}\bar{x}_k, (n-1)s_1^2, \ldots, (n-1)s_k^2)$ be the conditional expectation in the previous exercise. We may consider $\bar{f}$ to be a function of $\sqrt{n}\bar{x}_i$ and $\sqrt{(n-1)}\, s_i$, $i = 1, \ldots, k$. Let $\bar{\bar{f}}$ be the spherical average of this function over the last k variables $\sqrt{n-1}\, s_i$, $i = 1, \ldots, k$. Show that $\bar{\bar{f}}$ depends on $\bar{x}_i$, $i = 1, \ldots, k$ and SSE.

4. Apply the results of Exercises 2 and 3 to show that $\bar{x}_1, \ldots, \bar{x}_k$, SSE, is a set of sufficient statistics. (*Hint:* See the discussion of sufficiency in Chapter 6, and, in particular, the proof of Theorem 6-1.)

13-2. ANALYSIS-OF-VARIANCE TEST AND ASSOCIATED DISTRIBUTION THEORY

In this section we present the specific form of the classical test of the hypothesis (13-2) against the alternative (13-3). This test is known as the "one-way analysis-of-variance test." It is called the "one-way" test because just one factor is being tested at various levels; for example, the factor of fertilizer is being applied in various amounts. When two factors—such as fertilizer and water—are being tested, the corresponding test is called a *two-way analysis of variance*. In general, when m factors are tested, the test is called the m-way analysis of variance test. We shall consider only the one-way test; the distribution theory for the multiple-way test is much more complicated. The test is based on a comparison of two different Estimators of the unknown variance σ^2. Under the null hypothesis (13-2) that the means are equal—that is, the populations are identical—an unbiased Estimator of σ^2 is the average sum of squares about the average of all the coordinates $\bar{x}$: $\sum_{i=1}^{k} \sum_{j=1}^{n} (x_{ij} - \bar{x})^2/(nk-1)$. [This is the Sample variance where nk takes the place of n (Chapter 7).] Under the alternative hypothesis that the means are different, the variance has to be estimated independently within each class by means of the average sum of squares within the class about the class average. The Estimator for the ith class is $\sum_{i=1}^{n} (x_{ij} - \bar{x}_i)^2/(n-1)$. These are averaged over all the classes to get a single Estimator, namely, $\text{SSE}/k(n-1)$. Let us compare these two Estimators. The former may be expressed as

$$\frac{\sum_{i=1}^{k}\sum_{j=1}^{n} \bar{x}_{ij}^2 - nk\bar{x}^2}{kn-1}.$$

(For the proof apply the original decomposition (4-12) to the double sum.) By (13-5), (13-7), and (13-8) this Estimator is equal to

$$\frac{\text{SSE} + \text{SSB}}{kn-1};$$

thus, its ratio to the second Estimator is

$$\frac{k(n-1)}{kn-1}\left(1 + \frac{\text{SSB}}{\text{SSE}}\right). \tag{13-9}$$

Under the null hypothesis, the two Estimators are both unbiased Estimators of the same parameter σ^2 so that they are of the "same order of magnitude"; however, under the alternative hypothesis the first Estimator tends to be larger because the various averages $\bar{x}_i$ are estimating different parameters μ_i, $i = 1, \ldots, k$ so that SSB, defined in (13-8), tends to be larger. The sum of squares SSE is not influenced by the equality or inequality of the means so that it is of the same order of magnitude under both the null and alternative hypotheses. For this reason the ratio (13-9) of the first Estimator to the second tends to be larger under the alternative hypothesis. This suggests that the ratio is an appropriate statistic upon which to base the test of the null hypothesis. By (13-9), this ratio varies directly with SSB/SSE, or equivalently, with the statistic

$$\frac{\text{SSB}/(k-1)}{\text{SSE}/k(n-1)}. \tag{13-10}$$

The analysis-of-variance test procedure is to accept or reject the null hypothesis accordingly as the ratio (13-10) is small or large. The point cutting the positive axis into "small" and "large" values of the ratio (13-10) is determined by the level of significance ϵ and by the distribution of the ratio under the null hypothesis. For any level of significance ϵ, let

$$F(\epsilon, k-1, k(n-1)) \tag{13-11}$$

be the upper ϵ-percentile of the F-distribution with $k - 1$ and $k(n - 1)$ degrees of freedom; this distribution was derived in Example 5-5, where its density was expressed with k and $n - k$ in place of $k - 1$ and $k(n - 1)$, respectively. The test procedure is to reject the null hypothesis if the observed value of the statistic (13-10) exceeds the level (13-11), and to accept in the other case; in other words, the rejection region is the set of points in nk-dimensional space upon which the function (13-10) exceeds the number (13-11).

Table IV, Appendix contains values of the percentile (13-11) for $\epsilon = .01$ and .05 and selected values of k and n. The quantities SSB and SSE appearing in the ratio (13-10) are not computed directly from the indicated function of the coordinates because the latter, when calculated, will contain rounding errors due to the means, compounded by squaring and summing. An alternative equivalent formula for the ratio which can be computed to any degree of accuracy is

$$\frac{k(n-1)}{k-1} \cdot \frac{k \sum_{i=1}^{k} \left(\sum_{j=1}^{n} x_{ij} \right)^2 - \left(\sum_{i=1}^{k} \sum_{j=1}^{n} x_{ij} \right)^2}{kn \sum_{i=1}^{k} \sum_{j=1}^{n} x_{ij}^2 - k \sum_{i=1}^{k} \left(\sum_{j=1}^{n} x_{ij} \right)^2}. \tag{13-12}$$

The verification is based on the sum of squares decompositions (Exercise 3). Note that this ratio depends only on the three sums

$$\sum_{i=1}^{k} \sum_{j=1}^{n} x_{ij}, \qquad \sum_{i=1}^{k} \sum_{j=1}^{n} x_{ij}^2, \qquad \sum_{i=1}^{k} \left(\sum_{j=1}^{n} x_{ij} \right)^2.$$

Example 13-1. Suppose that we have a Sample of $n = 3$ observations from each of $k = 3$ populations, and wish to test the null hypothesis at the level of significance $\epsilon = .05$. The following observations are recorded:

$$\begin{array}{lll} X_{11} = -1 & X_{21} = 0 & X_{31} = -2 \\ X_{12} = 0 & X_{22} = 1 & X_{32} = -1 \\ X_{13} = +1 & X_{23} = 1 & X_{33} = 0. \end{array}$$

The three sums involved in the ratio (13-12) are calculated:

$$\sum_{i=1}^{k}\sum_{j=1}^{n} x_{ij} = -1, \qquad \sum_{i=1}^{k}\sum_{j=1}^{n} x_{ij}^2 = 9, \qquad \sum_{i=1}^{k}\left(\sum_{j=1}^{n} x_{ij}\right)^2 = 17,$$

so that the ratio has the value 5. The percentile F (.05, 2, 6) is equal to 5.14. Since the ratio does not exceed the percentile, the null hypothesis is accepted. ■

In the following theorems we derive the distribution of the statistic (13-10). Under the null hypothesis it has the F-distribution with $k - 1$ and $k(n - 1)$ degrees of freedom for all parameter values μ and σ; thus, the test above has the level of significance ϵ as claimed. Under the alternative hypothesis the ratio has a more general distribution—the noncentral F-distribution with the same number of degrees of freedom. The latter is derived from the noncentral chi-square distribution in a manner similar to the way the F-distribution is derived from the ordinary chi-square distribution.

THEOREM 13-1

Consider a Sample of nk observations, n from each of k normal populations with means $\mu + \tau_i$, $i = 1, \ldots, k$ *and common standard deviation* σ, *where* $\tau_1 + \cdots + \tau_k = 0$. *The three statistics*

$$SSE/\sigma^2, \qquad SSB/\sigma^2, \qquad \text{and} \qquad nk(\bar{x} - \mu)^2/\sigma^2, \tag{13-13}$$

are independent. The first has the chi-square distribution with $k(n - 1)$ *degrees of freedom, the second has the noncentral chi-square distribution with* $k - 1$ *degrees of freedom and noncentrality parameter*

$$\tau = \left(\sum_{i=1}^{k} \tau_i^2\right)^{1/2} \Big/ \sigma,$$

and the third has the chi-square distribution with one degree of freedom.

Proof: By changing the variables of integration in the multiple integral defining the joint distribution of the three statistics (13-13) (from x_{ij} to $(x_{ij} - \mu - \tau_i)/\sigma$), we see that this joint distribution is the same as that of the three statistics

$$\sum_{i=1}^{k}\sum_{j=1}^{n}(x_{ij}-\bar{x}_i)^2, \qquad n\sum_{i=1}^{k}(x_i-\bar{x}+\tau_i/\sigma)^2, \qquad nk\bar{x}^2, \quad (13\text{-}14)$$

in sampling from nk standard normal populations (Exercise 4). We shall prove the assertion of the theorem by showing that the joint distribution of the statistics (13-14) is the same as that of the statistics

$$\sum_{i=1}^{k}\sum_{j=2}^{n}x_{ij}^2, \qquad (x_{21}+\tau)^2+\sum_{i=3}^{k}x_{i1}^2, \qquad x_{11}^2. \qquad (13\text{-}15)$$

The reader can verify that this is equivalent to the assertion of the theorem (Exercise 5).

In order to prove the equivalence of the joint distributions of (13-14) and (13-15) in sampling from standard normal populations, we construct a succession of rotations transforming the functions (13-14) into the corresponding functions (13-15), and then invoke Theorems 5-2 and 5-3. More specifically we find rotations which when applied in the given order carry a point $\mathbf{x}$ onto $\mathbf{x}'$ in such a way that the following relations hold among their coordinates:

$$\begin{aligned}\sum_{i=1}^{k}\sum_{j=1}^{n}(x_{ij}-\bar{x}_i)^2 &= \sum_{i=1}^{k}\sum_{j=2}^{n}x_{ij}'^2 \\ n\sum_{i=1}^{k}\left(\bar{x}_i-\bar{x}+\frac{\tau_i}{\sigma}\right)^2 &= (x_{21}'+\tau)^2+\sum_{i=3}^{k}x_{i1}'^2 \qquad (13\text{-}16) \\ nk\bar{x}^2 &= x_{11}'^2.\end{aligned}$$

There are $k+2$ rotations in this succession. The first k act independently on each of the k sets of n variables x_{ij}, $j=1,\ldots,n$, leaving the others fixed. The ith rotation is the *diagonal* rotation on the variables x_{ij}, $j=1,\ldots,n$; and leaves the others fixed, $i=1,\ldots,k$. Let $\mathbf{x}'$ be the image of $\mathbf{x}$ under the succession of diagonal rotations; then, by Eqs. 5-8, the coordinates are related by

$$x_{i1}'=\sqrt{n}\,\bar{x}_i, \qquad \sum_{j=2}^{n}x_{ij}'^2=\sum_{j=1}^{n}(x_{ij}-\bar{x}_i)^2, \qquad i=1,\ldots,k.$$

From these we obtain the following equations for the expressions on the left-hand sides of Eqs. 13-16:

$$\sum_{i=1}^{k}\sum_{j=1}^{n}(x_{ij}-\bar{x}_i)^2=\sum_{i=1}^{k}\sum_{j=2}^{n}x_{ij}'^2$$

$$n \sum_{i=1}^{k} \left(\bar{x}_i - \bar{x} + \frac{\tau_i}{\sigma}\right)^2 = \sum_{i=1}^{k} \left(x'_{i1} - \frac{1}{k} \sum_{j=1}^{k} x'_{j1} + \frac{\tau_i}{\sigma}\right)^2 \quad (13\text{-}17)$$

$$nk\bar{x}^2 = k\left(\sum_{i=1}^{k} x'_{i1}/k\right)^2.$$

The function of $\mathbf{x}'$ appearing in the first equation in (13-17) depends only on the coordinates x'_{ij}, $i = 1, \ldots, k$, $j = 2, \ldots, n$; thus it is invariant under any rotation not involving these coordinates; therefore, if $\mathbf{x}''$ is the image of $\mathbf{x}'$ under such a rotation then the first relation in (13-17) is unchanged if $\mathbf{x}'$ is replaced by $\mathbf{x}''$. Consider the diagonal rotation on the variables x'_{i1}, $i = 1, \ldots, k$ which leaves the variables x'_{ij}, $i = 1, \ldots, k, j = 2, \ldots, n$ unchanged: the coordinates of $\mathbf{x}'$ and its image $\mathbf{x}''$ are related by the equations

$$x''_{11} = \sqrt{k}\left(\sum_{i=1}^{k} x'_{i1}/k\right), \qquad \sum_{i=2}^{k} x''^2_{i1} = \sum_{i=1}^{k} \left(x'_{i1} - \frac{1}{k} \sum_{j=1}^{k} x'_{j1}\right)^2. \quad (13\text{-}18)$$

Now apply the decomposition (4-12) to the sum of squares on the right-hand side of the second equation in (13-17), with $x'_{i1} + \tau_i/\sigma$ in place of x_i:

$$\sum_{i=1}^{k} \left(x'_{i1} - \frac{1}{k} \sum_{j=1}^{k} x'_{j1} + \frac{\tau_i}{\sigma}\right)^2 = \sum_{i=1}^{k} \left(x'_{i1} + \frac{\tau_i}{\sigma}\right)^2 - k\left(\sum_{i=1}^{k} x'_{i1}/k\right)^2.$$

(13-19)

Let $(\tau''_1, \ldots, \tau''_k)$ be the image of $(\tau_1, \ldots, \tau_k)$ under the diagonal rotation; then, by Theorem 4-3 (on the image of the hyperplane $\{\tau: \tau_1 + \cdots + \tau_k = 0\}$), we have

$$\tau''_1 = 0. \quad (13\text{-}20)$$

The first term on the right-hand side of 13-19) is unchanged when x''_{i1} and τ''_i are substituted for x'_{i1} and τ_i, respectively, because a rotation preserves distance; thus, by (13-20), we have

$$\sum_{i=1}^{k} \left(x'_{i1} + \frac{\tau_i}{\sigma}\right)^2 = x''^2_{11} + \sum_{i=2}^{k} \left(x''_{i1} + \frac{\tau''_i}{\sigma}\right)^2.$$

From this and from (13-18) and (13-20) it follows that the right-hand side of (13-19) is equal to

$$\sum_{i=2}^{k} \left(x''_{i1} + \frac{\tau''_i}{\sigma}\right)^2.$$

From this and from (13-18) we infer that Eqs. 13-17 are equivalent to

$$\sum_{i=1}^{k}\sum_{j=1}^{n}(x_{ij}-\bar{x}_i)^2=\sum_{i=1}^{k}\sum_{j=2}^{n}x_{ij}''^2$$

$$n\sum_{i=1}^{k}\left(\bar{x}_i-\bar{x}+\frac{\tau_i}{\sigma}\right)^2=\sum_{i=2}^{k}\left(x_{i1}''+\frac{\tau_i''}{\sigma}\right)^2 \qquad (13\text{-}21)$$

$$nk\bar{x}^2=x_{11}''^2.$$

The final rotation, carrying $\mathbf{x}''$ onto $\mathbf{x}'''$, transforms the variables x_{i1}'', $i=2,\ldots,k$ in accordance with the rotation carrying the point $(\tau_2'',\ldots,\tau_k'')$ onto the point on the first coordinate axis, $\left(\left(\sum_{i=2}^{k}\tau_i''^2\right)^{1/2},0,\ldots,0\right)$, and it leaves the other variables unchanged. The coordinates of $\mathbf{x}''$ and $\mathbf{x}'''$ are related by

$$x_{ij}''=x_{ij}''',\quad i=j=1,\quad\text{and}\quad i=1,\ldots,k,\quad j=2,\ldots,n.$$

$$\sum_{i=2}^{k}\left(x_{i1}''+\frac{\tau_i''}{\sigma}\right)^2=\left(x_{21}'''+\sqrt{\sum_{i=2}^{k}\frac{\tau_i''^2}{\sigma}}\right)^2+\sum_{i=3}^{k}x_{i1}'''^2. \qquad (13\text{-}22)$$

By (13-20) and the invariance of distance under rotation we have

$$\sum_{i=2}^{k}\tau_i''^2=\sum_{i=1}^{k}\tau_i''^2=\sum_{i=1}^{k}\tau_i^2=\tau^2\sigma^2.$$

From this and the relations (13-21) and (13-22) we obtain the set of equations (13-16), with $\mathbf{x}'''$ in place of $\mathbf{x}'$. This is the desired rotation. ■

This theorem provides most of the proof of the result on the distribution of the ratio (13-10).

THEOREM 13-2

The ratio

$$\frac{SSB/(k-1)}{SSE/k(n-1)}$$

has the noncentral F-distribution with $k-1$ and $k(n-1)$ degrees of freedom, and noncentrality parameter τ, defined by

$$\tau^2 = \sum_{i=1}^{k} \frac{\tau_i^2}{\sigma^2}.$$

This distribution has a density function equal to 0 for $y \leq 0$, and to

$$\int_0^\infty u\psi_{k-1}(uy;\tau)\psi_{k(n-1)}(u)\,du, \qquad y > 0, \qquad (13\text{-}23)$$

where $\psi_{k-1}(u;\tau)$ is the noncentral chi-square density defined by (11-22). Under the null hypothesis (13-2) the parameter τ is equal to 0, and the distribution is the ordinary F-distribution with $k-1$ and $k(n-1)$ degrees of freedom, defined in Example 5-5.

Proof: This is a consequence of Theorem 13-1 and Formula (5-15) for the distribution function of the quotient of two independent statistics in terms of their individual distributions. The density is obtained by differentiation under the sign of integration. We omit the details of the argument as they are similar to the corresponding ones for Examples 5-4 and 5-5 (Exercise 6). The derivation for the special case $\tau = 0$, when both sums of squares have ordinary chi-square distributions, has already been done in Example 5-5. ■

Theorem 13-2 not only shows that the one-way analysis of variance test with the "cutoff point" (13-11) has the claimed level ϵ of significance, but also provides a formula for the power of the test against any alternative $(\tau_1, \ldots, \tau_k)$ with some $\tau_j \neq 0$. This is the integral of the noncentral F-density over the half-line from the point (13-11) to $+\infty$. This integral has been tabulated for various values of τ, n, and k. Even though the integral is not expressible in closed analytic form it is still of theoretical value. It will be used in the next section to prove a certain optimal property of the analysis of variance test.

EXERCISES

1. The following are observations from $k = 4$ populations; here $n = 3$:

10, 13, 13   7, 10, 10
6, 7, 14   11, 16, 15

Test at the level $\epsilon = .05$ whether or not the populations are all the same.

2. Repeat Exercise 1 with the following data, and $k = 5, n = 4$:

7, 4, 7, 6	5, 4, 8, 3
6, 11, 10, 9	7, 7, 10, 12
9, 3, 8, 4	

3. Verify that (13-12) is equivalent to (13-10).

4. Show that the joint distribution of the three statistics (13-13) is the same as that of (13-14) when, in the latter case, the sampling is from a standard normal population.

5. What previous theorems imply that the statistics (13-15) are independent and have the distributions indicated in the statement of Theorem 13-1?

6. Furnish the details of the proof of Theorem 13-2; in particular, justify the differentiation under the sign of integration used to obtain the formula (13-23). (*Hint:* Exercise 3, Sec. 5-3.)

7. In the particular case $n = 2$ show that SSB may be written as $(n/2)(\bar{x}_1 - \bar{x}_2)^2$.

8. Put $s_i^2 = \sum_{j=1}^{n} (x_{ij} - \bar{x}_i)^2, i = 1, 2$. Prove that the four statistics $\sqrt{n}\bar{x}_1$, $\sqrt{n}\bar{x}_2$, $(n - 1)s_1^2$ and $(n - 1)s_2^2$ are independent. Find the joint distribution.

9. Using the result of the previous exercise, derive the distribution of

$$\frac{\sqrt{n}(\bar{x}_1 - \bar{x}_2)}{(\mathrm{SSE}/(n - 1))^{1/2}}.$$

(This is a special case of Theorem 13-2.)

10. Applying the results of the previous three exercises show that the analysis of variance test for $k = 2$ is equivalent to a test based on the t-distribution.

13-3. THEORETICAL JUSTIFICATION OF ANALYSIS-OF-VARIANCE TEST: INVARIANCE AND AVERAGE POWER (OPTIONAL)

The ratio (13-10), upon which is based the analysis-of-variance test, has the following invariance property. Its value is unchanged if each variable x_{ij} is multiplied by a common constant $a \neq 0$ and a common constant b is added; in other words, the ratio has the same value for the point $\mathbf{x}$ (with coordinates x_{ij}) as for the point $a\mathbf{x} + b\mathbf{1}$ (with coordinates $ax_{ij} + b$). This is useful in numerical calculations: if the observed values are very large in magnitude, a common constant may be subtracted from (or added to) each value to reduce the magnitude of the numbers and make the computation of (13-10) simpler.

The analysis-of-variance test was devised by R. A. Fisher in the 1920's. It has been valuable in many applications. Biologists, engineers, psychologists, and many other specialists use it in testing. The student of mathematical statistics might ask if the test has any optimal mathematical properties: is it a best test in any sense? In earlier chapters we proved the optimality of certain procedures; for example, our Estimators were efficient, and certain tests were most powerful, most powerful invariant or unbiased, or had largest average power. When an absolutely best test did not exist we defined a certain class of tests (unbiased, invariant) and showed that a given test was the most powerful in the class. The current hypothesis testing problem stated in (13-2) and (13-3) invites no best procedure because the simpler homogeneity testing problem in Sec. 11-5 does not.

In defining the optimal property of the analysis of variance test, we shall combine the notion of invariance, described above and in Sec. 9-4, with the concept of average power, described in Sec. 11-5. Our result is: *Among all invariant tests, the analysis of variance test has largest average power over spheres in the parameter space of points* $\boldsymbol{\tau} = (\tau_1, \ldots, \tau_k)$.

The null hypothesis (13-2) is really composite because μ and σ are arbitrary. As in Chapter 9, a test is characterized by a rejection region M in the nk-dimensional Sample space. The level of significance of an arbitrary test of the null hypothesis is not fixed because the integral

$$\int_M \cdots \int L(\mathbf{x}; \boldsymbol{\mu}, \sigma)\, d\mathbf{x}$$

depends in general on μ and σ even when $\boldsymbol{\tau} = \mathbf{0}$. As in Sec. 9-4, we shall restrict our considerations to invariant tests; such tests will be shown to have the property that the level of significance is the same for all μ and σ. The concept of invariance is slightly more general than that in Sec. 9-4: here a test is invariant if its rejection region is closed both under scalar multiplication and addition of a common scalar constant; more exactly, the test is invariant if its rejection region M has the property that $\mathbf{x}$ belongs to M *if and only if*

$$a\mathbf{x} + b\mathbf{1} \quad \text{belongs to } M, \quad a > 0, \quad -\infty < b < \infty .$$

As in Sec. 9-4 we have to add another technical condition on the region for the purpose of proving the main theorem at this level. A continuous function f on R^{nk} is called invariant if for every pair of real numbers $a > 0$ and b,

$$f(a\mathbf{x} + b\mathbf{1}) = f(\mathbf{x}), \qquad \mathbf{x} \in R^{nk}. \tag{13-24}$$

Let $\mathfrak{U}_0(M)$ be the subclass of $\mathfrak{U}(M)$ (the upper class of continuous functions for the set M, defined in Chapter 3) consisting of the invariant functions; then M is called *invariant* if not only it is invariant under scalar multiplication and addition, but the power function

$$\int_M \cdots \int L \, d\mathbf{x}$$

is approximable by the integrals

$$\int_{R^{nk}} \cdots \int f L \, d\mathbf{x}$$

of *invariant* functions f in $\mathfrak{U}(M)$; more formally:

$$\int_M \cdots \int L \, d\mathbf{x} = \operatorname*{glb}_{f \in \mathfrak{U}_0(M)} \int_{R^{nk}} \cdots \int f(\mathbf{x}) \, L \, d\mathbf{x} \tag{13-25}$$

for all parameter values $\boldsymbol{\mu}$ and σ, where L is given by (13-1). This condition signifies that the greatest lower bound in (3-35) over the class $\mathfrak{U}(M)$ is the same as over the subclass $\mathfrak{U}_0(M)$. The test is then called *invariant*.

The formulation of the testing problem is similar to that in Chapter 9. The first step is to replace integrals over invariant rejection regions by integrals of invariant functions; then the latter integrals are reduced by sufficiency and invariance.

LEMMA 13-1

Put

$$L(\mathbf{x}; \mu, \boldsymbol{\tau}, \sigma) = (2\pi\sigma^2)^{-nk/2} e^{-(1/2\sigma^2)\sum_{i=1}^{k}\sum_{j=1}^{n}(x_{ij}-\mu-\tau_i)^2}$$

If f is an invariant function, then

$$\int_{R^{nk}}\cdots\int f(\mathbf{x})\,L(\mathbf{x}; \mu, \boldsymbol{\tau}, \sigma)\,d\mathbf{x} = \int_{R^{nk}}\cdots\int f(\mathbf{x})\,L(\mathbf{x}; 0, \boldsymbol{\tau}/\sigma, 1)\,d\mathbf{x}, \tag{13-26}$$

where $\boldsymbol{\tau}/\sigma = (1/\sigma)\boldsymbol{\tau}$, for all μ and $\sigma > 0$.

Proof: Change the variables of integration from x_{ij} to $(x_{ij} - \mu)/\sigma$, and apply the invariance of f:

$$f(\sigma\mathbf{x} + \mu\mathbf{1}) = f(\mathbf{x}). \quad \blacksquare$$

COROLLARY

The level of significance of an invariant test of the null hypothesis (13-2) is the same for all μ and $\sigma > 0$; thus, we may without ambiguity refer to a fixed level of significance of an invariant test.

Proof: By Lemma 13-1, for any invariant function f, the integral

$$\int_{R^{nk}}\cdots\int f(\mathbf{x})\,L(\mathbf{x}; \mu, \mathbf{0}, \sigma)\,d\mathbf{x}$$

(here $\boldsymbol{\tau} = \mathbf{0}$) does not depend on μ or σ; thus, by the condition (13-25), neither does the integral

$$\int_{M}\cdots\int L(\mathbf{x}; \mu, \mathbf{0}, \sigma)\,d\mathbf{x},$$

which is the probability of error of the first kind. $\blacksquare$

LEMMA 13-2

Let f be an invariant function on R^{nk} such that $0 \leq f \leq 1$; then there exists a similar function $\bar{f}$ on R^{nk}, depending on $\mathbf{x}$ only through the $k + 1$ functions

$$\sqrt{n}\bar{x}_i, \quad i = 1, \ldots, k, \quad \sum_{i=1}^{k} \sum_{j=1}^{n} (x_{ij} - x_i)^2 \tag{13-27}$$

such that

$$\int \cdots \int_{R^{nk}} f(\mathbf{x}) L\left(\mathbf{x}; 0, \frac{\boldsymbol{\tau}}{\sigma}, 1\right) d\mathbf{x} = \int \cdots \int_{R^{nk}} \bar{f}(\mathbf{x}) L\left(\mathbf{x}; 0, \frac{\boldsymbol{\tau}}{\sigma}, 1\right) d\mathbf{x}, \tag{13-28}$$

for all $\boldsymbol{\tau}$ *and* $\sigma > 0$.

Proof: By successive decompositions of the sums of squares, first over index j and then over index i, as in the identity (13-5), the sum in the exponent of L may be decomposed as

$$\sum_{i=1}^{k} \sum_{j=1}^{n} \left(x_{ij} - \frac{\tau_i}{\sigma}\right)^2$$

$$= \sum_{i=1}^{k} \sum_{j=1}^{n} (x_{ij} - \bar{x}_i)^2 + n \sum_{i=1}^{k} \left(\bar{x}_i - \bar{x} - \frac{\tau_i}{\sigma}\right)^2 + nk\bar{x}^2. \tag{13-29}$$

(This may also be considered to be a special case of (13-5) with $x_{ij} - \tau_i/\sigma$ in place of x_{ij}.) Let $\mathbf{x}'$ be the image of $\mathbf{x}$ under the succession of the first k rotations in the proof of Theorem 13-1: the coordinates are related by the set of equations (13-17), and so the sum of squares (13-29) is, in terms of the coordinates of $\mathbf{x}'$, equal to

$$\sum_{i=1}^{k} \sum_{j=2}^{n} x_{ij}'^2 + n \sum_{i=1}^{k} \left(x_{i1}' - \frac{1}{k} \sum_{j=1}^{k} x_{j1}' + \frac{\tau_i}{\sigma}\right)^2 + nk\left(\sum_{i=1}^{k} x_{i1}'/k\right)^2. \tag{13-30}$$

Put $f^*(\mathbf{x}') = f(\mathbf{x})$; then the integral on the left-hand side of (13-28) is, by virtue of the equality of (13-29) and (13-30), equal to

$$\int \cdots \int_{R^{nk}} f^*(\mathbf{x}')(2\pi)^{-nk/2} \exp\left\{-\frac{1}{2}\left[\sum_{i=1}^{k} \sum_{j=2}^{n} x_{ij}'^2\right.\right.$$

$$\left.\left. + n \sum_{i=1}^{k} \left(x_{i1}' - \frac{1}{k} \sum_{j=1}^{k} x_{j1}' - \frac{\tau_i}{\sigma}\right)^2 + nk\left(\sum_{i=1}^{k} x_{i1}'/k\right)^2\right]\right\} d\mathbf{x}. \tag{13-31}$$

By the invariance of integration under rotation (Theorem 3-4) the integral is unchanged when the primes are removed from the variables in the integrand. Now for fixed x_{i1}, $i = 1, \ldots, k$, integrate

$$f^*(\mathbf{x})e^{-1/2\sum_{i=1}^{k}\sum_{j=2}^{n}x_{ij}^2}$$

with respect to x_{ij}, $i = 1, \ldots, k$, $j = 2, \ldots, n$. The integral is unchanged when f^*, as a function of the latter variables, is replaced by its spherical average, which we denote by

$$\bar{f}\left(x_{11}, \ldots, x_{1k}, \sum_{i=1}^{k}\sum_{j=2}^{n} x_{ij}^2\right)$$

(Eq. 6-5). By the same invariance of integration the primes may now be put back on the variables in the integrand. In terms of the original variables we have, by the relations preceding (13-17),

$$\bar{f}\left(x'_{11}, \ldots, x'_{1k}, \sum_{i=1}^{k}\sum_{j=1}^{n} x'^2_{ij}\right) = \bar{f}\left(\sqrt{n}\bar{x}_1, \ldots, \sqrt{n}\bar{x}_k, \sum_{i=1}^{k}\sum_{j=1}^{n} (x_{ij} - \bar{x}_i)^2\right); \tag{13-32}$$

thus the integral on the left-hand side of (13-28) is unchanged when f is replaced by the latter function $\bar{f}$. In this way there is constructed a correspondence between the function f and the above spherical average $\bar{f}$:

$$f(\mathbf{x}) \rightarrow \bar{f}\left(\sqrt{n}\bar{x}_1, \ldots, \sqrt{n}\bar{x}_k, \sum_{i=1}^{k}\sum_{j=1}^{n} (x_{ij} - \bar{x}_i)^2\right).$$

We prove the invariance of $\bar{f}$ as a function of $\mathbf{x}$:

$$\bar{f}\left(\sqrt{n}\bar{x}_1, \ldots, \sqrt{n}\bar{x}_k, \sum_{i=1}^{k}\sum_{j=1}^{n} (x_{ij} - \bar{x}_i)^2\right)$$

$$= \bar{f}\left(\sqrt{n}a\bar{x}_1 + b, \ldots, \sqrt{n}a\bar{x}_k + b, a^2\sum_{i=1}^{k}\sum_{j=1}^{n} (x_{ij} - \bar{x}_i)^2\right), \tag{13-33}$$

$a \neq 0$, *b real number.*

To show this we note that from the definition of $f^*(\mathbf{x})$ and the invariance of f we get:

$$f^*((a\mathbf{x} + b\mathbf{1})') = f(a\mathbf{x} + b\mathbf{1}) = f(\mathbf{x}) = f^*(\mathbf{x}'). \tag{13-34}$$

Now $(a\mathbf{x} + b\mathbf{1})' = a\mathbf{x}' + b\mathbf{1}'$; furthermore, by the definition of the diagonal rotation (for each of the k groups of n coordinates), the coordinates of indices $1, n + 1, \ldots, (k - 1)n + 1$ of $\mathbf{1}'$ are equal to $\sqrt{n}$ and the others are equal to 0. Equation (13-34) signifies that f^*, as a function of the coordinates of $\mathbf{x}'$, is invariant. This invariance is preserved when f^* is averaged over x'_{ij}, $i = 1, \ldots, k, j = 2, \ldots, n$:

$$\bar{f}\left(x'_{11}, \ldots, x'_{k1}, \sum_{i=1}^{k} \sum_{j=2}^{n} x'^2_{ij}\right)$$

$$= \bar{f}\left(ax'_{11} + b, \ldots, ax'_{k1} + b, a^2 \sum_{i=1}^{k} \sum_{j=2}^{n} x'^2_{ij}\right).$$

We obtain (13-33) by replacing the expressions for the coordinates of $\mathbf{x}'$ by the equivalent ones in terms of the coordinates of $\mathbf{x}$. ■

LEMMA 13-3

For any invariant function f, let $\bar{f}$ be the function constructed from it in accordance with the statement of Lemma 13-2. Put

$$P(\boldsymbol{\tau}) = \int_{R^{nk}} \cdots \int \bar{f}(\mathbf{x}) L\left(\mathbf{x}; 0, \frac{\boldsymbol{\tau}}{\sigma}, 1\right) d\mathbf{x}.$$

Then there exists a well-defined continuous function $g(u)$ of a real variable u such that the spherical average of $P(\boldsymbol{\tau})$ over a sphere of radius $\|\boldsymbol{\tau}\|$ is representable as

$$\bar{P}(\|\boldsymbol{\tau}\|^2) = \int_0^\infty g(u) H\left(u; \frac{\|\boldsymbol{\tau}\|}{\sigma}, \quad k - 1, \quad k(n - 1)\right) du \tag{13-35}$$

where H is the noncentral F-density with noncentrality parameter $\|\boldsymbol{\tau}\|/\sigma$, and $k - 1$ and $k(n - 1)$ degrees of freedom, defined by (13-23).

Proof: The $k + 1$ statistics

$$\sqrt{n}\bar{x}_1, \ldots, \sqrt{n}\bar{x}_k, \sum_{i=1}^{k} \sum_{j=1}^{n} (x_{ij} - \bar{x}_i)^2 \tag{13-36}$$

are independent, the first k having normal distributions with means τ_i/σ, $i = 1, \ldots, k$ and unit standard deviations, and the last having a chi-square distribution with $k(n-1)$ degrees of freedom. This is a consequence of the rotations constructed in the first part of the proof of Theorem 13-1—in particular, of the relations preceding (13-17). It follows that the multiple integral defining $P(\boldsymbol{\tau})$ may be simplified by applying Theorem 5-1 to the joint distribution of the statistics (13-36):

$$P(\boldsymbol{\tau}) = \int_{-\infty}^{\infty} \cdots \int_{-\infty}^{\infty} \int_{0}^{\infty} \bar{f}(u_1, \ldots, u_k, y)\, \psi_{k(n-1)}(y) \cdot \prod_{i=1}^{k} \phi\left(u_i - \frac{\tau_i}{\sigma}\right) dy\, du_1, \ldots, du_k. \tag{13-37}$$

For fixed y, let $\bar{\bar{f}}(u_1^2 + \cdots + u_k^2, y)$ be the average of $\bar{f}$, as a function of $u_1, \ldots, u_k$, over the sphere of radius $(u_1^2 + \cdots + u_k^2)^{1/2}$; then the spherical average of

$$\int_{0}^{\infty} \bar{f}(u_1, \ldots, u_k, y)\, \psi_{k(n-1)}(y)\, dy$$

is equal to

$$\int_{0}^{\infty} \bar{\bar{f}}(u_1^2 + \cdots + u_k^2, y)\, \psi_{k(n-1)}(y)\, dy$$

because the averaging operation may be interchanged with the process of integration over y, as in the proofs of Lemmas 11-3 and 11-4; furthermore, as a consequence the latter lemma, the spherical average of $P(\boldsymbol{\tau})$, defined by (13-37), is equal to

$$\bar{P}(\|\boldsymbol{\tau}\|^2) = \int_{-\infty}^{\infty} \cdots \int_{-\infty}^{\infty} \int_{0}^{\infty} \bar{\bar{f}}(u_1^2 + \cdots + u_k^2, y)\, \psi_{k(n-1)}(y) \cdot \prod_{i=1}^{k} \phi\left(u_i - \frac{\tau_i}{\sigma}\right) dy\, du_1 \cdots du_k.$$

In terms of the original variable $\mathbf{x}$ in R^{nk}, and by Theorem 5-1, the latter integral is equal to

$$\bar{P}(\|\boldsymbol{\tau}\|^2) = \int_{R^{nk}} \cdots \int \bar{\bar{f}}\left(n \sum_{i=1}^{k} \bar{x}_i^2, \sum_{i=1}^{k} \sum_{j=1}^{n} (x_{ij} - \bar{x}_i)^2\right) \quad L\left(\mathbf{x}; 0, \frac{\boldsymbol{\tau}}{\sigma}, 1\right) d\mathbf{x}. \tag{13-38}$$

The function $\bar{\bar{f}}$ inherits the invariance of $\bar{f}$ in the following way: for every $a > 0$ and every b,

$$\bar{\bar{f}}\left(na^2 \sum_{i=1}^{k} (\bar{x}_i - b)^2, a^2 \sum_{i=1}^{k} \sum_{j=1}^{n} (x_{ij} - \bar{x}_i)^2\right)$$
$$= \bar{\bar{f}}\left(n \sum_{i=1}^{k} \bar{x}_i^2, \sum_{i=1}^{k} \sum_{j=1}^{n} (x_{ij} - \bar{x}_i)^2\right) \qquad (13\text{-}39)$$

(Exercise 1). Put

$$a^2 = \left[\sum_{i=1}^{k} \sum_{j=1}^{n} (x_{ij} - \bar{x}_i)^2\right]^{-1}, \quad b = \bar{x};$$

then the invariance relation (13-39) implies:

$$\bar{\bar{f}}\left(n \sum_{i=1}^{k} \bar{x}_i^2, \sum_{i=1}^{k} \sum_{j=1}^{n} (x_{ij} - \bar{x}_i)^2\right) = \bar{\bar{f}}\left(1, \frac{n \sum_{i=1}^{k} (\bar{x}_i - \bar{x})^2}{\sum_{i=1}^{k} \sum_{j=1}^{n} (x_{ij} - \bar{x}_i)^2}\right); \qquad (13\text{-}40)$$

therefore, $\bar{\bar{f}}$ is a function of

$$\frac{n \sum_{i=1}^{k} (\bar{x}_i - \bar{x})^2}{\sum_{i=1}^{k} \sum_{j=1}^{n} (x_{ij} - \bar{x}_i)^2} = \frac{\text{SSB}}{\text{SSE}},$$

or, equivalently, of the ratio

$$\frac{\text{SSB}/k - 1}{\text{SSE}/k(n - 1)}.$$

Call this function $g = g(u)$:

$$g(u) = \bar{\bar{f}}\left(1, \frac{(k - 1)u}{k(n - 1)}\right).$$

Apply (13-40) to the integrand of (13-38), and then substitute g for $\bar{\bar{f}}$:

$$\bar{P}(\|\boldsymbol{\tau}\|^2) = \int_{R^{nk}} \cdots \int g\left(\frac{\text{SSB}/k - 1}{\text{SSE}/k(n - 1)}\right) L\left(\mathbf{x}; 0, \frac{\boldsymbol{\tau}}{\sigma}, 1\right) d\mathbf{x}.$$

By Theorem 13-2, the ratio in the argument of g has the noncentral

F-distribution; thus, by Theorem 4-1, the multiple integral representing $\bar{P}(\|\tau\|^2)$ is reducible to the single integral (13-35). ■

Our goal is similar to that of Theorem 11-7—finding an upper bound on the spherical average. The following result is analogous to Lemma 11-6:

LEMMA 13-4
Let $H(w, \tau)$ be the noncentral F-density with noncentrality parameter τ, and with $k - 1$ and $k(n - 1)$ degrees of freedom; then the ratio

$$H(w, \tau)/H(w; 0), \quad w > 0,$$

is a nondecreasing function of w for $\tau > 0$.

Proof: Recall that the noncentral chi-square density with $k - 1$ degrees of freedom and noncentrality parameter τ is given by (11-22) as

$$\psi_{k-1}(y; \tau) = \int_0^y (1/2)(y - u)^{-1/2}\psi_{k-2}(u)$$

$$\cdot[\phi(\sqrt{y - u} - \tau) + \phi(\sqrt{y - u} + \tau)]\, du, \quad y > 0.$$

This is of the form

$$\psi_{k-1}(y; \tau) = C\psi_{k-1}(y)\, G(y; \tau), \tag{13-41}$$

where C is a positive constant independent of y, and

$$G(y; \tau) = \int_0^1 (1 - u)^{-1/2} u^{(n - 3)/2} \cosh(\tau\sqrt{y(1 - u)})\, du$$

(Exercise 2; recall that $\cosh x = (e^x + e^{-x})/2$). In accordance with the definition (13-23) of the noncentral F-density, and the representation (13-41), we have

$$H(w, \tau) = C\int_0^\infty y\psi_{k-1}(yw)\, G(yw; \tau)\, \psi_{k(n-1)}(y)\, dy.$$

The ratio $H(w, \tau)/H(w, 0)$ is, by a change of variable of integration from y to yw, equal to

$$\frac{\int_0^\infty y\psi_{k-1}(y)\,G(y;\tau)\,\psi_{k(n-1)}(y/w)\,dy}{\int_0^\infty y\psi_{k-1}(y)\,\psi_{k(n-1)}(y/w)\,dy}. \tag{13-42}$$

Put

$$q(y;w) = \frac{y\psi_{k-1}(y)\,\psi_{k(n-1)}(y/w)}{\int_0^\infty t\psi_{k-1}(t)\,\psi_{k(n-1)}(t/w)\,dt};$$

then the ratio (13-42) is equal to

$$\int_0^\infty G(y;\tau)\,q(y;w)\,dy. \tag{13-43}$$

We want to show that this is a nondecreasing function of w; for this purpose we invoke Lemma 9-5. Let w_0 and w_1 be two numbers such that

$$0 < w_0 < w_1;$$

and put $p_i(y) = q(y, w_i)$, $i = 0, 1$. It follows from the definitions of $q(y; w)$ and of the ordinary chi-square density (4-8), that p_i satisfies the conditions in the hypothesis of Lemma 9-5:

$$\int_0^\infty p_i(y)\,dy = 1, \quad i = 1,2$$

$$p_1(y)/p_0(y) \quad \text{is nondecreasing for} \quad y > 0$$

(Exercise 3). (As in the proof of Lemma 9-6, the variable y is restricted to positive values.) The function $\cosh x$ increases for $x > 0$; thus $G(y;\tau)$ increases with y for $y > 0$ and $\tau > 0$; finally, by Lemma 9-5:

$$\int_0^\infty G(y,\tau)\,q(y;w_0)\,dy \le \int_0^\infty G(y;\tau)\,q(y;w_1)\,dy,$$

and so the ratio (13-42) is a nondecreasing function of w. ■

LEMMA 13-5
Let $f(\mathbf{x})$, $\mathbf{x} \in R^{nk}$, be an invariant function such that $0 \le f \le 1$; put

$$P(\boldsymbol{\tau}) = \int_{R^{nk}} \cdots \int f(\mathbf{x}) \cdot L(\mathbf{x}; 0, \boldsymbol{\tau}/\sigma, 1)\, dx,$$

$$P(\mathbf{0}) = \epsilon,$$

and let F_ϵ be the percentile (13-11) of the F-distribution. Then the spherical average $\bar{P}(\|\boldsymbol{\tau}\|^2)$ of P is at most equal to the probability that the ratio

$$\frac{SSB/(k-1)}{SSE/k(n-1)}$$

exceeds F_ϵ, that is

$$\int \cdots \int_{\left\{\mathbf{x}:\, n\sum_{i=1}^{k}(\bar{x}_i - \bar{x})^2/k-1 > F_\epsilon \sum_{i=1}^{k}\sum_{j=1}^{n}(x_{ij}-\bar{x}_i)^2/k(n-1)\right\}} L(\mathbf{x}; 0, \boldsymbol{\tau}/\sigma, 1)\, d\mathbf{x}. \quad (13\text{-}44)$$

Proof: By Lemma 13-3 the spherical average is representable as the integral (13-35) of a continuous function g with respect to the noncentral F-density. For $\boldsymbol{\tau} = \mathbf{0}$, P and $\bar{P}$ are equal. These facts imply that the bounding of $\bar{P}(\|\boldsymbol{\tau}\|^2)$ subject to $\bar{P}(0) = \epsilon$ is equivalent to the bounding of the integral (13-35) subject to its being equal to ϵ for $\boldsymbol{\tau} = \mathbf{0}$. Put

$$p_0(w) = H(w, 0), \qquad p_1(w) = H\left(w, \frac{\|\boldsymbol{\tau}\|}{\sigma}\right),$$

and invoke the Neyman-Pearson lemma (Lemma 9-1); the conditions are satisfied on account of Lemma 13-4, and so the integral (13-35) is not more than

$$\int_{F_\epsilon}^{\infty} H\left(w, \frac{\|\boldsymbol{\tau}\|}{\sigma}\right) dw.$$

The integral (13-35) is equal to the multiple integral (13-44) because the ratio (13-10) has the density $H(w, \|\boldsymbol{\tau}\|/\sigma)$ (Theorem 13-2). ■

THEOREM 13-3

Among all invariant tests of size ϵ the analysis of variance test has the maximum average power over every sphere $\{\boldsymbol{\tau}: \|\boldsymbol{\tau}\| = r\}$, $r > 0$.

Proof: The proof is similar to that of Theorem 11-8, which states an analogous result for the homogeneity test. Let M be the rejection region of an invariant test of size ϵ, and let $f(\mathbf{x})$ be an arbitrary function in the subclass $\mathfrak{U}_0(M)$ of invariant members of the upper class $\mathfrak{U}(M)$. By Lemma 13-1 we may assume without loss of generality that $\mu = 0$, that the mean of the ith population is τ_i/σ, and that their common standard deviation is equal to 1. From the relation (3-35) we get

$$\int_{R^{nk}} \cdots \int f(\mathbf{x})\, L\left(\mathbf{x}; 0, \frac{\boldsymbol{\tau}}{\sigma}, 1\right) d\mathbf{x} \geq \int_{M} \cdots \int L\left(\mathbf{x}; 0, \frac{\boldsymbol{\tau}}{\sigma}, 1\right) d\mathbf{x} \qquad (13\text{-}45)$$

for all $\boldsymbol{\tau}$ and $\sigma > 0$; thus, in particular:

$$\epsilon' = \int_{R^{nk}} \cdots \int f(\mathbf{x})\, L(\mathbf{x}; 0, \mathbf{0}, 1)\, d\mathbf{x} \geq \epsilon.$$

The inequality (13-45) is preserved under spherical averaging. The rest of the proof is, after appropriate simple modifications, the same as that of Theorem 11-8: The spherical average of the left-hand side of (13-45) is dominated by the expression (13-44) with ϵ' in place of ϵ; furthermore, ϵ' may be taken arbitrarily close to ϵ; finally, the F-distribution is continuous (Exercise 4). It has already been shown that the analysis-of-variance test has a rejection region which is invariant under scalar multiplication and addition of $b\mathbf{1}$, and that the power function is given by (13-44). The proof that the test satisfies the technical condition (13-25) is very similar to the verification of the corresponding conditions in the case of unbiased and invariant tests in Chapter 9; the details can be inferred from the proof of Theorem 9-2 (Exercise 5). ■

EXERCISES

1. Verify Eq. 13-39.
2. Verify Eq. 13-41.
3. Show that the functions p_0 and p_1 in the proof of Lemma 13-4 actually satisfy the conditions of Lemma 9-1.

4. Following the proof of Theorem 11-8, give the details of the proof of Theorem 13-3 which immediately follow Eq. 13-45.

5. Show that the rejection region of the analysis-of-variance test satisfies the technical condition (13-25).

6. Consider k normal populations with means $\mu_1, \ldots, \mu_k$ and common standard deviation σ. Prove that for any $c_1, \ldots, c_k$ the ratio

$$\frac{\sum_{i=1}^{k} c_i(\bar{x}_i - \mu_i)/\|\mathbf{c}\|}{\left[\sum_{i=1}^{k}\sum_{j=1}^{n}(x_{ij} - \bar{x}_i)^2/k(n-1)\right]^{1/2}}$$

has the t-distribution with $k(n - 1)$ degrees of freedom.

7. Let $t_{\epsilon/2}$ be the upper $\epsilon/2$-percentile of the t-distribution with $k(n - 1)$ degrees of freedom. Prove that

$$\sum_{i=1}^{k} c_i\bar{x}_i \pm \|\mathbf{c}\|\, t_{\epsilon/2}\left[\frac{\sum_{i=1}^{k}\sum_{j=1}^{n}(x_{ij} - x_i)^2}{k(n-1)}\right]^{1/2}$$

is a confidence interval for $\sum_{i=1}^{k} c_i\mu_i$ of coefficient $1 - \epsilon$.

8. Consider k normal populations with means $\mu_1, \ldots, \mu_k$ and common standard deviation σ. Prove that the ratio

$$\frac{n\sum_{i=1}^{k}(\bar{x}_i - \mu_i)^2/k}{\sum_{i=1}^{k}\sum_{j=1}^{n}(x_{ij} - \bar{x}_i)^2/k(n-1)}$$

has the F-distribution with k and $k(n - 1)$ degrees of freedom.

9. (*Multiple comparisons*) By the result of Exercise 7 we can construct a confidence interval for a given linear combination of the means; however, the resulting probability statement is valid only for that fixed linear combination. It is of more interest to get a confidence interval which serves simultaneously for all linear combinations of the means. *Prove:* The probability that the coordinates of the Sample point satisfy the inequality

$$\left|\sum_{i=1}^{k} c_i\bar{x}_i - \sum_{i=1}^{k} c_i\mu_i\right| \leq \|\mathbf{c}\|\left[kF_{\epsilon/2}\sum_{i=1}^{k}\sum_{j=1}^{n}(x_{ij} - x_i)^2/k(n-1)\right]^{1/2}$$

simultaneously for all $c_1, \ldots, c_k$ is at least equal to $1 - \epsilon$. [*Hint:* Apply the Cauchy-Schwarz inequality

$$\left|\sum_{i=1}^{k} c_i(\bar{x}_i - \mu_i)\right|^2 \leq \sum_{i=1}^{k} c_i^2 \cdot \sum_{i=1}^{k} (\bar{x}_i - \mu_i)^2$$

(Sec. 3-1, Exercise 9), and the result of Exercise 8 above.]

10. Suppose $\mu_i = \mu + \tau_i$ where $\tau_1 + \cdots + \tau_k = 0$. Show that

$$\sum_{i=1}^{k} c_i \bar{x}_i \pm \|\mathbf{c}\| \left[F_{\epsilon/2} \sum_{i=1}^{k} \sum_{j=1}^{n} (x_{ij} - \bar{x}_i)^2/(n-1)\right]^{1/2}$$

is a confidence interval for $\sum_{i=1}^{k} c_i \tau_i$ for *all* $\mathbf{c}$ such that $c_1 + \cdots + c_k = 0$, and that the confidence coefficient is at least equal to $1 - \epsilon$.

13-4. ESTIMATION OF PARAMETERS IN THE ONE-WAY ANALYSIS-OF-VARIANCE MODEL (OPTIONAL)

The first three sections concerned testing the equality of the means of several normal populations. Now we consider the *estimation* of the means not only for normal but general populations.

Let us begin with the case of n observations from each of k normal populations with means $\mu_1, \ldots, \mu_k$, respectively, and common standard deviation σ. No relations among the means are assumed. It is apparent that "information" about μ_i is furnished only by that part of the Sample from the ith population. By a simple extension of the proof of Theorem 11-2 it can be shown that $\bar{x}_i$, the mean of the observations from the ith population, is the efficient Estimator of μ_i (Exercise 1). If the means are put in the form

$$\mu_i = \mu + \tau_i$$

where $\tau_1 + \cdots + \tau_k = 0$, then it can be shown that $\bar{x}$, the average of all the observations, is the efficient Estimator of μ, and that $\bar{x}_i - \bar{x}$ is the efficient Estimator of τ_i; this follows, as in Theorem 12-3, from the corresponding relations for the means:

$$\mu = \frac{1}{n}\sum_{i=1}^{n} \mu_i, \quad \tau_i = \mu_i - \frac{1}{n}\sum_{j=1}^{n} \mu_j.$$

If the populations have distributions which are unknown

except for the fact that they have the same standard deviation σ, then the efficient Estimators for the normal case are best linear unbiased Estimators in the general case. Let us use the method of Sec. 12-5 to show that $\bar{x}_h$ is a best linear unbiased Estimator of μ_h. First of all it is clear from (11-46) and (11-48) that $\bar{x}_h$ is unbiased, and that its variance is equal to σ^2/n. Now let

$$\sum_{i,j} c_{ij} x_{ij}$$

be an arbitrary linear unbiased Estimator of μ_h; then

$$\sum_{i,j} c_{ij} \mu_i = \mu_h, \qquad \textit{for all} \quad \mu_1, \ldots, \mu_k.$$

This implies:

$$\sum_j c_{ij} = 1 \quad \text{if} \quad i = h; \quad = 0 \quad \text{if} \quad i \neq h;$$

therefore

$$\bar{c}_h = 1/n, \quad \bar{c}_i = 0 \quad \text{if} \quad i \neq h.$$

Put c_{ij} in place of x_{ij} in (13-6), and employ the given values of $\bar{c}_i$ above:

$$\sum_{i,j} c_{ij}^2 \geq n \sum_{i=1}^{k} \bar{c}_i^2 = 1/n.$$

This inequality implies that the variance of the linear Estimator, which is $\sigma^2 \Sigma c_{ij}^2$, is at least equal to σ^2/n; therefore $\bar{x}_h$ is a best one because it has that variance.

In both the normal and general cases the quantity SSE/ $k(n-1)$ is an unbiased Estimator of σ^2; indeed, the single average sum of squares

$$\frac{1}{n-1} \sum_{j=1}^{n} (x_{ij} - \bar{x}_i)^2$$

is an unbiased Estimator with respect to the ith population, so that the average of these Estimators over the k populations is also unbiased (Exercise 2).

We conclude this section with a remark on the validity of the one-way analysis-of-variance test when the underlying populations are not normal. If n is large and if the central limit theorem holds, then, as in the case of confidence intervals based on the t-

distribution (Sec. 8-4), the test based on the statistic (13-10) and the F-distribution is robust. To explain this, let us briefly show why the distribution of the ratio

$$\frac{n \sum_{i=1}^{k} (\bar{x}_i - \bar{x})^2/(k - 1)}{\text{SSE}/k(n - 1)}$$

is, for large values of n, almost the same *for all populations obeying the central limit theorem and the condition* (8-28). The denominator is an unbiased Estimator of σ^2 which is likely to be near σ^2 if n is very large; therefore, the distribution of the ratio is approximately the same as that of

$$\frac{\sum_{i=1}^{k} (\sqrt{n}\bar{x}_i - \sqrt{n}\bar{x})^2}{\sigma^2(k - 1)}$$

The statistics $\sqrt{n}\bar{x}_i$ are from different populations, and $\sqrt{n}\bar{x}$ is their average; furthermore, by the central limit theorem, the distribution of $\sqrt{n}\bar{x}_i$ is approximately normal with standard deviation σ; consequently, the distribution of the ratio above is similar to that of s^2/σ^2 in a Sample of k observations from a single normal population with standard deviation σ—the scaled chi-square distribution with the density $(k - 1)\psi_{k-1}((k - 1)u)$.

EXERCISES

1. Show that $\bar{x}_i$ is the efficient Estimator of μ_i in the normal case.

2. Let $f_1(\mathbf{x}), \ldots, f_k(\mathbf{x})$ be k unbiased Estimators, based on n observations, of a parameter θ. Prove that the average

$$\frac{1}{k}(f_1(\mathbf{x}) + \cdots + f_k(\mathbf{x}))$$

is also an unbiased Estimator.

13-5. A NUMERICAL EXAMPLE (OPTIONAL)

We shall use the data of Table 7-1 to illustrate the analysis of variance test for $k = 3$ and $n = 5$; here, of course, the null hypothesis holds—namely, the three populations are standard normal. The first three Samples in the table are considered to be from three populations, with five observations from each; then the ratio (13-12) is computed. This is repeated for the fourth, fifth, and sixth Samples, then for the seventh, eighth, and ninth, and so on up to the twenty-fourth; the last Sample is not used. The three sums appearing in the F-ratio (13-12) are displayed along with the latter in Table 13-1.

TABLE 13-1

F-Ratios for Seven Analysis-of-Variance Tests for $k = 3$ and $n = 5$.

Sample	$\left(\sum_{i,j} x_{ij}\right)^2$	$\sum_i \left(\sum_j x_{ij}\right)^2$	$\sum_{i,j} x_{ij}^2$	F-Ratio
1, 2, 3	45.590	24.128	12.920	1.324
4, 5, 6	6.518	5.332	9.966	.426
7, 8, 9	33.570	15.315	11.830	.565
10, 11, 12	3.176	11.718	15.148	.999
13, 14, 15	3.056	3.418	8.652	.361
16, 17, 18	2.338	1.347	13.673	1.061
19, 20, 21	.005	43.915	29.721	2.517
22, 23, 24	17.716	16.499	13.924	1.197

In Table IV, Appendix, we find that the upper .05-percentile of the F-distribution with $k - 1 = 2$ and $k(n - 1) = 12$ degrees of freedom is 8.51; thus each of the computed ratios falls below this percentile; hence, the null hypothesis is accepted in every one of the seven cases.

EXERCISE (Optional)

1. Using the normal deviates on the top of the second page of Table VI, Appendix, (instead of the first page as in Table 13-1) calculate seven F-ratios as in the previous example.

13-6. APPLICATION TO EDUCATIONAL TESTING (OPTIONAL)

We give an example illustrating the use of the analysis of variance in testing various educational methods. In recent years the number of students enrolled in courses in analytic geometry and calculus has increased. Three different ways of handling the large number of students has been proposed:

1. An experienced, talented instructor teaches a large number of students on closed circuit TV.
2. An instructor of similar ability teaches a large number of students in one classroom.
3. The students are divided in smaller sections of normal size and taught by individual instructors of varying ability.

The results of an experiment in this area were reported by Lancaster and Erskine in "*Achievement in Small Class, Large Class, and TV Instruction in College Mathematics*," *J. Engineering Education*, Vol. 52 (1962), pp. 583–598. Freshmen engineering students were assigned to a TV class, a large section, and various small sections in analytic geometry. The numbers of students in each category and the average final examination scores are given in the table below:

Type of Instruction	Number of Students	Average Final Examination Grade
TV	102	74.45
Large section	58	81.97
Twenty-two small sections	459	70.90
Total	619	72.53

We think of the students in the three categories as three normal populations, whose variances are assumed to be equal; thus $k = 3$. For simplicity of exposition our discussion of the analysis of variance was restricted to the case where the numbers of observations n from each population are equal; however, as in this particular example, the numbers are often not equal. The analysis-of-variance test has a more general form for unequal numbers of observations. Let $n_1, \ldots, n_k$ be the numbers from the respective populations. We put

$$\text{SSB} = \sum_{i=1}^{k} n_i(\bar{x}_i - \bar{x})^2$$

$$\text{SSE} = \sum_{i=1}^{k} \sum_{j=1}^{n_i} (x_{ij} - \bar{x}_i)^2,$$

where, as before, $\bar{x}_i$ is the ith group average and $\bar{x}$ is the overall average. By direct extension of the foregoing theory it can be shown that the ratio

$$\frac{\text{SSB}/(k - 1)}{\text{SSE}/\sum_i n_i - k}$$

has, under the null hypothesis, the F-distribution with $k - 1$ and $\sum_i n_i - k$ degrees of freedom; thus the statistic may be used for the test of homogeneity. In the example above we have

$$n_1 = 102, \quad n_2 = 58, \quad n_3 = 459;$$

thus $k - 1 = 2$ and $\Sigma n_i - k = 616$. The ratio above was found to be equal to 12.59. The table of the F-distribution (Table IV, Appendix) indicates that the upper .01-point on this distribution is about 4.61, which is much less than 12.59; thus (at the level of significance $\epsilon = .01$) we reject the null hypothesis that the instruction for the three groups is the same. The data indicate that the best results are obtained by a skilled instructor in a large class.

BIBLIOGRAPHY

The earliest book on the analysis of variance is R. A. Fisher, *Statistical Methods for Research Workers*, 12th ed. (London: Oliver and Boyd, 1954).

Two comprehensive works on the analysis of variance and its extensions are Henry Scheffe, *The Analysis of Variance* (New York: Wiley, 1959), and Oscar Kempthorne, *The Design and Analysis of Experiments* (New York: Wiley, 1952).

CHAPTER 14

Correlation and the Bivariate Normal Distribution

14-1. DEFINITION AND EMPIRICAL BACKGROUND OF THE BIVARIATE NORMAL DISTRIBUTION

In order to describe bivariate populations we recall the population of labeled balls in Chapter 1; there each ball was tagged by a single numerical value x. A bivariate population is one in which each ball is labeled by a *pair* of numbers (x, y). Let $x_1, x_2, \ldots$ and $y_1, y_2, \ldots$ be the sets of x and y labels, and let p_{ij} be the proportion of balls having the labels (x_i, y_j), $i = 1, 2, \ldots, j = 1, 2, \ldots$. Suppose a ball is drawn at random from the population, as described in Chapter 1; and let X and Y stand for the actual pair of labels x and y on the ball drawn. The pair (X, Y) is a *pair* of random variables; it is called a random pair or a random vector. The probability that it assumes the particular value (x_i, y_j) is defined as p_{ij}—the proportion of balls so labeled. The system of numbers (p_{ij}) is called the joint probability distribution of (X, Y). For each i, the proportion of balls whose x-label is x_i is the sum over j of the proportions whose pairs of values are (x_i, y_j); we call this proportion $p_{i\cdot}$:

$$p_{i\cdot} = \sum_j p_{ij}.$$

It represents the probability that the label X is equal to x_i. The system $(p_{i\cdot}: i = 1, 2, \ldots)$ is called the *marginal distribution of* X. The marginal distribution of Y is similarly defined: the probability that Y is equal to y_j is

$$p_{\cdot j} = \sum_i p_{ij}.$$

We introduce the *standard bivariate normal population with correlation coefficient* ρ. As in Chapter 2 imagine an increasing sequence of finite populations such that the sizes of the successive populations increase to infinity. For any set A in the xy-plane, let $P_n(A)$ stand for the probability that the random vector (X, Y) selected from the nth population assumes some value in the set A: it is the proportion of balls whose label pairs belong to A:

$$P_n(A) = \sum_{(x_i, y_j) \in A} p_{ij}.$$

Let $I = [a, b]$ and $J = [c, d]$ be intervals on the x- and y-axes, respectively, and let $I \times J$ be the rectangle $\{(x, y)\colon a \leq x \leq b, c \leq y \leq d\}$. A standard bivariate normal population is the "limit" of the increasing sequence of finite populations in the following sense:

The limit of $P_n(I \times J)$, $n \to \infty$, exists for every pair of intervals I and J. The limit, denoted $P(I \times J)$, is representable as the double integral

$$P(I \times J) = \int_a^b \int_c^d \frac{1}{2\pi\sqrt{1 - \rho^2}} e^{-[1/2(1-\rho^2)](x^2 - 2\rho xy + y^2)}\, dy\, dx,$$

where the parameter ρ satisfies $-1 < \rho < 1$. The latter is called the *correlation coefficient*.

The integrand above,

$$\phi(x, y; \rho) = (2\pi)^{-1}(1 - \rho^2)^{-1/2} \exp\left[\frac{x^2 - 2\rho xy + y^2}{2(1 - \rho^2)}\right], \tag{14-1}$$

is called the *standard bivariate normal density* with correlation coefficient ρ. Note that for $\rho = 0$ we have

$$\phi(x, y; 0) = \phi(x)\phi(y), \tag{14-2}$$

where the functions on the right-hand side are the usual standard normal densities of one variable.

The general bivariate normal population with means μ_1 and μ_2, standard deviations σ_1 and σ_2, and correlation coefficient ρ is defined in exactly the same way as the standard bivariate population except that $\phi(x, y; \rho)$ is replaced by

$$\frac{1}{\sigma_1 \sigma_2} \phi\left(\frac{x - \mu_1}{\sigma_1}, \frac{x - \mu_2}{\sigma_2}; \rho\right). \tag{14-3}$$

As in Sec. 2-2 many of the calculations for the standard density can be directly extended to the more general density.

Many populations studied in science and engineering can be thought of as bivariate populations and have approximately bivariate normal distributions. In Sec. 2-5 it was shown that measurements of certain biological characteristics have approximately normal distributions. It is also true that measurements of pairs often have bivariate normal distributions. As an example, consider the height and weight gains of every person in a given group over a specified period: each such pair of measurements is considered a label (x, y) corresponding to a member of the group or population. A classical example of a bivariate normal population is the measured heights of fathers and sons; it is of interest in studying heredity. Another example is furnished by the pairs of scores arising from a pair of tests taken by a population of students. Not all populations are normal. General bivariate populations are discussed in Sec. 14-5.

One of the main reasons for studying bivariate populations is to measure the association or *correlation* between the two characteristics in the population. It is clear from simple physical considerations that there is an association between the height and weight of a typical member of a population. The geneticist studies heights of fathers and sons to determine hereditary association. The psychologist studies scores on pairs of tests to measure the relation between two kinds of skills. The correlation coefficient ρ is a quantitative description of such an association.

In order to see how ρ measures association between two characteristics, let us analyze the bivariate normal density for various values of ρ. It is enough to consider just the standard bivariate normal density because the general density (14-3) is obtained from the standard one by linear transformation of the variables; thus we consider $\phi(x, y; \rho)$ for various values of ρ, $-1 < \rho < 1$. As a function of two variables x and y, the density is not representable by a two-dimensional graph. The domain of $\phi(x, y; \rho)$, for fixed ρ, is the xy-plane; its values at the various points may be conceived as the corresponding heights of a mound over the plane. The functional relation between the point (x, y) in the plane and the corresponding height $\phi(x, y; \rho)$ may be graphically represented by means of the device used in altitude maps—the lines of equal altitude. By analogy, we describe the bivariate density by means of the "curves of equal density"; these are the loci of points (x, y) in the plane satisfying the condition

$$\phi(x, y; \rho) = b \tag{14-4}$$

for various values $b > 0$. As defined in (14-1) the function ϕ depends on (x, y) only through the quadratic function of x and y in the exponent; thus Eq. 14-4 is equivalent to a corresponding equation

$$(1 - \rho^2)^{-1}(x^2 - 2\rho xy + y^2) = c \tag{14-5}$$

for the quadratic function. The locus of points satisfying this equation is an ellipse. If $\rho > 0$, the major axis is along the line $x = y$; if $\rho < 0$, it is along $x = -y$. The lengths of the major and minor axes are $2\sqrt{c(1 + |\rho|)}$ and $2\sqrt{c(1 - |\rho|)}$, respectively (Exercise 2). If $\rho = 0$, then the ellipse becomes a circle of radius $\sqrt{c}$. The graph of the ellipse for positive, negative and zero values of ρ is displayed in Fig. 14-1. As ρ tends to either 1 or -1, the length of the major axis tends to $2\sqrt{2c}$ and that of the minor axis converges to 0. This indicates that the ellipse collapses onto the major axis line.

The density is symmetric about the line $x = y$ and about the line $x = -y$. This means that if (x_1, y_1) and (x_2, y_2) are points on a line perpendicular to the line $x = y$ (or $x = -y$), and equidistant from the latter line, then $\phi(x_1, y_1; \rho) = \phi(x_2, y_2; \rho)$. This follows from the fact that the lines of equal density form an ellipse, which is symmetric about the major and minor axes.

The density has a maximum at the origin, and decreases as the ellipse of equal density expands: the ellipse corresponding to a curve of equal density b_1 contains in its interior the one of density b_2 if $b_1 < b_2$. A positive correlation—a value of $\rho > 0$—signifies that the size of the x-label of a typical member of the bivariate population is likely to be large if the y-label is, and conversely. The reason for this is that the ellipses of equal density have their major axes along the line $x = y$ when $\rho > 0$, so that large values of x on the ellipses tend to have correspondingly large y-values, and conversely. In the extreme case $\rho = 1$, the ellipses degenerate into line segments along the line $x = y$; thus every x-label is equal to the accompanying y-value. If $\rho < 0$, the major axes of the ellipses of equal density are along the line $x = -y$ so that small x-values are associated with large y-values, and, conversely, large

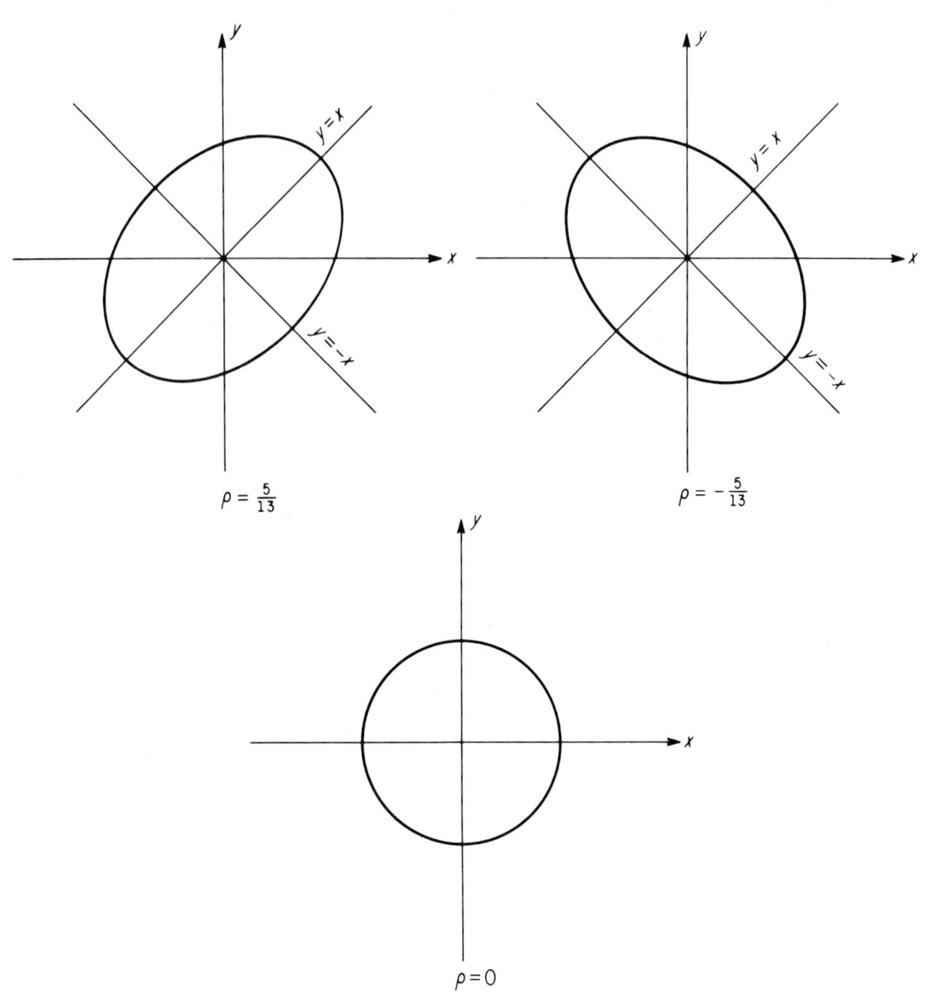

Fig. 14-1. Curves of equal density for positive, negative, and zero values of ρ.

x-values with small y-values. If $\rho = 0$, the density (14-2) is invariant under rotations about the origin; in this case, the x- and y-labels of the members of the population are "unrelated" because, for any value of x, the corresponding y-value is just as likely to be small as large. In this way the parameter ρ measures the correlation between two numerical labels; for this reason it is called the correlation coefficient.

An example of a bivariate normal population is furnished by the diameters and weights of ears of maize. The frequencies for 327 ears are given in Table 14-1.

TABLE 14-1
Frequency of Occurrence of Ears of Maize Having Each Diameter and Weight.*

Y = Weight, Grams	X = Diameter, Millimeters																f_y
	36	37	38	39	40	41	42	43	44	45	46	47	48	49	50	51	
320													1				1
310												1					1
300													3				3
290											1		1				2
280										1	2	1			4		8
270									1	1	1		2	1		1	7
260							1	3	2	7		3	3				19
250								1	2		3	2	2	1			11
240							3	1	4	3	5	6			1		23
230						1	1	1	4	4	3	4		2			20
220						1	2	6	7	3	1	1	1		1		23
210						4	1	7	5	4	3	3			1		28
200				1	1	3	2	6	6	4	3	1	2				29
190						1	2	5	5	11	2	5	1				32
180				2		4	4	6	3	2	2	4	1				28
170			1	1		2	5	5	1	4			2				21
160				1	5		1	2	4	2	3				1		19
150			1		1	2	1	4	1	2	1	1					14
140					3	3	2	1		1		1					11
130						2	1										3
120			1	1	5			2									9
110						2					1						3
100					1			1									2
90			1				1		2								4
80					1	1	1										3
70				1													1
60					1												1
50	1																1
f_x	1	0	4	7	18	26	28	51	47	49	31	33	19	4	8	1	327

*Reprinted by permission from STATISTICAL METHODS, 6th edition, by George W. Snedecor and William C. Cochran, ©1967 by the Iowa State University Press, Ames Iowa.

EXERCISES

1. Let (x', y') be the image of (x, y) under the rotation carrying the diagonal line onto the x-axis—that is, carrying $(1, 1)$ onto $(\sqrt{2}, 0)$. Write the left-hand side of (14-5) in terms of the coordinates x' and y'.

2. Use the result of Exercise 1 to verify that the lengths of the major and minor axes are as indicated following (14-5).

3. *Prove:* The double integral of $\phi(x, y; \rho)$ over the interior of the ellipse (14-5) is equal to $1 - e^{-c/2}$. (*Hint:* Use the rotation in Exercise 1.) Note that the integral is independent of ρ.

14-2. ANALYTIC PROPERTIES OF THE BIVARIATE NORMAL DISTRIBUTION

The introduction of the bivariate density in the previous section was mathematically arbitrary; we gave no explanation for its particular functional form. In this section we shall show how it arises in sampling from ordinary (univariate) normal populations; then we shall relate it to the standard univariate normal density, and derive some integral formulas.

The bivariate density $\phi(x, y; \rho)$ arises as the joint density of certain linear functions of the coordinates of the Sample point in sampling from a normal population. The main identity used in the proof of this and other facts about the bivariate density is

$$\phi(x, y; \rho) = \phi(x)\,\phi\left(\frac{y - x\rho}{\sqrt{1 - \rho^2}}\right)\frac{1}{\sqrt{1 - \rho^2}} \tag{14-6}$$

(Exercise 1).

THEOREM 14-1

In a Sample of two observations from a standard normal population, the joint density of the two statistics

$$f_1(x_1, x_2) = x_1 \quad \text{and} \quad f_2(x_1, x_2) = \rho x_1 + \sqrt{1 - \rho^2}\,x_2 \tag{14-7}$$

at the point (x, y) is $\phi(x, y; \rho)$.

Proof: By definition (Sec. 5-1) the joint distribution of these two statistics at the point (x, y) is the integral of $\phi(x_1)\,\phi(x_2)$ over the portion of the x_1x_2—plane which is to the left of the vertical line $x_1 = x$, and below (and left) of the line $y = x_1\rho + \sqrt{1 - \rho^2}x_2$; this region is portrayed in Fig. 14-2.

The double integral is computed by fixing x_1, integrating over x_2 from $-\infty$ to $(y - x_1\rho)/\sqrt{1 - \rho^2}$, and then integrating over x_1 from $-\infty$ to x:

$$\int_{-\infty}^{x} \int_{-\infty}^{(y - x_1\rho)/\sqrt{1-\rho^2}} \phi(x_1)\,\phi(x_2)\,dx_2\,dx_1.$$

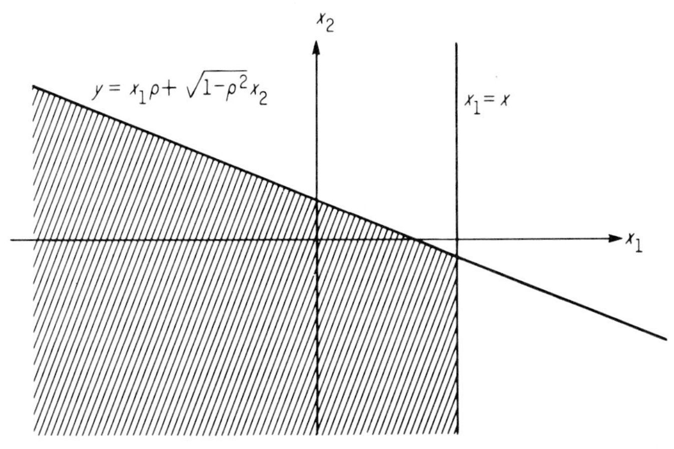

Fig. 14-2. Domain of integration for the joint distribution of the linear functions.

Change the variable of integration in the inner integral from x_2 to $(x_2 - x_1\rho)/\sqrt{1-\rho^2}$; the double integral becomes

$$\int_{-\infty}^{x}\int_{-\infty}^{y}\phi(x_1)\,\phi\left(\frac{x_2 - x_1\rho}{\sqrt{1-\rho^2}}\right)\frac{1}{\sqrt{1-\rho^2}}\,dx_2\,dx_1,$$

which, by (14-6), is

$$\int_{-\infty}^{x}\int_{-\infty}^{y}\phi(x_1, x_2; \rho)\,dx_2\,dx_1.$$

The second-order mixed derivative of the joint distribution is equal to $\phi(x, y; \rho)$. ■

The function $\phi(x, y; \rho)$ is our first example of a bivariate density which is not a product of two univariate densities except when $\rho = 0$. This means that the statistics (14-7), which have this joint density, are not independent for $\rho \neq 0$. (The concept of independence is defined in Sec. 5-1.) The effect of ρ on the dependence between the statistics (14-7) is illustrated by Theorem 14-1: ρ is a measure of linear association. If $\rho > 0$, then f_2 varies linearly and positively with $f_1 = x_1$; furthermore, if ρ is nearly 1, then f_2 and f_1 are nearly equal so that the association between the two is very strong. If $\rho = 0$, then f_1 and f_2 are identical with x_1 and x_2, respectively, so that there is no association between them. If $\rho < 0$, then f_2 varies linearly and negatively with f_1.

An immediate consequence of Theorem 14-1 is that the marginal distributions associated with $\phi(x, y; \rho)$ are standard normal. (The marginal distribution is defined in Sec. 5-1.) This follows from the fact that f_1 has a standard normal distribution; and, as a linear function of x_1 and x_2, the statistic f_2 has a normal distribution with mean 0 and standard deviation $(\rho^2 + (1 - \rho^2))^{1/2} = 1$ (Theorem 11-4).

The parameter ρ satisfies the equation

$$\int_{-\infty}^{\infty} \int_{-\infty}^{\infty} xy\,\phi(x, y; \rho)\,dx\,dy = \rho; \tag{14-8}$$

this follows from Eq. 14-6 (Exercise 2). A more general version of (14-8) is

$$\int_{-\infty}^{\infty} \int_{-\infty}^{\infty} (x - \mu_1)(y - \mu_2)\,\phi\left(\frac{x - \mu_1}{\sigma_1}, \frac{y - \mu_2}{\sigma_2}; \rho\right) dx\,dy = \rho\sigma_1\sigma_2 \tag{14-9}$$

(Exercise 3).

Formula (14-6) shows that the standard bivariate normal density is the product of a "marginal density" and a "conditional density," that is, the product of (i) $\phi(x)$, the standard normal, and (ii) the function

$$\frac{1}{\sqrt{1 - \rho^2}}\,\phi\left(\frac{y - x\rho}{\sqrt{1 - \rho^2}}\right)$$

which, for each x, is a normal density with mean $x\rho$ and standard deviation $\sqrt{1 - \rho^2}$.

The second density is called the "conditional density of y given x." By Eq. 14-6 this conditional density is equal to

$$\frac{\phi(x, y; \rho)}{\phi(x)}.$$

In terms of the original bivariate normal population this ratio represents the relative proportion (divided by dy) of the population whose second label falls in $[y, y + dy]$, relative to those whose first label falls in $[x, x + dx]$. It is remarkable that this conditional density is also a normal one. The mean $x\rho$ of the conditional density is called the "conditional mean of y given x," and $1 - \rho^2$ is called the *conditional variance*. The line $y = \rho x$ in the plane is called the regression line of y on x. The conditional density has the following geometric significance in terms of the mound over the xy-plane which represents the bivariate normal density. If the mound is cut by a plane perpendicular to the xy-plane and parallel to the y-axis, then the intersection of this plane with the surface of the mound is a curve with the shape of the conditional density: if the cutting plane passes through the point x_0 on the x-axis, it is the conditional density given x_0. It is interesting that the variance of the conditional density is the same for all x (Fig. 14-3).

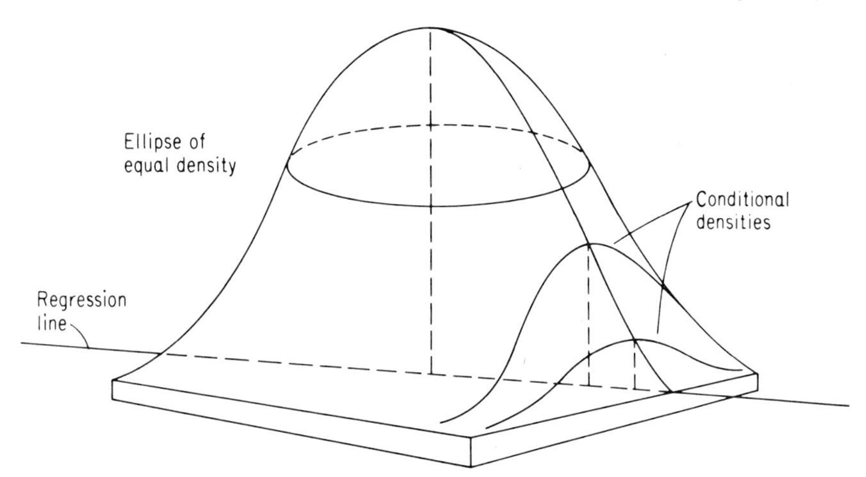

Fig. 14-3. The surface of the bivariate normal mound and the conditional densities.

The equation for the general bivariate normal density corresponding to (14-6) is

$$\frac{1}{\sigma_1}\frac{1}{\sigma_2}\phi\left(\frac{x-\mu_1}{\sigma_1}, \frac{y-\mu_2}{\sigma_2}; \rho\right) = \frac{1}{\sigma_1}\phi\left(\frac{x_1-\mu_1}{\sigma_1}\right)\cdot\frac{1}{\sigma_2\sqrt{1-\rho^2}}\cdot$$
$$\phi\left(\frac{y-\mu_2-(\sigma_2\rho/\sigma_1)(x-\mu_1)}{\sigma_2\sqrt{1-\rho^2}}\right).$$

In this case the conditional density is

$$\frac{1}{\sigma_2\sqrt{1-\rho^2}}\cdot\phi\left(\frac{y-\mu_2-(\sigma_2\rho/\sigma_1)(x-\mu_1)}{\sigma_2\sqrt{1-\rho^2}}\right)$$

which, for fixed x, is a normal density with mean

$$\mu_2 + (\sigma_2\rho/\sigma_1)(x-\mu_1)$$

and variance $\sigma_2^2(1 - \rho^2)$. As a function of x, the mean is called the regression of y on x. The geometric interpretation of the conditional density is the same as in the standard bivariate case.

We point out a difference between the linear regression model in Chapter 12 and the present bivariate normal model—a difference in terms of applications. The former is used in the analysis of the relation between a *nonrandom* independent variable and a *random* dependent variable, for example, the relation between *time* and a *random* economic time series. The bivariate normal model is used in analyzing the relation between *two random* variables, for example, the diameter and weight of an ear of corn (Table 14-1).

EXERCISES

1. By elementary algebra verify the identity (14-6).

2. Verify (14-8) by applying (14-6) and evaluating the double integral by integrating over y and then over x.

3. By changing the variables of integration, derive (14-9) from (14-8).

4. Let (X_1, X_2) be a Sample from two normal populations with means μ_1 and μ_2 and standard deviations σ_1 and σ_2, respectively. Show that the statistics

$$f_1 = ax_1 + bx_2 \qquad \text{and} \qquad f_2 = cx_1 + dx_2$$

have a bivariate normal distribution with means

$$a\mu_1 + b\mu_2, \qquad c\mu_1 + d\mu_2,$$

standard deviations

$$(a^2\sigma_1^2 + b^2\sigma_2^2)^{1/2}, \qquad (c^2\sigma_1^2 + d^2\sigma_2^2)^{1/2},$$

and correlation coefficient

$$\frac{ac\sigma_1^2 + bd\sigma_2^2}{(a^2\sigma_1^2 + b^2\sigma_2^2)^{1/2}(c^2\sigma_1^2 + d^2\sigma_2^2)^{1/2}},$$

provided the latter is strictly between -1 and $+1$. (*Hint:* Extend the proof of Theorem 14-1.)

5. *Prove:*

$$\int_{-\infty}^{\infty}\int_{-\infty}^{\infty} x\phi(x, y; \rho)\, dx\, dy = 0.$$

6. *Prove:*

$$\int_{-\infty}^{\infty} \int_{-\infty}^{\infty} x^2 \phi(x, y; \rho)\, dx\, dy = 1.$$

7. Let f_1 and f_2 be the two statistics on the standard normal Sample space defined by (14-7). Show that the two statistics

$$af_1(x_1, x_2) + b \qquad \text{and} \qquad cf_2(x_1, x_2) + d$$

have a bivariate normal distribution with the same correlation coefficient ρ. (*Hint:* Use Theorem 14-1 and extend its proof.)

8. Show that the marginal distributions associated with (14-3) have the densities

$$\frac{1}{\sigma_1} \phi\left(\frac{x - \mu_1}{\sigma_1}\right) \qquad \text{and} \qquad \frac{1}{\sigma_2} \phi\left(\frac{y - \mu_2}{\sigma_2}\right),$$

respectively.

14-3. SAMPLING FROM A BIVARIATE NORMAL POPULATION; THE SAMPLE CORRELATION COEFFICIENT

The theory of sampling from finite populations, described in Chapter 1, can be extended to sampling from bivariate populations. The Sample of n observations is replaced by a Sample of n pairs of observations

$$(X_1, Y_1), \ldots, (X_n, Y_n).$$

The Sample space is the $2n$-dimensional space of points represented by multiplets of real numbers $(x_1, y_1, \ldots, x_n, y_n)$. We shall denote a point in this space by an ordered pair of points $(\mathbf{x}, \mathbf{y})$ in R^n:

$$(\mathbf{x}, \mathbf{y}) = (x_1, y_1, \ldots, x_n, y_n).$$

The likelihood function (1-8) is replaced by a more general product over the probabilities p_{ij} of two indices. We shall not elaborate on the sampling theory for finite populations but go directly to the sampling theory for bivariate normal populations. The likelihood function for the Sample space associated with n pairs of observations from the population with the density (14-3) is the product of n such functions:

$$L(\mathbf{x}, \mathbf{y}; \mu_1, \mu_2, \sigma_1, \sigma_2, \rho) = \prod_{i=1}^{n} \frac{1}{\sigma_1 \sigma_2} \phi\left(\frac{x_i - \mu_1}{\sigma_1}, \frac{y_i - \mu_2}{\sigma_2}; \rho\right).$$

By the particular form of $\phi(x, y; \rho)$, it is equal to

$$(2\pi\sigma_1\sigma_2(1-\rho^2)^{1/2})^{-n}\exp\left(-\frac{1}{2(1-\rho^2)}\left[\sum_{i=1}^{n}\left(\frac{x_i-\mu_1}{\sigma_1}\right)^2\right.\right.$$
$$\left.\left.-\frac{2\rho}{\sigma_1\sigma_2}\sum_{i=1}^{n}(x_i-\mu_1)(y_i-\mu_2)+\sum_{i=1}^{n}\left(\frac{y_i-\mu_2}{\sigma_2}\right)^2\right]\right). \qquad (14\text{-}10)$$

The parameters may be unknown or partially known, and are estimated or tested on the basis of a Sample from the bivariate population. The first four parameters—the means μ_1 and μ_2 and the standard deviations σ_1 and σ_2—are also parameters in the marginal distributions (Exercise 8, Sec. 14-2); hence, they can be estimated or tested on the basis of just the x or y coordinates of the Sample point, so that the previous univariate theory covers their analysis. The parameter of prime interest in a bivariate population is the correlation coefficient ρ because it measures the association between the two characteristics of the population.

The following five statistics are used in the theory of inference for the bivariate normal population:

$$\bar{x} = (x_1 + \cdots + x_n)/n, \qquad \bar{y} = (y_1 + \cdots + y_n)/n$$

$$s_1^2 = \sum_{i=1}^{n}(x_i - \bar{x})^2/(n-1), \qquad s_2^2 = \sum_{i=1}^{n}(y_i - \bar{y})^2/(n-1),$$

and, finally, the "Sample correlation coefficient,"

$$r = \frac{\sum_{i=1}^{n}(x_i-\bar{x})(y_i-\bar{y})}{\left\{\sum_{i=1}^{n}(x_i-\bar{x})^2\cdot\sum_{i=1}^{n}(y_i-\bar{y})^2\right\}^{1/2}}. \qquad (14\text{-}11)$$

As in the univariate case, $\bar{x}$, $\bar{y}$, s_1^2, and s_2^2 are used for estimation and testing of the population means and standard deviations. The statistic r is the usual Estimator of ρ. It can be shown that these five statistics form a set of "sufficient statistics" (Exercise 7).

We briefly explain the use of r as an Estimator of ρ. Just as $\bar{x}$ and s^2 are Sample analogues of the population mean and variance, so is r the analogue of the population correlation coefficient. This may be seen as follows. Divide the numerator and denominator of the fraction (14-11) by n; then r is the ratio of the average of $(x_i - \bar{x})(y_i - \bar{y})$ to the product of the square roots of the averages of $(x_i - \bar{x})^2$ and $(y_i - \bar{y})^2$, respectively. Compare this to the equation (14-9) for ρ: the latter is the ratio of the integral of $(x - \mu_1)(y - \mu_2)\phi$ to the square roots of the integrals of $(x - \mu_1)^2\phi$ and $(y - \mu_2)^2\phi$.

The derivation of the distribution of r, except for $\rho = 0$, is too involved to be included here. For this reason it is not possible for us to demonstrate any optimal properties of r; however, we shall just indicate why it is a good Estimator of ρ when the number n of observations is large. The following generalization of the decomposition (4-12) holds: *For any pair of sets of real numbers* $x_1, \ldots, x_n$ *and* $y_1, \ldots, y_n$ *the relations*

$$\sum_{i=1}^{n} (x_i - \bar{x})(y_i - \bar{y}) = \sum_{i=1}^{n} x_i(y_i - \bar{y}) = \sum_{i=1}^{n} x_i y_i - n\bar{x}\bar{y} \qquad (14\text{-}12)$$

hold. The proof is similar to that of (4-12) (Exercise 1). It follows that r may be expressed as

$$r = \frac{\frac{1}{n}\sum_{i=1}^{n} x_i y_i - \bar{x}\bar{y}}{\sqrt{(n-1)/n}\, s_1 s_2}. \qquad (14\text{-}13)$$

The functions $\bar{x}$, $\bar{y}$, s_1^2, and s_2^2 are unbiased Estimators of μ_1, μ_2, σ_1^2 and σ_2^2 with variances equal to multiples of $1/n$ or $1/(n-1)$ (Sec. 7-3 and 7-4); thus if n is very large, they are very "likely" to be close to the true parameter values (Exercise 8). The function

$$\frac{1}{n}\sum_{i=1}^{n} x_i y_i \qquad (14\text{-}14)$$

is an unbiased Estimator of $\rho\sigma_1\sigma_2 + \mu_1\mu_2$ with variance equal to a constant multiple of $1/n$ (Exercise 9); thus, it is also likely to be close to $\rho\sigma_1\sigma_2 + \mu_1\mu_2$ if n is very large. It follows that, for large n, the ratio (14-13) is likely to be near to the value of the corresponding function of the parameters

$$\frac{(\rho\sigma_1\sigma_2 + \mu_1\mu_2) - \mu_1\mu_2}{\sigma_1\sigma_2} = \rho.$$

EXERCISES

1. Verify the identity (14-12).

2. By applying the decompositions (4-12) and (14-12) prove that the co-

efficient of $-1/2(1-\rho^2)$ in the exponent of the likelihood function (14-10) may be expressed as

$$\sum_{i=1}^{n}\left(\frac{x_i-\bar{x}}{\sigma_1}\right)^2-\frac{2\rho}{\sigma_1\sigma_2}\sum_{i=1}^{n}(x_i-\bar{x})(y_i-\bar{y})+\sum_{i=1}^{n}\left(\frac{y_i-\bar{y}}{\sigma_2}\right)^2$$
$$+\,n\left(\frac{\bar{x}-\mu_1}{\sigma_1}\right)^2-\frac{2\rho n}{\sigma_1\sigma_2}(\bar{x}-\mu_1)(\bar{y}-\mu_2)+n\left(\frac{\bar{y}-\mu_2}{\sigma_2}\right)^2.$$

3. Let $\mathbf{x}'$ and $\mathbf{y}'$ be the images of two points $\mathbf{x}$ and $\mathbf{y}$ in R^n under a diagonal rotation. By applying the invariance of $\sum_{i=1}^{n} x_i y_i$ under rotation (Exercise 5, Sec. 11-2), the identity (14-12), and the relations (5-8) between the coordinates of $\mathbf{x}$ and $\mathbf{y}$ and $\mathbf{x}'$ and $\mathbf{y}'$, respectively, prove that

$$\frac{\sum_{i=1}^{n}(x_i-\bar{x})(y_i-\bar{y})}{\left[\sum_{i=1}^{n}(x_i-\bar{x})^2\cdot\sum_{i=1}^{n}(y_i-\bar{y})^2\right]^{1/2}}=\frac{\sum_{i=2}^{n}x_i'y_i'}{\left[\sum_{i=2}^{n}x_i'^2\cdot\sum_{i=2}^{n}y_i'^2\right]^{1/2}}.$$

4. For a continuous function $f(\mathbf{x},\mathbf{y})$, $\mathbf{x},\mathbf{y}\in R^n$, let f^* be defined as

$$f^*(\mathbf{x}',\mathbf{y}')=f(\mathbf{x},\mathbf{y}),$$

where $\mathbf{x}'$ and $\mathbf{y}'$ are as in Exercise 3. Show that f^* is a continuous function of the $2n$ variables

$$x_1',\ y_1',\ x_i'+y_i',\ x_i'-y_i',\qquad i=2,\ldots,n$$

that is, there exists a continuous function f_1^* of these variables such that

$$f^*(\mathbf{x}',\mathbf{y}')=f_1^*(x_1',y_1',\ x_i'+y_i',\ x_i'-y_i',\qquad i=2,\ldots,n).$$

5. For fixed x_1', y_1', let $f_2^*\left(x_1',y_1',\sum_{i=2}^{n}(x_i'+y_i')^2,\sum_{i=2}^{n}(x_i'-y_i')^2\right)$ be the iterated spherical average of f_1^* over the two sets of variables $x_i'+y_i'$, $i=2,\ldots,n$, and $x_i'-y_i'$, $i=2,\ldots,n$. Show that f_2^* is also a function of

$$x_1',\ y_1',\ \sum_{i=2}^{n}x_i'^2,\ \sum_{i=2}^{n}y_i'^2,\ \sum_{i=2}^{n}x_i'y_i'.$$

(*Hint:* Use Exercise 4, Sec. 11-2.)

6. Write f_2^*, a function of the five variables in Exercise 5, as a function of the original variables **x** and **y**. Show that it is a function of the five statistics $\bar{x}$, $\bar{y}$, s_1^2, s_2^2, and r. This is the conditional expectation of f given these statistics.

7. Prove the analogue of Theorem 6-1 for the conditional expectation constructed in Exercise 6. This result signifies that the five statistics form a sufficient set.

8. Using the result of Exercise 5, Sec. 7-1, show that an unbiased Estimator is very likely to be close to the parameter value if the variance is inversely proportional to the number of observations and the latter is large.

9. Prove that the statistic (14-14) is an unbiased Estimator and that its variance is a multiple of $1/n$.

10. Applying the Cauchy-Schwarz inequality (Exercise 9, Sec. 3-1), prove that $|r| \leq 1$.

14-4. TESTING THE HYPOTHESIS $\rho = 0$ (OPTIONAL)

An important problem in the analysis of bivariate populations is to determine whether or not the correlation is 0; in other words, to find out if the two characteristics of the population are linearly unrelated or not. As shown in the previous section, the statistic r is an appropriate Estimator of ρ; thus, if a Sample is drawn from the population and the computed value of r is close to 0, we infer that $\rho = 0$; on the other hand, if r is far from 0, we infer that $\rho \neq 0$. This is the basis of the classical procedure for testing the null hypothesis $\rho = 0$ against the two-sided composite alternative $\rho \neq 0$. Let $t = t_{n-2,\alpha/2}$ be the upper $\alpha/2$-percentile of the t-distribution with $n - 2$ degrees of freedom. The rejection region of this test is the set of points in the $2n$-dimensional space where

$$r^2 > \frac{t^2/(n-2)}{1 + t^2/(n-2)}; \tag{14-15}$$

in other words, the null hypothesis $\rho = 0$ is rejected if r^2 exceeds the ratio on the right-hand side of (14-15). The inequality (14-15) is equivalent to

$$\frac{\sqrt{n-2}\,|r|}{\sqrt{1-r^2}} > t; \tag{14-16}$$

thus the probability that the Sample point falls in the rejection region can be computed from the distribution of the function of r,

$$\frac{\sqrt{n-2}\,r}{\sqrt{1-r^2}}. \tag{14-17}$$

As a function of $\mathbf{x}$ and $\mathbf{y}$, r is invariant under multiplication of the coordinates of $\mathbf{x}$ by a common positive constant, and the addition of a common constant; the same is true for $\mathbf{y}$. This means that the value of r at the point $(\mathbf{x}, \mathbf{y})$ is the same as its value at the point $(a\mathbf{x} + b\mathbf{1}, c\mathbf{y} + d\mathbf{1})$, $a > 0$, $c > 0$; this follows immediately from the definition of r (Exercise 1). This fact implies that the distribution of r is the same for all μ_1, μ_2, σ_1, and σ_2, and so depends only on the value of ρ (Exercise 4); thus the same is true of the statistic (14-17). It follows that the probability of an error of the first kind for the test with the rejection region (14-16) has the same value for all μ_1, μ_2, σ_1, and σ_2, and so the probability may be computed by considering only the special case where $\mu_1 = \mu_2 = 0$ and $\sigma_1 = \sigma_2 = 1$. We shall not derive the power function because, as indicated in the previous section, the distribution of r is very complicated when $\rho \neq 0$; however, we shall show that the statistic (14-17) has, under the null hypothesis $\rho = 0$, the t-distribution with $n - 2$ degrees of freedom so that the test with the rejection region (14-16) has the level of significance α.

THEOREM 14-2

When $\rho = 0$ the statistic (14-17) has the t-distribution with $n - 2$ degrees of freedom.

Proof: By the previous remarks it suffices to consider the particular case $\mu_1 = \mu_2 = 0$, $\sigma_1 = \sigma_2 = 1$; thus when $\rho = 0$ the likelihood function (14-10) is

$$\prod_{i=1}^{n} \phi(x_i) \prod_{i=1}^{n} \phi(y_i). \tag{14-18}$$

The distribution of the statistic (14-17) at the point u is the integral of (14-18) over the set in $2n$-dimensional space where

$$\frac{\sqrt{n-2}\,r}{\sqrt{1-r^2}} \leq u. \tag{14-19}$$

We shall identify this integral by transforming (14-17). By the identity (14-12), r may be written as

$$r = \frac{\sum_{i=1}^{n} x_i(y_i - \bar{y})}{\left[\sum_{i=1}^{n} (x_i - \bar{x})^2 \cdot \sum_{i=1}^{n} (y_i - \bar{y})^2\right]^{1/2}}.$$

Put $t_i = y_i - \bar{y}$, and $b = \sum_{i=1}^{n} x_i t_i \Big/ \sum_{i=1}^{n} t_i^2$; thus

$$r = \frac{b\,\|\mathbf{t}\|}{\left[\sum_{i=1}^{n} (x_i - \bar{x})^2\right]^{1/2}},$$

and so

$$\frac{\sqrt{n-2}\,r}{\sqrt{1-r^2}} = \frac{b\,\|\mathbf{t}\|}{\left[\frac{1}{n-2}\left(\sum_{i=1}^{n} (x_i - \bar{x})^2 - b^2\|\mathbf{t}\|^2\right)\right]^{1/2}}. \tag{14-20}$$

The numbers $t_1, \ldots, t_n$ satisfy the condition (12-3):

$$\sum_{i=1}^{n} t_i = \sum_{i=1}^{n} (y_i - \bar{y}) = \sum_{i=1}^{n} y_i - n\bar{y} = 0;$$

thus, by the sum of squares decomposition (12-5), with $x_i - \bar{x}$ in place of x_i, we have

$$\sum_{i=1}^{n} (x_i - \bar{x})^2 = \sum_{i=1}^{n} (x_i - \bar{x} - bt_i)^2 + b^2\|\mathbf{t}\|^2.$$

From this and (14-20) we get

$$\frac{\sqrt{n-2}\,r}{\sqrt{1-r^2}} = \frac{b\,\|\mathbf{t}\|}{\left[\sum_{i=1}^{n} (x_i - \bar{x} - bt_i)^2/(n-2)\right]^{1/2}};$$

therefore the region (14-19) is identical with the region where

$$\frac{b\,\|\mathbf{t}\|}{\left[\sum_{i=1}^{n}(x_i - \bar{x} - bt_i)^2/(n-2)\right]^{1/2}} \le u, \tag{14-21}$$

where $t_i = (y_i - \bar{y})$.

The multiple integral of (14-18) over the region (14-21) will be computed by fixing $y_1, \ldots, y_n$, integrating over $x_1, \ldots, x_n$, and then integrating over the former variables. For fixed $y_1, \ldots, y_n$ consider the integral of (14-18) with respect to $x_1, \ldots, x_n$ over the set (14-21): the product $\prod_{i=1}^{n} \phi(y_i)$ is fixed and may be factored out of the integral, while the numbers $t_1, \ldots, t_n$ are also fixed by the values of the y's. It follows that the integral with respect to the x's is the same as that defining the distribution of

$$\frac{b\,\|\mathbf{t}\|}{\left[\sum_{i=1}^{n}(x_i - \bar{x} - bt_i)^2/(n-2)\right]^{1/2}}$$

for fixed $t_1, \ldots, t_n$ and with respect to the likelihood function $\prod_{i=1}^{n} \phi(x_i)$. This distribution has been shown in Theorem 12-5 (for $\beta = 0$) to be the t-distribution with $n - 2$ degrees of freedom. The latter distribution does not depend on the particular values $t_1, \ldots, t_n$; thus, the multiple integral of $x_1, \ldots, x_n$ for fixed $y_1, \ldots, y_n$ does not depend on the latter, and so the integral

$$\underset{\{\text{Region (14-21) with } \mathbf{y} \text{ fixed}\}}{\int \cdots \int} \prod_{i=1}^{n} \phi(x_i)\, d\mathbf{x} \tag{14-22}$$

has the same value for all $\mathbf{y}$. Now multiply by $\prod_{i=1}^{n} \phi(y_i)$ and integrate with respect to $y_1, \ldots, y_n$: since (14-22) does not depend on $\mathbf{y}$, it may be factored out of the integral, and the integral with respect to $\mathbf{y}$ is

$$\int_{-\infty}^{\infty} \int_{-\infty}^{\infty} \prod_{i=1}^{n} \phi(y_i)\, d\mathbf{y} = 1.$$

It follows that the multiple integral of (14-18) over (14-21) is equal to the n-fold integral (14-22), which has just been shown to be the t-distribution with $n - 2$ degrees of freedom. ■

EXERCISES

1. A function $f(\mathbf{x}, \mathbf{y}), \mathbf{x}, \mathbf{y} \in R^n$ is called *invariant if*

$$f(a\mathbf{x} + b\mathbf{1}, c\mathbf{y} + d\mathbf{1}) = f(\mathbf{x}, \mathbf{y})$$

for all $a > 0$ and $c > 0$. Prove that r is invariant.

2. Let $f(\mathbf{x}, \mathbf{y})$ be a function depending on $\mathbf{x}$ and $\mathbf{y}$ only through the five variables

$$\bar{x},\ \bar{y},\ s_1^2,\ s_2^2,\quad \text{and} \quad r.$$

Prove: If $\bar{f}$ is also invariant then it necessarily depends on $\mathbf{x}$ and $\mathbf{y}$ through r alone.

3. Let $\bar{f}(\mathbf{x}, \mathbf{y})$ be a continuous invariant function on R^{2n}. Show that there exists a continuous function $f(\mathbf{x}, \mathbf{y})$ depending on $(\mathbf{x}, \mathbf{y})$ through r alone such that

$$\int \cdots \int_{R^{2n}} f L \, d\mathbf{x}\, d\mathbf{y} = \int \cdots \int_{R^{2n}} \bar{f} L \, d\mathbf{x}\, d\mathbf{y}$$

where L is defined by (14-10). This result signifies that r is "sufficient among invariant statistics."

4. Prove that the distribution of r depends only on ρ and not on the other four parameters. [*Hint:* Write the multiple integral defining the distribution, change the variables of integration from x_i and y_i to $(x_i - \mu_1)/\sigma_1$ and $(y_i - \mu_2)/\sigma_2$, respectively, $i = 1, \ldots, n$, and apply the result of Exercise 1.]

14-5. THE GENERAL BIVARIATE POPULATION (OPTIONAL)

There is little organized analytical work on bivariate populations other than the normal. A bivariate population with a density $h(x, y)$ is defined in the same way as the bivariate normal except that the special normal density ϕ is replaced by a more general one. The properties of a bivariate density $h(x, y)$ are exactly those of the joint density of a pair of statistics, discussed in Sec. 5-1: it is nonnegative, and

$$\int_{-\infty}^{\infty} \int_{-\infty}^{\infty} h(x, y)\, dx\, dy = 1.$$

The marginal density of x is obtained by integration over y:

$$\int_{-\infty}^{\infty} h(x, y)\, dy.$$

In the normal case the correlation coefficient appears as an important parameter in the bivariate density. The equations (14-8) (14-9) suggest a definition of correlation in the general case. Let μ_1, μ_2, σ_1, and σ_2 represent the means and standard deviations of the marginal distributions:

$$\mu_1 = \int_{-\infty}^{\infty} x \left(\int_{-\infty}^{\infty} h(x, y)\, dy \right) dx$$

$$\mu_2 = \int_{-\infty}^{\infty} y \left(\int_{-\infty}^{\infty} h(x, y)\, dx \right) dy$$

$$\sigma_1^2 = \int_{-\infty}^{\infty} (x - \mu_1)^2 \left(\int_{-\infty}^{\infty} h(x, y)\, dy \right) dx$$

$$\sigma_2^2 = \int_{-\infty}^{\infty} (y - \mu_2)^2 \left(\int_{-\infty}^{\infty} h(x, y)\, dx \right) dy.$$

Generalizing (14-9) we define the correlation coefficient ρ of the population with the bivariate density $h(x, y)$:

$$\rho = (1/\sigma_1 \sigma_2) \int_{\infty}^{\infty} \int_{\infty}^{\infty} (x - \mu_1)(y - \mu_2) h(x, y)\, dx\, dy. \quad (14\text{-}23)$$

As in the normal lase, ρ satisfies the inequalities $-1 \leq \rho \leq 1$ (Exercise 1).

We recall from Sec. 5-1 that two statistics are called independent if their joint distribution (or density) factors into a product of the marginals. Extending this to bivariate populations, we say that the two characteristics x and y of the population with bivariate density $h(x, y)$ are independent if

$$h(x, y) = h_1(x) h_2(y), \quad (14\text{-}24)$$

where h_1 and h_2 are the marginal densities:

$$h_1(x) = \int_{-\infty}^{\infty} h(x, y)\, dy, \quad h_2(y) = \int_{-\infty}^{\infty} h(x, y)\, dx.$$

Equation 14-24 signifies that the relative proportion of the population whose second label is equal to y, relative to those whose first label is x, is the same for all x; in other words, the first label by itself provides no information about the second one. Equation 14-2 states that if the population is normal and if $\rho = 0$, then the conditions (14-24) holds and so the two characteristics are independent; conversely, the analysis of the bivariate normal density in Sec. 14-1 shows that the bivariate density factors only if $\rho = 0$ (Exercise 2); therefore the condition $\rho = 0$ is equivalent to (14-24) in the normal case.

For a general bivariate density $h(x, y)$, the condition (14-24) implies that $\rho = 0$ (Exercise 3). The converse is not true: there are bivariate populations with $\rho = 0$ but not satisfying (14-24); thus, for a general population, the lack of correlation between two characteristics does not imply independence. We shall give an example of a bivariate density such that $\rho = 0$ but (14-24) is not satisfied. Put

$$h(x, y) = \frac{\exp\left(-\sqrt{x^2 + y^2}\right)}{\sqrt{x^2 + y^2}}. \tag{14-25}$$

This is nonnegative; its double integral is 1:

$$\int_{-\infty}^{\infty} \int_{-\infty}^{\infty} h(x, y)\, dx\, dy = \frac{1}{2\pi} \int_0^{2\pi} \int_0^{\infty} e^{-r}\, dr\, d\theta = 1.$$

The means μ_1 and μ_2 are equal to 0 (Exercise 4); thus by (14-23) the correlation coefficient is proportional to

$$\int_{-\infty}^{\infty} \int_{-\infty}^{\infty} xy \exp\left(-\sqrt{x^2 + y^2}\right)(x^2 + y^2)^{-1/2}\, dx\, dy, \tag{14-26}$$

which is equal to 0 (Exercise 4).

EXERCISES

1. Using an appropriate version of the Cauchy-Schwarz inequality prove that $-1 \le \rho \le 1$.
2. Show that if $\rho \ne 0$, then the bivariate normal density does not factor in accordance with (14-24).
3. Show, by iterated integration, that the condition (14-24) and the assumption of finite variances imply $\rho = 0$.
4. Without computation show why the means of the density (14-25) and the correlation are equal to 0.

14-6. A NUMERICAL EXAMPLE OF A STANDARD BIVARIATE NORMAL POPULATION WITH $\rho = 1/2$ (OPTIONAL)

Theorem 14-1 can be used to construct bivariate normal Samples from Table VI, Appendix. If (x_1, x_2) is a pair of normal deviates, then, by the above theorem, the pair $(x_1, \rho x_1 + \sqrt{1 - \rho^2} x_2)$ represents a pair from the standard bivariate density with correlation coefficient ρ. Fifty pairs of normal deviates $(x_1 x_2)$ were taken from the first two columns of the first page of Table VI, Appendix; the pairs are $(-1.082, -1.094)$, $(-.153, -.845)$, etc. These were transformed into the pairs

$$(s_1, (.5)\, x_1 + (.8660)\, x_2).$$

(Note that $\sqrt{1 - (.5)^2} = .8660$). The points in the xy-plane corresponding to these pairs are plotted in Fig. 14-4. The points tend to run along the diagonal, which is a manifestation of the positive correlation.

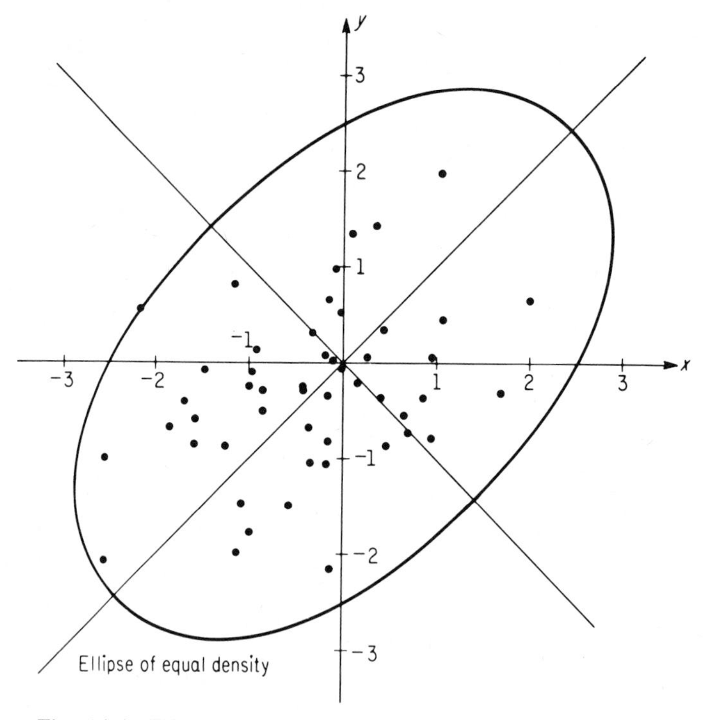

Fig. 14-4. Fifty random points in the plane with a standard bivariate normal distribution with $\rho = 1/2$.

EXERCISES

1. Show that $(x_1, -\rho x_1 + \sqrt{1 - \rho^2} x_2)$ represents a pair from the standard bivariate normal density with correlation $-\rho$.

2. (Optional) Using the same pairs of deviates as in Fig. 14-4, construct a Sample of fifty from a standard bivariate density with correlation $-.6$.

BIBLIOGRAPHY

Many applications of the bivariate normal distribution are given by G. Udny Yule and M. G. Kendall, *An Introduction to the Theory of Statistics*, 14th ed. (New York: Hafner, 1950).

Appendix

TABLE I
Area Under the Normal Curve

$$A = \frac{1}{\sqrt{2\pi}} \int_0^z e^{-(1/2)t^2}\, dt$$

z	.00	.01	.02	.03	.04	.05	.06	.07	.08	.09
.0	.0000	.0040	.0080	.0120	.0160	.0199	.0239	.0279	.0319	.0359
.1	.0398	.0438	.0478	.0517	.0557	.0596	.0636	.0675	.0714	.0753
.2	.0793	.0832	.0871	.0910	.0948	.0987	.1026	.1064	.1103	.1141
.3	.1179	.1217	.1255	.1293	.1331	.1368	.1406	.1443	.1480	.1517
.4	.1554	.1591	.1628	.1664	.1700	.1736	.1772	.1808	.1844	.1879
.5	.1915	.1950	.1985	.2019	.2054	.2088	.2123	.2157	.2190	.2224
.6	.2257	.2291	.2324	.2357	.2389	.2422	.2454	.2486	.2517	.2549
.7	.2580	.2611	.2642	.2673	.2704	.2734	.2764	.2794	.2823	.2852
.8	.2881	.2910	.2939	.2967	.2995	.3023	.3051	.3078	.3106	.3133
.9	.3159	.3186	.3212	.3238	.3264	.3289	.3315	.3340	.3365	.3389
1.0	.3413	.3438	.3461	.3485	.3508	.3531	.3554	.3577	.3599	.3621
1.1	.3643	.3665	.3686	.3708	.3729	.3749	.3770	.3790	.3810	.3830
1.2	.3849	.3869	.3888	.3907	.3925	.3944	.3962	.3980	.3997	.4015
1.3	.4032	.4049	.4066	.4082	.4099	.4115	.4131	.4147	.4162	.4177
1.4	.4192	.4207	.4222	.4236	.4251	.4265	.4279	.4292	.4306	.4319
1.5	.4332	.4345	.4357	.4370	.4382	.4394	.4406	.4418	.4429	.4441
1.6	.4452	.4463	.4474	.4484	.4495	.4505	.4515	.4525	.4535	.4545
1.7	.4554	.4564	.4573	.4582	.4591	.4599	.4608	.4616	.4625	.4633
1.8	.4641	.4649	.4656	.4664	.4671	.4678	.4686	.4693	.4699	.4706
1.9	.4713	.4719	.4726	.4732	.4738	.4744	.4750	.4756	.4761	.4767
2.0	.4772	.4778	.4783	.4788	.4793	.4798	.4803	.4808	.4812	.4817
2.1	.4821	.4826	.4830	.4834	.4838	.4842	.4846	.4850	.4854	.4857
2.2	.4861	.4864	.4868	.4871	.4875	.4878	.4881	.4884	.4887	.4890
2.3	.4893	.4896	.4898	.4901	.4904	.4906	.4909	.4911	.4913	.4916
2.4	.4918	.4920	.4922	.4925	.4927	.4929	.4931	.4932	.4934	.4936
2.5	.4938	.4940	.4941	.4943	.4945	.4946	.4948	.4949	.4951	.4952
2.6	.4953	.4955	.4956	.4957	.4959	.4960	.4961	.4962	.4963	.4964
2.7	.4965	.4966	.4967	.4968	.4969	.4970	.4971	.4972	.4973	.4974
2.8	.4974	.4975	.4976	.4977	.4977	.4978	.4979	.4979	.4980	.4981
2.9	.4981	.4982	.4982	.4982	.4984	.4984	.4985	.4985	.4986	.4986
3.0	.4987	.4987	.4987	.4988	.4988	.4989	.4989	.4989	.4990	.4990
3.1	.4990	.4991	.4991	.4991	.4992	.4992	.4992	.4992	.4993	.4993
3.2	.4993	.4993	.4994	.4994	.4994	.4994	.4994	.4995	.4995	.4995
3.3	.4995	.4995	.4995	.4996	.4996	.4996	.4996	.4996	.4996	.4997
3.4	.4997	.4997	.4997	.4997	.4997	.4997	.4997	.4997	.4997	.4998
3.5	.4998	.4998	.4998	.4998	.4998	.4998	.4998	.4998	.4998	.4998
3.6	.4998	.4998	.4999	.4999	.4999	.4999	.4999	.4999	.4999	.4999
3.7	.4999	.4999	.4999	.4999	.4999	.4999	.4999	.4999	.4999	.4999
3.8	.4999	.4999	.4999	.4999	.4999	.4999	.4999	.4999	.4999	.4999
3.9	.5000	.5000	.5000	.5000	.5000	.5000	.5000	.5000	.5000	.5000
4.0	.5000	.5000	.5000	.5000	.5000	.5000	.5000	.5000	.5000	.5000
z	.00	.01	.02	.03	.04	.05	.06	.07	.08	.09

TABLE II
Percentage Points of the χ^2-Distribution*

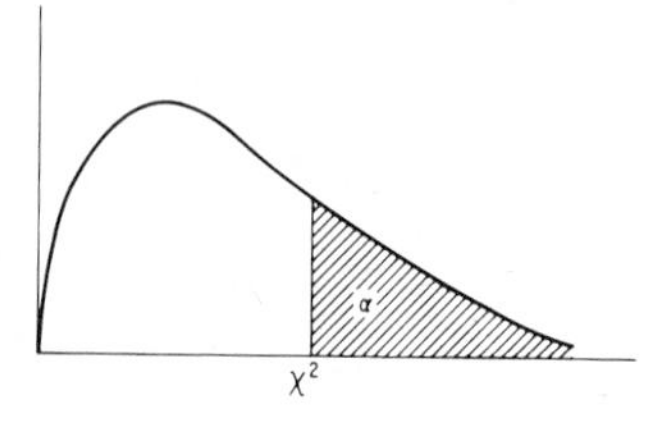

d.f. \ α	0·995	0·990	0·975	0·950	0·900	0·750	0·500
1	392704.10^{-10}	157088.10^{-9}	982069.10^{-9}	393214.10^{-8}	0·0157908	0·1015308	0·454936
2	0·0100251	0·0201007	0·0506356	0·102587	0·210721	0·575364	1·38629
3	0·0717218	0·114832	0·215795	0·351846	0·584374	1·212534	2·36597
4	0·206989	0·297109	0·484419	0·710723	1·063623	1·92256	3·35669
5	0·411742	0·554298	0·831212	1·145476	1·61031	2·67460	4·35146
6	0·675727	0·872090	1·23734	1·63538	2·20413	3·45460	5·34812
7	0·989256	1·239043	1·68987	2·16735	2·83311	4·25485	6·34581
8	1·34441	1·64650	2·17973	2·73264	3·48954	5·07064	7·34412
9	1·73493	2·08790	2·70039	3·32511	4·16816	5·89883	8·34283
10	2·15586	2·55821	3·24697	3·94030	4·86518	6·73720	9·34182
11	2·60322	3·05348	3·81575	4·57481	5·57778	7·58414	10·3410
12	3·07382	3·57057	4·40379	5·22603	6·30380	8·43842	11·3403
13	3·56503	4·10692	5·00875	5·89186	7·04150	9·29907	12·3398
14	4·07467	4·66043	5·62873	6·57063	7·78953	10·1653	13·3393
15	4·60092	5·22935	6·26214	7·26094	8·54676	11·0365	14·3389
16	5·14221	5·81221	6·90766	7·96165	9·31224	11·9122	15·3385
17	5·69722	6·40776	7·56419	8·67176	10·0852	12·7919	16·3382
18	6·26480	7·01491	8·23075	9·39046	10·8649	13·6753	17·3379
19	6·84397	7·63273	8·90652	10·1170	11·6509	14·5620	18·3377
20	7·43384	8·26040	9·59078	10·8508	12·4426	15·4518	19·3374
21	8·03365	8·89720	10·28293	11·5913	13·2396	16·3444	20·3372
22	8·64272	9·54249	10·9823	12·3380	14·0415	17·2396	21·3370
23	9·26043	10·19567	11·6886	13·0905	14·8480	18·1373	22·3369
24	9·88623	10·8564	12·4012	13·8484	15·6587	19·0373	23·3367
25	10·5197	11·5240	13·1197	14·6114	16·4734	19·9393	24·3366
26	11·1602	12·1981	13·8439	15·3792	17·2919	20·8434	25·3365
27	11·8076	12·8785	14·5734	16·1514	18·1139	21·7494	26·3363
28	12·4613	13·5647	15·3079	16·9279	18·9392	22·6572	27·3362
29	13·1211	14·2565	16·0471	17·7084	19·7677	23·5666	28·3361
30	13·7867	14·9535	16·7908	18·4927	20·5992	24·4776	29·3360
40	20·7065	22·1643	24·4330	26·5093	29·0505	33·6603	39·3353
50	27·9907	29·7067	32·3574	34·7643	37·6886	42·9421	49·3349
60	35·5345	37·4849	40·4817	43·1880	46·4589	52·2938	59·3347
70	43·2752	45·4417	48·7576	51·7393	55·3289	61·6983	69·3345
80	51·1719	53·5401	57·1532	60·3915	64·2778	71·1445	79·3343
90	59·1963	61·7541	65·6466	69·1260	73·2911	80·6247	89·3342
100	67·3276	70·0649	74·2219	77·9295	82·3581	90·1332	99·3341

*From E. S. Pearson and H. O. Hartley (eds), *Biometrika Tables for Statisticians*, Vol. 1, 3rd ed. (London: Cambridge University Press, 1966), by permission.

TABLE II (*continued*)

d.f. \ α	0·250	0·100	0·050	0·025	0·010	0·005	0·001
1	1·32330	2·70554	3·84146	5·02389	6·63490	7·87944	10·828
2	2·77259	4·60517	5·99146	7·37776	9·21034	10·5966	13·816
3	4·10834	6·25139	7·81473	9·34840	11·3449	12·8382	16·266
4	5·38527	7·77944	9·48773	11·1433	13·2767	14·8603	18·467
5	6·62568	9·23636	11·0705	12·8325	15·0863	16·7496	20·515
6	7·84080	10·6446	12·5916	14·4494	16·8119	18·5476	22·458
7	9·03715	12·0170	14·0671	16·0128	18·4753	20·2777	24·322
8	10·2189	13·3616	15·5073	17·5345	20·0902	21·9550	26·125
9	11·3888	14·6837	16·9190	19·0228	21·6660	23·5894	27·877
10	12·5489	15·9872	18·3070	20·4832	23·2093	25·1882	29·588
11	13·7007	17·2750	19·6751	21·9200	24·7250	26·7568	31·264
12	14·8454	18·5493	21·0261	23·3367	26·2170	28·2995	32·909
13	15·9839	19·8119	22·3620	24·7356	27·6882	29·8195	34·528
14	17·1169	21·0641	23·6848	26·1189	29·1412	31·3194	36·123
15	18·2451	22·3071	24·9958	27·4884	30·5779	32·8013	37·697
16	19·3689	23·5418	26·2962	28·8454	31·9999	34·2672	39·252
17	20·4887	24·7690	27·5871	30·1910	33·4087	35·7185	40·790
18	21·6049	25·9894	28·8693	31·5264	34·8053	37·1565	42·312
19	22·7178	27·2036	30·1435	32·8523	36·1909	38·5823	43·820
20	23·8277	28·4120	31·4104	34·1696	37·5662	39·9968	45·315
21	24·9348	29·6151	32·6706	35·4789	38·9322	41·4011	46·797
22	26·0393	30·8133	33·9244	36·7807	40·2894	42·7957	48·268
23	27·1413	32·0069	35·1725	38·0756	41·6384	44·1813	49·728
24	28·2412	33·1962	36·4150	39·3641	42·9798	45·5585	51·179
25	29·3389	34·3816	37·6525	40·6465	44·3141	46·9279	52·618
26	30·4346	35·5632	38·8851	41·9232	45·6417	48·2899	54·052
27	31·5284	36·7412	40·1133	43·1945	46·9629	49·6449	55·476
28	32·6205	37·9159	41·3371	44·4608	48·2782	50·9934	56·892
29	33·7109	39·0875	42·5570	45·7223	49·5879	52·3356	58·301
30	34·7997	40·2560	43·7730	46·9792	50·8922	53·6720	59·703
40	45·6160	51·8051	55·7585	59·3417	63·6907	66·7660	73·402
50	56·3336	63·1671	67·5048	71·4202	76·1539	79·4900	86·661
60	66·9815	74·3970	79·0819	83·2977	88·3794	91·9517	99·607
70	77·5767	85·5270	90·5312	95·0232	100·425	104·215	112·317
80	88·1303	96·5782	101·879	106·629	112·329	116·321	124·839
90	98·6499	107·565	113·145	118·136	124·116	128·299	137·208
100	109·141	118·498	124·342	129·561	135·807	140·169	149·449

TABLE III
Percentage Points of the t-Distribution*

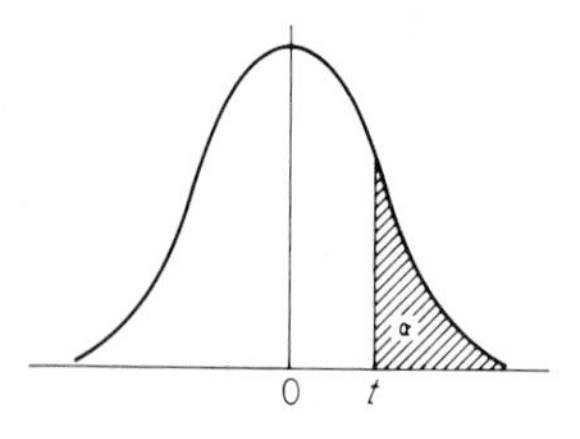

d.f.	$\alpha = 0{\cdot}4$	0·25	0·1	0·05	0·025	0·01	0·005	0·0025	0·001	0·0005
1	0·325	1·000	3·078	6·314	12·706	31·821	63·657	127·32	318·31	636·62
2	·289	0·816	1·886	2·920	4·303	6·965	9·925	14·089	22·327	31·598
3	·277	·765	1·638	2·353	3·182	4·541	5·841	7·453	10·214	12·924
4	·271	·741	1·533	2·132	2·776	3·747	4·604	5·598	7·173	8·610
5	0·267	0·727	1·476	2·015	2·571	3·365	4·032	4·773	5·893	6·869
6	·265	·718	1·440	1·943	2·447	3·143	3·707	4·317	5·208	5·959
7	·263	·711	1·415	1·895	2·365	2·998	3·499	4·029	4·785	5·408
8	·262	·706	1·397	1·860	2·306	2·896	3·355	3·833	4·501	5·041
9	·261	·703	1·383	1·833	2·262	2·821	3·250	3·690	4·297	4·781
10	0·260	0·700	1·372	1·812	2·228	2·764	3·169	3·581	4·144	4·587
11	·260	·697	1·363	1·796	2·201	2·718	3·106	3·497	4·025	4·437
12	·259	·695	1·356	1·782	2·179	2·681	3·055	3·428	3·930	4·318
13	·259	·694	1·350	1·771	2·160	2·650	3·012	3·372	3·852	4·221
14	·258	·692	1·345	1·761	2·145	2·624	2·977	3·326	3·787	4·140
15	0·258	0·691	1·341	1·753	2·131	2·602	2·947	3·286	3·733	4·073
16	·258	·690	1·337	1·746	2·120	2·583	2·921	3·252	3·686	4·015
17	·257	·689	1·333	1·740	2·110	2·567	2·898	3·222	3·646	3·965
18	·257	·688	1·330	1·734	2·101	2·552	2·878	3·197	3·610	3·922
19	·257	·688	1·328	1·729	2·093	2·539	2·861	3·174	3·579	3·883
20	0·257	0·687	1·325	1·725	2·086	2·528	2·845	3·153	3·552	3·850
21	·257	·686	1·323	1·721	2·080	2·518	2·831	3·135	3·527	3·819
22	·256	·686	1·321	1·717	2·074	2·508	2·819	3·119	3·505	3·792
23	·256	·685	1·319	1·714	2·069	2·500	2·807	3·104	3·485	3·767
24	·256	·685	1·318	1·711	2·064	2·492	2·797	3·091	3·467	3·745
25	0·256	0·684	1·316	1·708	2·060	2·485	2·787	3·078	3·450	3·725
26	·256	·684	1·315	1·706	2·056	2·479	2·779	3·067	3·435	3·707
27	·256	·684	1·314	1·703	2·052	2·473	2·771	3·057	3·421	3·690
28	·256	·683	1·313	1·701	2·048	2·467	2·763	3·047	3·408	3·674
29	·256	·683	1·311	1·699	2·045	2·462	2·756	3·038	3·396	3·659
30	0·256	0·683	1·310	1·697	2·042	2·457	2·750	3·030	3·385	3·646
40	·255	·681	1·303	1·684	2·021	2·423	2·704	2·971	3·307	3·551
60	·254	·679	1·296	1·671	2·000	2·390	2·660	2·915	3·232	3·460
120	·254	·677	1·289	1·658	1·980	2·358	2·617	2·860	3·160	3·373
∞	·253	·674	1·282	1·645	1·960	2·326	2·576	2·807	3·090	3·291

*From E. S. Pearson and H. O. Hartley (eds.), *Biometrika Tables for Statisticians*, Vol. 1, 3rd ed. (London: Cambridge University Press, 1966), by permission.

TABLE IV

Upper 5 Percentage Points of the *F*-Distribution*

n \ m	1	2	3	4	5	6	7	8	9	10	12	15	20	24	30	40	60	120	∞
1	161·4	199·5	215·7	224·6	230·2	234·0	236·8	238·9	240·5	241·9	243·9	245·9	248·0	249·1	250·1	251·1	252·2	253·3	254·3
2	18·51	19·00	19·16	19·25	19·30	19·33	19·35	19·37	19·38	19·40	19·41	19·43	19·45	19·45	19·46	19·47	19·48	19·49	19·50
3	10·13	9·55	9·28	9·12	9·01	8·94	8·89	8·85	8·81	8·79	8·74	8·70	8·66	8·64	8·62	8·59	8·57	8·55	8·53
4	7·71	6·94	6·59	6·39	6·26	6·16	6·09	6·04	6·00	5·96	5·91	5·86	5·80	5·77	5·75	5·72	5·69	5·66	5·63
5	6·61	5·79	5·41	5·19	5·05	4·95	4·88	4·82	4·77	4·74	4·68	4·62	4·56	4·53	4·50	4·46	4·43	4·40	4·36
6	5·99	5·14	4·76	4·53	4·39	4·28	4·21	4·15	4·10	4·06	4·00	3·94	3·87	3·84	3·81	3·77	3·74	3·70	3·67
7	5·59	4·74	4·35	4·12	3·97	3·87	3·79	3·73	3·68	3·64	3·57	3·51	3·44	3·41	3·38	3·34	3·30	3·27	3·23
8	5·32	4·46	4·07	3·84	3·69	3·58	3·50	3·44	3·39	3·35	3·28	3·22	3·15	3·12	3·08	3·04	3·01	2·97	2·93
9	5·12	4·26	3·86	3·63	3·48	3·37	3·29	3·23	3·18	3·14	3·07	3·01	2·94	2·90	2·86	2·83	2·79	2·75	2·71
10	4·96	4·10	3·71	3·48	3·33	3·22	3·14	3·07	3·02	2·98	2·91	2·85	2·77	2·74	2·70	2·66	2·62	2·58	2·54
11	4·84	3·98	3·59	3·36	3·20	3·09	3·01	2·95	2·90	2·85	2·79	2·72	2·65	2·61	2·57	2·53	2·49	2·45	2·40
12	4·75	3·89	3·49	3·26	3·11	3·00	2·91	2·85	2·80	2·75	2·69	2·62	2·54	2·51	2·47	2·43	2·38	2·34	2·30
13	4·67	3·81	3·41	3·18	3·03	2·92	2·83	2·77	2·71	2·67	2·60	2·53	2·46	2·42	2·38	2·34	2·30	2·25	2·21
14	4·60	3·74	3·34	3·11	2·96	2·85	2·76	2·70	2·65	2·60	2·53	2·46	2·39	2·35	2·31	2·27	2·22	2·18	2·13
15	4·54	3·68	3·29	3·06	2·90	2·79	2·71	2·64	2·59	2·54	2·48	2·40	2·33	2·29	2·25	2·20	2·16	2·11	2·07
16	4·49	3·63	3·24	3·01	2·85	2·74	2·66	2·59	2·54	2·49	2·42	2·35	2·28	2·24	2·19	2·15	2·11	2·06	2·01
17	4·45	3·59	3·20	2·96	2·81	2·70	2·61	2·55	2·49	2·45	2·38	2·31	2·23	2·19	2·15	2·10	2·06	2·01	1·96
18	4·41	3·55	3·16	2·93	2·77	2·66	2·58	2·51	2·46	2·41	2·34	2·27	2·19	2·15	2·11	2·06	2·02	1·97	1·92
19	4·38	3·52	3·13	2·90	2·74	2·63	2·54	2·48	2·42	2·38	2·31	2·23	2·16	2·11	2·07	2·03	1·98	1·93	1·88
20	4·35	3·49	3·10	2·87	2·71	2·60	2·51	2·45	2·39	2·35	2·28	2·20	2·12	2·08	2·04	1·99	1·95	1·90	1·84
21	4·32	3·47	3·07	2·84	2·68	2·57	2·49	2·42	2·37	2·32	2·25	2·18	2·10	2·05	2·01	1·96	1·92	1·87	1·81
22	4·30	3·44	3·05	2·82	2·66	2·55	2·46	2·40	2·34	2·30	2·23	2·15	2·07	2·03	1·98	1·94	1·89	1·84	1·78
23	4·28	3·42	3·03	2·80	2·64	2·53	2·44	2·37	2·32	2·27	2·20	2·13	2·05	2·01	1·96	1·91	1·86	1·81	1·76
24	4·26	3·40	3·01	2·78	2·62	2·51	2·42	2·36	2·30	2·25	2·18	2·11	2·03	1·98	1·94	1·89	1·84	1·79	1·73
25	4·24	3·39	2·99	2·76	2·60	2·49	2·40	2·34	2·28	2·24	2·16	2·09	2·01	1·96	1·92	1·87	1·82	1·77	1·71
26	4·23	3·37	2·98	2·74	2·59	2·47	2·39	2·32	2·27	2·22	2·15	2·07	1·99	1·95	1·90	1·85	1·80	1·75	1·69
27	4·21	3·35	2·96	2·73	2·57	2·46	2·37	2·31	2·25	2·20	2·13	2·06	1·97	1·93	1·88	1·84	1·79	1·73	1·67
28	4·20	3·34	2·95	2·71	2·56	2·45	2·36	2·29	2·24	2·19	2·12	2·04	1·96	1·91	1·87	1·82	1·77	1·71	1·65
29	4·18	3·33	2·93	2·70	2·55	2·43	2·35	2·28	2·22	2·18	2·10	2·03	1·94	1·90	1·85	1·81	1·75	1·70	1·64
30	4·17	3·32	2·92	2·69	2·53	2·42	2·33	2·27	2·21	2·16	2·09	2·01	1·93	1·89	1·84	1·79	1·74	1·68	1·62
40	4·08	3·23	2·84	2·61	2·45	2·34	2·25	2·18	2·12	2·08	2·00	1·92	1·84	1·79	1·74	1·69	1·64	1·58	1·51
60	4·00	3·15	2·76	2·53	2·37	2·25	2·17	2·10	2·04	1·99	1·92	1·84	1·75	1·70	1·65	1·59	1·53	1·47	1·39
120	3·92	3·07	2·68	2·45	2·29	2·17	2·09	2·02	1·96	1·91	1·83	1·75	1·66	1·61	1·55	1·50	1·43	1·35	1·25
∞	3·84	3·00	2·60	2·37	2·21	2·10	2·01	1·94	1·88	1·83	1·75	1·67	1·57	1·52	1·46	1·39	1·32	1·22	1·00

*From E. S. Pearson and H. O. Hartley (eds.), *Biometrika Tables for Statisticians*, Vol. 1, 3rd ed. (London: Cambridge University Press, 1966), by permission.

TABLE IV (*continued*)
Upper 1 Percentage Points of the F-Distribution

n \ m	1	2	3	4	5	6	7	8	9	10	12	15	20	24	30	40	60	120	∞
1	4052	4999·5	5403	5625	5764	5859	5928	5981	6022	6056	6106	6157	6209	6235	6261	6287	6313	6339	6366
2	98·50	99·00	99·17	99·25	99·30	99·33	99·36	99·37	99·39	99·40	99·42	99·43	99·45	99·46	99·47	99·47	99·48	99·49	99·50
3	34·12	30·82	29·46	28·71	28·24	27·91	27·67	27·49	27·35	27·23	27·05	26·87	26·69	26·60	26·50	26·41	26·32	26·22	26·13
4	21·20	18·00	16·69	15·98	15·52	15·21	14·98	14·80	14·66	14·55	14·37	14·20	14·02	13·93	13·84	13·75	13·65	13·56	13·46
5	16·26	13·27	12·06	11·39	10·97	10·67	10·46	10·29	10·16	10·05	9·89	9·72	9·55	9·47	9·38	9·29	9·20	9·11	9·02
6	13·75	10·92	9·78	9·15	8·75	8·47	8·26	8·10	7·98	7·87	7·72	7·56	7·40	7·31	7·23	7·14	7·06	6·97	6·88
7	12·25	9·55	8·45	7·85	7·46	7·19	6·99	6·84	6·72	6·62	6·47	6·31	6·16	6·07	5·99	5·91	5·82	5·74	5·65
8	11·26	8·65	7·59	7·01	6·63	6·37	6·18	6·03	5·91	5·81	5·67	5·52	5·36	5·28	5·20	5·12	5·03	4·95	4·86
9	10·56	8·02	6·99	6·42	6·06	5·80	5·61	5·47	5·35	5·26	5·11	4·96	4·81	4·73	4·65	4·57	4·48	4·40	4·31
10	10·04	7·56	6·55	5·99	5·64	5·39	5·20	5·06	4·94	4·85	4·71	4·56	4·41	4·33	4·25	4·17	4·08	4·00	3·91
11	9·65	7·21	6·22	5·67	5·32	5·07	4·89	4·74	4·63	4·54	4·40	4·25	4·10	4·02	3·94	3·86	3·78	3·69	3·60
12	9·33	6·93	5·95	5·41	5·06	4·82	4·64	4·50	4·39	4·30	4·16	4·01	3·86	3·78	3·70	3·62	3·54	3·45	3·36
13	9·07	6·70	5·74	5·21	4·86	4·62	4·44	4·30	4·19	4·10	3·96	3·82	3·66	3·59	3·51	3·43	3·34	3·25	3·17
14	8·86	6·51	5·56	5·04	4·69	4·46	4·28	4·14	4·03	3·94	3·80	3·66	3·51	3·43	3·35	3·27	3·18	3·09	3·00
15	8·68	6·36	5·42	4·89	4·56	4·32	4·14	4·00	3·89	3·80	3·67	3·52	3·37	3·29	3·21	3·13	3·05	2·96	2·87
16	8·53	6·23	5·29	4·77	4·44	4·20	4·03	3·89	3·78	3·69	3·55	3·41	3·26	3·18	3·10	3·02	2·93	2·84	2·75
17	8·40	6·11	5·18	4·67	4·34	4·10	3·93	3·79	3·68	3·59	3·46	3·31	3·16	3·08	3·00	2·92	2·83	2·75	2·65
18	8·29	6·01	5·09	4·58	4·25	4·01	3·84	3·71	3·60	3·51	3·37	3·23	3·08	3·00	2·92	2·84	2·75	2·66	2·57
19	8·18	5·93	5·01	4·50	4·17	3·94	3·77	3·63	3·52	3·43	3·30	3·15	3·00	2·92	2·84	2·76	2·67	2·58	2·49
20	8·10	5·85	4·94	4·43	4·10	3·87	3·70	3·56	3·46	3·37	3·23	3·09	2·94	2·86	2·78	2·69	2·61	2·52	2·42
21	8·02	5·78	4·87	4·37	4·04	3·81	3·64	3·51	3·40	3·31	3·17	3·03	2·88	2·80	2·72	2·64	2·55	2·46	2·36
22	7·95	5·72	4·82	4·31	3·99	3·76	3·59	3·45	3·35	3·26	3·12	2·98	2·83	2·75	2·67	2·58	2·50	2·40	2·31
23	7·88	5·66	4·76	4·26	3·94	3·71	3·54	3·41	3·30	3·21	3·07	2·93	2·78	2·70	2·62	2·54	2·45	2·35	2·26
24	7·82	5·61	4·72	4·22	3·90	3·67	3·50	3·36	3·26	3·17	3·03	2·89	2·74	2·66	2·58	2·49	2·40	2·31	2·21
25	7·77	5·57	4·68	4·18	3·85	3·63	3·46	3·32	3·22	3·13	2·99	2·85	2·70	2·62	2·54	2·45	2·36	2·27	2·17
26	7·72	5·53	4·64	4·14	3·82	3·59	3·42	3·29	3·18	3·09	2·96	2·81	2·66	2·58	2·50	2·42	2·33	2·23	2·13
27	7·68	5·49	4·60	4·11	3·78	3·56	3·39	3·26	3·15	3·06	2·93	2·78	2·63	2·55	2·47	2·38	2·29	2·20	2·10
28	7·64	5·45	4·57	4·07	3·75	3·53	3·36	3·23	3·12	3·03	2·90	2·75	2·60	2·52	2·44	2·35	2·26	2·17	2·06
29	7·60	5·42	4·54	4·04	3·73	3·50	3·33	3·20	3·09	3·00	2·87	2·73	2·57	2·49	2·41	2·33	2·23	2·14	2·03
30	7·56	5·39	4·51	4·02	3·70	3·47	3·30	3·17	3·07	2·98	2·84	2·70	2·55	2·47	2·39	2·30	2·21	2·11	2·01
40	7·31	5·18	4·31	3·83	3·51	3·29	3·12	2·99	2·89	2·80	2·66	2·52	2·37	2·29	2·20	2·11	2·02	1·92	1·80
60	7·08	4·98	4·13	3·65	3·34	3·12	2·95	2·82	2·72	2·63	2·50	2·35	2·20	2·12	2·03	1·94	1·84	1·73	1·60
120	6·85	4·79	3·95	3·48	3·17	2·96	2·79	2·66	2·56	2·47	2·34	2·19	2·03	1·95	1·86	1·76	1·66	1·53	1·38
∞	6·63	4·61	3·78	3·32	3·02	2·80	2·64	2·51	2·41	2·32	2·18	2·04	1·88	1·79	1·70	1·59	1·47	1·32	1·00

TABLE V
Random Digits*

13962	70992	65172	28053	02190	83634	66012	70305	66761	88344
43905	46941	72300	11641	43548	30455	07686	31840	03261	89199
00504	48658	38051	59408	16508	82979	92002	63606	41078	86326
61274	57238	47267	35303	29066	02140	60867	39847	50968	96719
43753	21159	16239	50595	62509	61207	86816	29902	23395	72640
83503	51662	21636	68192	84294	38754	84755	34053	94582	29215
36807	71420	35804	44862	23577	79551	42003	58684	09271	68396
19110	55680	18792	41487	16614	83053	00812	16749	45347	88199
82615	86984	03290	87971	60022	35415	20852	02909	99476	45568
05621	26584	36493	63013	68181	57702	49510	75304	38724	15712
06936	37293	55875	71213	83025	46063	74665	12178	10741	58362
84981	60458	16194	92403	80951	80068	47076	23310	74899	87929
66354	88441	96191	04794	14714	64749	43097	83976	83281	72038
49602	94109	36460	62353	00721	66980	82554	90270	12312	56299
78430	72391	96973	70437	97803	78683	04670	70667	58912	21883
33331	51803	15934	75807	46561	80188	78984	29317	27971	16440
62843	84445	56652	91797	45284	25842	96246	73504	21631	81223
19528	15445	77764	33446	41204	70067	33354	70680	66664	75486
16737	01887	50934	43306	75190	86997	56561	79018	34273	25196
99389	06685	45945	62000	76228	60645	87750	46329	46544	95665
36160	38196	77705	28891	12106	56281	86222	66116	39626	06080
05505	45420	44016	79662	92069	27628	50002	32540	19848	27319
85962	19758	92795	00458	71289	05884	37963	23322	73243	98185
28763	04900	54460	22083	89279	43492	00066	40857	86568	49336
42222	40446	82240	79159	44168	38213	46839	26598	29983	67645
43626	40039	51492	36488	70280	24218	14596	04744	89336	35630
97761	43444	95895	24102	07006	71923	04800	32062	41425	66862
49275	44270	52512	03951	21651	53867	73531	70073	45542	22831
15797	75134	39856	73527	78417	36208	59510	76913	22499	68467
04497	24853	43879	07613	26400	17180	18880	66083	02196	10638
95468	87411	30647	88711	01765	57688	60665	57636	36070	37285
01420	74218	71047	14401	74537	14820	45248	78007	65911	38583
74633	40171	97092	79137	30698	97915	36305	42613	87251	75608
46662	99688	59576	04887	02310	35508	69481	30300	94047	57096
10853	10393	03013	90372	89639	65800	88532	71789	59964	50681
68583	01032	67938	29733	71176	35699	10551	15091	52947	20134
75818	78982	24258	93051	02081	83890	66944	99856	87950	13952
16395	16837	00538	57133	89398	78205	72122	99655	25294	20941
53892	15105	40963	69267	85534	00533	27130	90420	72584	84576
66009	26869	91829	65078	89616	49016	14200	97469	88307	92282
45292	93427	92326	70206	15847	14302	60043	30530	57149	08642
34033	45008	41621	79437	9874[illegible]	84455	66769	94729	17975	50[illegible]63
13364	09937	00535	88122	47278	90758	23542	35273	67912	97670
03343	62593	93332	09921	25306	57483	98115	33460	55304	43572
46145	24476	62507	19530	41257	97919	02290	40357	38408	50031
37703	51658	17420	30593	39637	64220	45486	03698	80220	12139
12622	98083	17689	56977	56603	93316	79858	52548	67364	72416
56043	00251	70085	28067	78135	53000	18138	40564	77086	49557
43401	35924	28308	55140	07515	53854	23023	70268	80435	24269
18053	53460	32125	81357	26935	67234	78460	47833	20498	35645

*From The Rand Corporation, *A Million Random Digits with 100,000 Normal Deviates* (New York: The Free Press, 1955), by permission.

Random Digits (*continued*)

42013	25126	49296	38839	98092	96100	44205	85129	46749	47707
66261	56987	46342	70656	04614	26422	32479	41453	82281	65793
99218	43326	71220	47549	69609	05780	01070	70739	29282	98507
60246	70506	12969	83611	57725	10209	67627	07864	05937	31892
36193	05504	57510	24880	43433	20377	33928	54749	73464	40652
00531	71458	96341	59955	54799	63186	22416	45953	94761	58992
74737	91290	58472	75246	44996	62216	27970	50154	44759	77127
53061	21680	80352	70951	15425	12816	51622	32075	85276	14589
08614	40071	68920	64920	23340	05380	28335	25114	61683	57618
96656	96439	54118	52156	52621	13824	59450	01023	11607	67538
62095	88876	35524	81750	08680	17349	89230	03916	67328	28455
77889	92840	57213	80607	03600	58153	38089	99100	73014	31305
16918	23456	81198	75611	25074	48084	03581	64703	27349	74763
46268	35453	53423	93232	96635	01540	07102	35254	17330	03758
55888	65437	40317	22775	57810	31889	32922	03784	05018	37517
25423	35242	78919	09390	37512	13982	05485	86728	34392	30716
89078	42669	29644	78077	72494	16407	04518	27476	60810	25292
13192	42580	66330	68901	44233	62951	29750	39622	68054	11176
70931	04040	92715	84711	50352	92935	37042	56655	97889	87259
77128	77829	88384	34423	38977	88885	79735	22965	69877	98333
37100	95870	34423	68405	52871	15661	85110	84313	99862	10238
10308	26146	30767	59839	47405	18099	88450	98937	72828	61186
01822	07742	73964	01142	16459	36271	07333	02936	50903	53965
07393	92846	14103	35259	30109	60824	52938	31191	27402	55803
50735	71154	23048	03035	08385	61502	23161	94518	63237	16715
07419	04879	07820	33621	56625	85884	49636	57122	98079	14220
50087	54480	86002	35638	98404	27118	02877	02755	85210	29357
71772	45715	74504	54733	80412	18241	88087	40118	06232	93326
45714	59014	12114	18042	90519	52413	93567	24945	85998	17555
59220	80301	51824	76885	85646	16102	74986	65006	53715	19746
10764	53574	99564	08495	80459	85802	76505	36292	40744	19788
17678	48295	83757	65027	35491	46418	67883	72615	65192	84193
25137	13748	72617	05969	41288	08094	72739	95361	06969	94802
73983	90100	48318	64269	19759	81850	24789	55982	30526	22676
99383	86972	98698	53377	78399	75624	77124	78996	42364	04273
81463	95646	47319	48775	93940	51645	60109	18278	29681	77883
47218	25488	68262	21410	61186	89952	11764	79953	56457	95497
81348	41579	58825	22003	49490	30865	37291	68696	04454	88205
87900	38409	86023	52902	65419	63324	88553	23217	25332	21388
55171	74690	78509	82507	74217	61944	45093	20674	95483	23757
72273	88384	07084	32642	58395	87752	57079	29525	06486	47759
31315	92691	65161	01439	80177	42240	87554	50096	89808	01816
14727	40472	08302	00582	81954	46590	43884	17637	88473	20390
83748	71311	64473	73025	90187	37261	38389	62432	31440	59825
88824	06478	50725	76185	50468	38948	23682	11328	77542	88781
15674	10172	11670	56256	01649	07564	38895	01649	51479	65397
27554	23402	41542	24214	93388	67693	29967	67477	04991	12243
37635	96312	73891	16729	11725	76610	39574	99565	36190	32008
07476	42396	95670	49941	51549	07010	53100	48997	33188	29414
66005	72644	19812	65399	48215	09132	25707	73648	60225	30702

Random Digits (*continued*)

52534	82857	47299	48748	72888	31705	13867	62620	68513	79149
88602	31385	63436	43035	10815	07119	35964	30360	63656	60661
91400	63007	97273	32057	91951	46330	12893	86602	03710	56516
80753	34577	58421	51026	28514	06233	70431	42107	97639	90818
35885	86909	61780	30921	33258	12960	84640	11301	05409	32044
93990	99581	70584	49811	43933	14954	49446	70379	27056	95378
09614	57360	60566	71263	39899	98163	43014	09500	47266	43147
06856	28724	93988	77589	22748	32910	14649	52967	34718	90217
79287	76856	85097	14422	94957	02732	73579	31711	52065	41789
63180	07117	50802	57008	72837	88961	86536	54036	01747	59887
09472	93904	79062	09699	06683	39134	22416	66211	31214	81730
15627	05471	52308	78528	25876	82840	20825	22134	94528	53834
31677	96908	03890	21314	64503	80250	90814	92145	73283	60898
65093	27865	69792	50778	30718	32243	52658	83178	67131	95642
01874	79890	73607	21145	65168	05079	55896	97865	93480	92630
70378	46175	35976	11715	13058	69755	45829	86870	86083	32591
27728	79685	87000	81766	13756	29480	63335	37586	03953	15324
39542	43035	24562	93335	00849	57574	20158	04989	41399	84132
15083	24701	87498	56370	41122	66845	35060	75773	07576	75952
38208	12759	03955	72612	18824	70458	73336	17085	22792	20258
15282	40571	74608	54511	01418	63856	60915	16317	60816	49287
36061	06128	71126	47235	87258	98985	91898	29157	53154	15095
66755	02682	33789	88898	29899	03361	80922	66412	36269	15562
36558	16521	10515	82211	47412	59691	15381	93119	31477	83235
48090	82667	87063	61246	10231	39794	18043	04686	49861	88349
63465	99694	98810	44399	90063	42883	40735	95238	44447	54255
96395	05290	86695	34823	17766	89280	86992	56403	48722	89742
82253	29278	46716	17803	16278	67634	46388	46708	48283	64313
22288	14812	38440	61537	68293	68072	81028	50708	40678	75296
35481	12317	84403	27489	14825	68187	86422	66783	00748	81409
05628	23186	49976	51151	27859	53810	50390	21892	36258	76361
23071	52184	85989	74678	20590	84137	25824	33136	56998	20918
77886	44504	48977	07431	29133	31306	19685	45540	78390	39250
99517	36561	13069	73593	96267	34628	01825	11804	67831	49082
55302	19261	76761	17529	76210	13943	25544	59446	50749	55514
44083	43083	43603	48769	75463	96623	84077	66487	06917	78976
98302	35491	15837	75768	25722	66625	73375	40050	29995	30804
56887	14082	29531	46695	63224	79911	69596	81584	80995	33835
67921	91143	53255	79854	08527	13399	74312	58701	71350	85986
29088	87873	31377	59474	68647	31290	64092	04332	60435	15943
75682	95072	81979	08783	39007	64506	38531	09770	69281	78108
17748	60659	16166	41789	99067	79611	16400	73810	00030	39514
75168	24241	75997	82268	18396	35742	22967	65975	60427	04488
75091	41347	29533	69180	76810	13149	36522	70276	69792	30860
93673	57951	74687	75747	48063	07832	99227	52971	09475	71006
69834	52374	59616	91240	18688	84379	31381	99735	95153	94818
96128	64581	02975	98263	59500	96218	18273	30260	04603	41966
98609	81146	56638	77827	28671	58102	92651	24204	25081	50459
49501	76986	33488	89038	45521	26386	07564	31116	06500	51750
84477	65814	16334	14684	25142	48115	52182	04834	70368	69631

Random Digits (*continued*)

77012	07384	73077	58917	90883	74805	32390	45582	23330	69756
99688	96577	36515	53327	27830	78936	03391	93326	41244	81366
20302	92144	50915	29248	36120	11778	86630	76882	41739	17986
87101	03409	76898	41137	43948	86332	51689	07086	70734	05550
81969	35233	89022	76526	43304	36463	83480	59540	47470	69287
34807	08126	85164	30912	62625	53139	16842	09266	52532	46581
80291	09797	06976	65531	94969	71925	26860	95874	35446	71293
00849	40816	97682	40753	67140	09518	54907	73276	11899	98687
34082	67803	78386	96338	90435	33261	72133	88780	99020	97004
48144	36678	28842	99910	91250	19469	81431	72494	37623	22987
71923	97743	68311	35887	84390	02619	02411	43127	51406	68460
97766	61888	77653	78082	32839	10269	29426	55488	01517	88955
18801	39524	79790	74616	77609	74732	61525	92718	25394	51080
09013	01087	80078	15951	91875	76013	13021	62817	71591	97002
13048	51611	74386	99985	44431	12773	90146	38524	22724	27506
41789	61818	51838	46690	75215	86387	93567	37939	30734	19879
58001	25588	78572	27581	72758	45851	25995	26819	21913	10179
75657	91363	06883	63767	13608	42773	63539	18529	31925	28668
59445	80665	88151	12051	63428	35053	01179	05254	68537	42581
44889	04294	07799	67874	79841	92780	71505	06464	76839	63347
81237	64397	68539	66659	52781	60654	84232	53804	65075	27220
10679	50334	11422	02765	29241	22632	91788	67316	68237	64200
58973	65467	38950	87853	16905	90671	57147	30893	06623	41475
49234	68411	12338	15153	83208	25784	40268	38287	73891	27376
36185	81850	31507	60906	23892	59827	67797	62799	91022	79253
02697	19279	56989	81942	80964	67913	08399	73386	94867	67321
55830	02431	93205	79197	74599	87585	74438	83704	21945	15137
23957	75773	68884	31936	81490	45020	55818	24592	84987	50474
06343	44395	48564	42993	65735	44594	57843	70724	92870	32617
39485	98566	02373	23953	30520	93264	31347	23349	38934	45635
48876	74880	49887	86365	18299	97041	31641	05984	88948	70140
13551	41581	09044	56419	16195	46709	05714	53352	19679	87224
74692	76567	86266	19132	45849	02131	74225	04155	43997	62830
29131	83732	75242	65232	08872	43279	00809	97046	28851	15337
38528	48380	94012	44693	53843	24203	04579	95053	43817	62309
45021	72055	37371	44018	98373	48304	90618	42059	97379	10602
06474	23471	13782	32396	97506	53770	03116	51076	66798	64104
83575	05580	81439	06635	02816	51003	78830	02038	76344	89083
17007	74305	01821	41041	42674	44803	36700	74785	25617	68049
37606	40561	54302	99589	22671	34258	26432	06732	00600	93367
17936	91024	84985	03289	07721	39729	45809	80821	39587	07684
34330	18439	69577	37771	68254	90970	10173	35598	82336	74141
69074	10171	17991	65383	51863	38535	01224	10097	14548	98917
16723	63250	14038	39436	46531	29073	50638	25920	04575	78448
97324	63679	10177	05637	62994	20462	61271	21702	84168	46679
02085	54343	26238	58376	39827	35866	33987	79586	29719	47766
56224	99229	54362	76943	13980	83508	22616	15862	84824	89884
96214	94020	06341	78763	62643	83854	16712	91383	15935	82958
67058	61329	23706	22885	44251	27633	82942	79253	40230	07629
95044	15713	92819	23532	40332	49307	57167	24892	63015	43145

Random Digits (*continued*)

88115	59359	88857	19687	34108	33685	47395	85450	74431	53889
28073	71635	12335	70915	34263	46352	13003	75724	91414	70539
32924	78792	18933	91796	63082	95063	10456	61652	81171	64157
79599	42519	05201	17045	39382	85426	78979	45796	77853	37515
67006	80140	74512	12408	64096	48444	88584	30026	86203	69963
53957	07994	42002	12668	50178	10182	60719	91408	52581	31505
49498	88978	17442	68378	54770	03452	35732	66049	71522	78488
18990	45671	72172	37095	63197	35698	27050	80879	22214	17387
56316	57505	67778	56592	42765	46357	21105	62470	27862	30463
90133	09824	25398	50303	93421	45316	67491	08712	49107	89776
53760	25900	43391	18319	08120	63323	15314	18931	06321	67593
24233	68519	57668	89044	21821	50937	98675	93242	03575	52357
11980	81333	87078	85924	87322	71992	37385	89058	03613	58247
72173	21019	66804	98240	86312	94275	65750	39800	89690	79511
55332	13184	41987	81401	08589	19763	15911	62758	90384	08462
57607	25649	41487	81745	25535	45433	38792	67984	14297	03353
69686	81509	19028	75140	03866	82198	64457	32441	90153	47008
34130	16327	81730	80949	11124	64870	04745	32499	40333	61195
85721	82231	09296	43409	80316	68520	05796	57927	49096	43765
40694	73282	88363	23493	66853	05007	88903	89201	38588	74264
28790	73555	39872	82425	67586	04692	58763	73089	62844	07701
01488	66532	81456	26070	54050	21033	34261	62446	04251	88138
95499	75165	80569	89454	25994	26350	81331	66392	22865	38179
23246	93026	49835	68011	46709	97573	41406	87120	23019	76957
54372	71661	98259	03429	90265	78363	79075	69264	01258	25346
50667	67447	12042	93977	84144	57085	92362	17519	44003	00728
39015	09690	22308	55079	46001	52100	62325	34030	86493	05479
30848	98916	84404	69121	62466	64858	20484	41417	70154	54470
03889	22848	43778	74537	75257	60289	91993	33513	47265	92628
94392	05327	45485	23672	49212	40702	27590	45199	41120	28897
68181	66796	01111	81046	03106	02797	46778	06137	74457	17698
54283	10083	48124	66874	40560	20775	77508	24814	42647	61096
48823	58942	76779	11831	71291	30891	75550	08406	35640	97312
24114	27526	36778	52558	03349	27244	32784	31944	04514	79805
58250	09433	49922	73517	05749	29521	11486	35378	02951	19991
53805	24698	90621	44865	30020	79181	27710	77898	18641	25822
67094	32424	78081	01866	06524	15797	50526	19737	46728	47806
96327	08369	01352	81145	76239	27756	92892	40249	12677	20351
96304	17442	26193	50470	40080	45962	27413	52596	74767	15419
14320	29318	35603	67402	11237	10278	34135	11550	08956	78074
17503	66734	92657	25594	48869	45477	07566	01028	98648	11852
78373	76435	19095	11690	87307	86055	69098	79807	32699	21694
78058	96964	06841	98848	17419	09237	80180	37075	15571	75381
74417	70127	90585	33464	32119	49159	97364	98803	41313	52271
05478	71773	34259	73504	17136	90957	04215	79673	59041	58532
95594	70376	56991	94958	73025	72241	18772	83987	66214	96335
86317	76771	66837	10831	54982	29193	64089	48005	22480	35522
71728	32787	76084	68568	38031	65199	70118	82010	63048	89269
39979	10146	27776	94907	33324	37254	91599	06189	53170	95295
15592	60403	42603	96773	89004	87192	57567	05696	98464	98800

Random Digits (*continued*)

70926	14068	90617	51352	05865	25126	21435	69981	06479	54942
38294	02507	86133	46888	21711	98619	05872	54301	87716	26002
51447	00598	36601	60566	32418	06444	18754	79000	29678	33535
33733	65492	09115	87007	43944	90683	36936	70086	26343	44499
98578	16717	53459	98243	68053	14785	94556	24145	73756	02966
98045	35229	65436	97032	90692	92839	48741	46015	60649	78986
11030	27678	40435	23421	30446	68355	41264	56845	25972	24239
06398	01268	40461	42165	76730	80605	64851	45178	40698	45023
66982	99502	99638	32288	81444	22186	99057	35949	63798	62111
49990	41133	65203	58173	33477	09119	34541	41143	46943	46256
22975	87537	49981	83262	55604	56229	84786	38692	16164	86948
56363	46172	55878	38017	28757	23228	36018	72170	51051	75235
25239	51964	68059	75456	39179	09456	92300	65626	48879	31402
59897	48512	00334	45937	19369	41725	17979	02825	84411	90936
07445	43199	11331	84333	24530	07944	18773	94012	01441	40655
52821	36525	86483	01485	27152	22479	34278	29029	82444	32543
04483	39213	14609	02255	65310	22945	48013	36887	71101	33008
85207	37425	35526	59376	85614	97070	30842	27193	62451	28179
36385	78567	05346	74610	26141	87177	19154	53851	05824	59090
55407	39812	30331	60144	17516	98353	69388	48352	58137	77898
46331	84545	74655	37810	71052	47561	38516	00995	29132	67466
04897	76974	58825	06219	35829	94137	86723	33959	58960	78467
66731	13895	78598	34046	33819	31321	28214	15552	79956	93786
43230	46708	41391	27181	83392	83917	50354	06317	82244	64964
66454	90538	67155	16330	47634	26177	91044	95720	98046	97875
15390	36462	55477	37908	88709	79835	02268	34899	43576	94090
46247	72875	26162	79014	84536	95251	08529	38081	20459	00237
55637	15060	33992	51861	89417	48441	81158	34236	95130	56722
02487	37297	82484	66076	52244	52682	24153	95903	16598	18839
42610	30603	43360	37782	74207	42717	30680	04798	49349	80247
76181	27103	98753	17717	75804	70789	38568	18708	96245	97479
24240	59688	23292	40949	74821	74395	15182	28469	39961	16523
81005	08991	34680	03514	44197	25382	59596	78425	98515	94769
56871	37208	97581	82371	26370	49617	59215	38569	14739	48414
33208	81662	43605	20558	79985	90844	17530	99091	10285	21540
43803	95466	93094	97246	31204	85156	37718	08525	91170	63419
39723	01384	22765	93642	60124	99086	45153	32542	47145	48575
62515	76499	84234	06259	51069	60919	55124	78419	58554	94540
45055	24644	30258	39939	53071	26932	18676	19285	62417	63764
69681	82442	84755	30753	38850	46942	61530	59202	11087	18121
83385	60015	99787	68177	43030	09512	56402	52055	69511	95801
04458	50325	13625	59771	52638	03817	31659	90880	61424	19064
35754	46420	40662	91784	10383	86003	18461	37487	07663	29044
10162	66613	00105	17031	85743	46022	84098	48084	05707	94180
89831	87771	78854	87869	25562	86955	25525	10040	73737	96766
55684	77321	42361	69034	44115	03720	97262	71890	16199	59265
52365	07152	28678	69439	55376	06525	59029	29933	27542	74515
33636	30300	46185	47790	80994	17002	12405	42203	99491	54380
38313	14021	53007	04659	93128	86269	29475	28220	71108	33211
12929	16404	65222	46174	52721	30713	10441	13115	06313	78985

TABLE VI
Normal Deviates*

1.082-	1.094-	.553	1.466-	.390	.857	.581-	.216-	2.016-	.472
.153-	.845-	.077-	1.559	.592	.760	1.193-	.016	.162-	1.009-
2.573-	.914-	.034-	.961	.304-	1.138	.238	1.795-	.974-	.564-
.286	.102-	.182	.090-	.069-	.138-	.377-	.573	.551-	1.270
.157-	.295-	.801-	.171-	.985-	.481-	.015	.910	.319	1.027
.974	.489-	.042	.778-	.785-	1.252-	.473	.599-	1.007	.890
.337-	1.024-	.566	.773-	.364-	.623	.129-	.514	.720-	1.124
.874-	1.083	.143	.571	.886-	1.297-	.215-	.514-	.970-	.025-
.939	1.452-	1.126	.416-	1.840-	1.302	1.174	2.638	.254	.191
1.587-	.053-	.625-	.051-	.345-	1.445-	.332	.042	.473	.169
1.481-	.949	.324-	.503	.148	1.483-	.560	2.747	.955	.445-
1.036	1.616-	.265	2.189-	.986-	.232	1.433	.761	.876	.163-
.422	.167	1.197	1.224	1.773	1.995-	1.665-	.199-	1.049-	.335-
.010-	.053-	1.564	1.160	.346-	.142-	.315	.034	2.446-	1.001-
.787	.828-	.595	.603-	1.877	.163	.537	.355-	2.015	.667-
.431-	.025-	.045	.536	.997-	.506-	.700	.346	.086-	.101
.328	1.477	1.013	.330	.168	.504-	1.392	.281-	.118-	1.308-
.859-	.160	1.018-	1.198	.764	1.447	.428-	.505	.222-	.699-
.965-	.454	2.078-	2.432	.023	.601-	.169-	.716-	.680	.418-
2.506	.291-	1.500-	1.044-	.102-	1.979-	.305	.195	.714	.404-
.354-	.588-	.300-	.298	.762	.839	.557-	.247	1.170-	1.217-
1.598-	.239	.115-	.226-	.151-	.490	.830	.534-	1.410	.514
1.079	.102-	.850-	1.253	1.461	.085	2.606-	1.104-	.562	1.216
.560-	1.396-	1.655-	2.010-	.112	1.043-	.769-	.680-	1.242	.627-
.196-	1.257	.242-	.142-	.925	.091-	.812	.344-	.147-	.731
.160-	1.096-	.252-	1.159-	.037	.070-	.345-	.489	.532	1.611
2.007	.383-	.165-	.234	1.739	1.617-	.703-	1.150	1.475-	2.268
.018-	.611	.144-	.196	.063	.441	1.756-	.004-	.784-	.515
1.003-	.307	.679	.521	.257-	.081-	.629	1.609-	.786	1.360
1.696-	.523	.679-	.176	.154-	2.007-	1.432-	.342	.631-	.777-
1.692	1.317-	1.146	.089	.506-	.178-	1.211	.306-	2.105-	.308-
2.192-	1.908	.650	2.505	.555	.017	1.062-	1.542-	.660	.263-
.663	.994-	.249-	.541	1.895	1.231	.811	.165	.504-	.281
.084-	1.193	2.735	.885-	.242	2.052-	1.656-	1.675	1.993-	.983-
1.234-	.278-	1.259-	1.156	1.879-	1.943-	1.303	.698-	1.261	.567-
.484	1.291-	2.314-	.929-	.466	1.089	.532-	.127-	.559	.800
.700	1.221-	.806	1.367-	.698-	.285-	1.480	.571-	2.037-	1.558
.978-	1.461-	.961-	.472	2.550-	.846-	.180	.302-	1.248	.766
.098	1.497	1.032-	1.286-	.228-	.417-	.503	.338	1.060	1.565
.413	.671-	.617-	.768	1.394	.835	.387	.012	1.260	.417
.118-	2.414-	1.508	.677-	1.427	.698	.530	.166-	1.001-	1.226-
.412-	.089-	.125-	2.033-	.212-	1.523-	.822	.435-	2.239	.177
1.109-	1.658-	1.285-	.391	2.552-	1.460	1.181	.191-	.068-	.117
1.833-	.294	.321	.142-	1.501	.435	1.286-	.302	.665-	.125
.097-	.069	.691-	.737	.319	.677	.859-	.335-	.221-	.361-
.313-	.543	.042	.978	.938-	.259-	.507-	1.435	.263-	.009
1.144-	1.534	2.071	.981	.927-	1.203	.176	.311	.166	1.000
.152	.343-	2.805-	1.986-	.527	.540-	.581-	.556-	.777	.169
.166-	.883	.360-	.013	.574-	.567-	.129	.247-	1.438-	.343
.913-	.687	.087	.636	.464-	.511-	.891	.214	.615-	.810

*From The Rand Corporation, *A Million Random Digits with 100,000 Normal Deviates* (New York: The Free Press, 1955), by permission.

Normal Deviates (*continued*)

.202-	1.303-	.671-	.140-	.018-	1.565	.284-	.622-	2.073	.481
.420	1.103-	.176	1.099	.092-	.482-	.543	.218-	1.683-	2.836
2.417	1.181	.168-	.238-	.560	1.847-	.061-	.578-	.513	2.014
.260	.580-	.539	.955	1.128-	.730	.979	1.812	.195	1.322-
.353-	.151-	1.598-	1.213-	.189	1.014-	.678-	.412-	.165-	.101
2.555-	.712-	.567	.085-	1.792	1.116	.252	1.676-	.121	.346
.666	.149-	1.359	.760-	.214	.446	.682	.584	.126-	.662-
.077	.526	.783-	1.950-	.854	.084	.552	.757-	1.018-	.528
1.365-	.027-	.251-	.273-	.494	.022-	.383	1.253-	.728-	.194
1.833	.154-	1.804	.414-	.103	.759	.054	.504-	.066	1.647
.308	2.537	1.220	1.250-	.371-	1.210-	.906	.604-	1.361-	.519-
.768	.132	1.464	.428-	.182	1.792-	.864	.483	1.799-	.349-
.957-	.265	.724-	.055	.885	.379-	.694	1.448-	.672-	.209
.094-	.957-	.373-	.792-	.086	.134-	1.493	.210-	1.830	.109-
.148	.539-	.397	.362-	.245-	1.194	.746-	.242	.197	1.375
.661-	.854-	.379-	.759-	.804	.282	1.317-	.219-	.318-	.580-
1.231	.337-	.185-	1.373-	.535-	.119	.775	.254-	.598	1.200
1.117-	.871-	.187-	.543-	.421	.311	.493	.574	.145-	2.332-
.551	.335	1.746-	.235	1.455	.251	1.024	.062	.009	.676
.743	1.076	.766	.052-	1.194	.517-	.401-	1.292	.280-	.540
.329-	.277	1.736	.175	.401-	.665	.479	1.322	.072	.867-
1.264-	.970	.639-	.761-	.502-	1.559-	.249	.119	.065-	.812-
2.092-	1.610	1.423-	1.071-	.642	.759-	2.276-	.133	.976-	1.506
1.447-	.154	1.464	.032	1.076-	.327	.378-	.055	.521-	1.400-
.018	.533	.558	.593	.737-	.189	1.876-	.140-	1.380-	.303-
1.445-	1.357	1.657-	.837-	1.417-	.548	.423-	.398	.167	.147
.002	1.537	.113	1.008-	1.080	.772-	.368-	.290-	2.146	.539-
.576	1.201-	.108-	.334	.659	1.192	.119	1.861	.856	.018-
.108-	.385-	.228	.166	1.169-	1.099	.914-	.462-	1.312	.266-
.233	1.043-	.852	.746-	.046	.395	.735	1.526-	1.065	1.450
1.239-	.155	.090	1.130	2.623	.811	1.372-	.647	.858	.740-
.928-	.802	.043-	.463-	.985	.395-	.386	.465	.372-	.278-
.670-	.821-	1.092-	1.062	.601	2.509	1.557-	.814-	.220-	.019-
.643	1.339	1.287	.446	.042-	.593	.366	.640	.850-	.847
2.503	.162-	1.125	1.241-	2.226	1.063	.085	.016	.786	.766-
.895	2.238-	1.711	.640	.067-	.088-	.031-	1.184	1.550	.417
.070-	1.367-	.659-	1.025-	.475	.059	.792-	.468	.284	.185-
.891	.903-	.213-	1.847	.223	1.640-	.772-	.324	.013-	1.757
1.170	.340-	.295-	.451	1.081	1.073-	.073	.477-	.397	1.282-
.130	.205	.665	.306	.790	.851-	.935	.502-	.650	.254
.591	1.342-	1.194	1.428	1.470-	1.202-	.450-	.668-	.212	1.161
.487-	.792-	1.453	1.465-	.390	.796	2.186-	.461	.848	.236-
1.048-	2.550-	.241-	.109-	1.385-	.066-	2.523-	1.270	.914	.157-
.984	.357	.563	1.177-	.371	.624-	.614	.566	1.292	.776
1.217	.976	1.516-	.737-	.018	.768-	.712	1.001-	.012	.456-
1.008-	.849-	1.272-	.903	1.192-	2.081-	.157	.708	2.132-	.297-
.596-	.219-	.726-	.417-	.214-	.625	.699-	.276	1.505	.672
.315-	.999-	1.788	.592	.640	.877	.965-	1.066	1.189-	.657
1.441-	1.171	.192-	.315-	1.714	1.131	.001-	.342-	.039	1.480
.413	.269-	.602	.085	.848-	.207-	.396	2.358-	.045-	.087-

Normal Deviates (*continued*)

.064	1.259-	.949	.595-	.068-	1.744	.031-	1.095	.609	.114-
1.206	.185-	.485-	2.179-	.343	.070	.175-	.670	1.236-	.781
1.366	2.142-	.332	.022	1.923	.327	.466-	.276	1.402	.025
.869	.561	.397-	.538-	.213	.525-	.026	.853-	.568-	.535
.362-	1.625-	1.122	.322	.300	.166	.498-	.022-	.433-	.159
1.554	1.453-	2.637	.711	.541	.275	.005-	.109	.153-	.176
1.304-	.985	.186	.163	.268	.577	.561	.766-	.256-	1.726
1.487-	1.119-	.561-	1.240	1.554	.309	.758	.230	.747-	.705
.816	.316	.734	.859-	1.041	.523	1.062-	.590	1.641	.448
.337	1.326-	1.467-	.361-	.020	.680-	.177	.623-	.608	.627-
1.312-	.222	1.547	1.091-	.809	.042-	1.299-	.118	1.500-	.215
1.010-	.043	1.601-	.047-	.058	.883	.790	.575	.647-	.017
.477-	.812-	1.868	.249	1.764-	.575	.796-	.411	.372	.096-
.388	.614-	.587-	.535-	.518	.061-	.020	.463	.504-	.422
2.080-	.352	.838	.862-	.631	1.502	.801-	1.783	.390	.661-
.535	.424	.096-	1.069	.359-	1.558-	1.189-	.629	1.124-	.311
.591-	.184-	.830	.674	1.126	1.545	.907	1.327	1.091-	.594
.265-	.449	.175-	1.676	.688-	.493-	1.501	1.504	2.387-	.876
1.033-	1.702-	.684-	1.114-	1.150	1.488-	.160	.593-	.224-	.456
.300-	.037	1.138-	.812-	1.756-	.095	.601	.700	.884-	1.459
1.024-	.284-	.239-	1.415	.662	.078	.113	.835-	2.192-	.800
.357-	1.126-	1.544-	1.132-	.557	.177	.069-	.444	1.139	1.342
.433	1.038	1.930-	1.181	.418-	.529-	1.221	1.222-	.527-	.867-
.344-	1.417-	.973-	1.435	1.253-	1.649	.923	2.646	.065-	.274
.048-	.909	.941	.324-	1.129	1.153-	.286	1.662	1.268-	.784-
.344	.450-	2.742	1.118-	2.261	2.516-	.141-	.854-	1.285	.332
1.799	.102-	1.617-	.730	1.112	.326	.390-	.681-	.924-	.364-
.925	.856	.545-	.291-	.082-	.378-	.923-	1.316-	.983-	.590-
.762-	.209	1.045-	.924	.294-	2.773	.293-	.149	.476	.692
.372-	2.184	1.159-	2.540-	1.011	2.382-	.598-	.465	1.044-	.037
1.587-	.329	.733-	.373-	.001-	.936	.029	1.290	.587-	1.106-
.736-	.047-	.055	.397-	1.080	1.740	.664	.635-	.821-	1.682-
.768	.369-	.138-	.860-	.026-	.150	1.444-	2.445-	.549-	1.576
2.588	.628	.343-	.179-	1.123-	.490	.631	1.471-	1.783	1.064
.133	.277	.868-	.548-	.731	.230-	.933-	.642	.713	.761-
.149-	1.567-	.174-	.101	.161-	1.574-	.031-	.492	.689	.000
2.121	.394	.372-	2.125-	1.001-	.533-	.699	1.325-	.652-	.430-
.174	1.547-	1.077-	.889-	.538-	1.360	.083-	1.403-	.338	.403-
.466	.030	1.350	.844	.082	.165	.836	1.277	1.370-	.429-
.551-	.824-	1.169	1.192	.485-	1.200-	.240	.423	.486	1.855-
.696	.065-	1.652	.440-	.915	.812-	1.354-	.872-	.279-	1.528-
.925-	.280	.270	1.322	.988-	1.381	.207-	.178	2.352	.084-
.680	.041	.699-	1.386	.706	.713	.926	.641-	.900-	.228
.677	1.434-	.219-	.407	.538-	2.494	.501	.206	.733	.543-
1.528	1.060-	.201	.190-	.665	.895	.698	2.367-	.049-	2.619-
.520	2.567	.060	.098-	.243	.013	1.356	.046	.889-	.788
1.766	1.002	.374	1.329	1.884-	.524	2.111	.020	.240	2.408-
2.200	1.607-	.883	2.853-	.167	.353	.766	1.464-	.563-	.766
.012-	.666	.738	.601	.426-	.343-	1.229	.316	.112-	1.468-
1.014	.007	.407	1.388	.981-	.152	1.050-	.786	.670-	1.617

Normal Deviates (*continued*)

.739	.619	1.448-	.394-	.615	.502	.225	1.772	1.334	1.156-
2.735	.960	1.822	.335-	.345-	.572	.084	.093-	.588-	.331
.831-	1.860	1.415	1.812	.023	2.016	.546-	1.333-	.355-	.357
1.131	.310	1.824-	.126	.736	.634-	.224-	.729-	.152-	.452-
.914	2.554	.379-	.347-	1.228	.538	.723	1.184	.169-	.045-
.391-	.149	1.397-	.517	1.044	.511-	.498-	1.038	.322	.748
.852	.100-	1.293-	1.864-	1.478-	.335	.400	.343-	1.642	.536-
2.387-	1.046	.232-	.379-	1.992	.863	.234-	2.018-	.444	1.793-
.410-	1.070	.462	.562	.785	2.211	1.791	.677-	1.307	.927-
.047-	.427-	.340-	.824-	.558-	.859-	3.123	.363	1.356	.238-
.581	2.111	2.003	.732-	.476	.463-	.361-	.745-	1.011	.524-
2.008	.842-	.303	.574-	.761	2.245	.775	.824	.444-	.071-
.885-	.188-	.266-	1.106-	.834	.306	.662	.415	.759	.171
1.340-	1.127-	2.295-	.585	.844	.091-	.996-	.513	1.397	.112
1.124-	1.350	.040	.405-	.655	.758-	3.624	2.310	.140-	.006
.207-	.371	.301	.408	.046	.995-	.083-	1.374	.681	.172-
.202	.176-	.656-	.280	.792	1.401-	.595-	.902	.606	.009
.695	.489-	1.363	.073	1.485-	.269	.352	.130	1.098-	.532-
.239	.349-	.866	.457-	1.183	.398-	1.172-	1.130-	.343	.774-
.128-	1.643-	1.717-	.897	1.419-	.640	.464	1.194	.836	.627
.887	1.549	.369	.345-	.483	.834	.431	.438-	.070	1.443-
1.244-	.211	.008	1.220	1.204-	.667	1.916-	1.068-	.546-	.476-
.227	2.070	.398	.072-	.281-	.190	1.168	.290-	.958-	1.823
.019-	.556	.479	.050-	1.158-	.532-	1.030-	.670	.962	1.965-
.354-	.454-	.910	1.146	.482-	.191-	.277	.078-	.710-	2.215
1.927-	.943-	.868-	.784-	.176	.372	.913	1.736	.877	.911-
.147	1.053	1.972-	.242-	1.491	.908-	.813	1.086-	.662	.268-
.708-	.148	.699	.476	.493	1.804	.469-	.183	.896	.493
1.527-	1.102-	.141-	1.383-	.036-	.901-	.177-	.268	.405	.364
.267-	1.313	2.188	1.300	1.982-	.046-	.708-	.325	.509-	.900-
.028-	.451-	.671	1.568	.003-	1.127-	1.097	1.386	.904-	2.134-
1.101-	1.323	.213-	.142-	1.338-	.224-	.162	1.265	.986-	.693
.665	1.235	.725	.713-	1.092	.835	.873-	1.199-	.104-	.428-
.550-	.763-	.984	.032	.682	.061	.392	1.424-	1.349-	.830-
.292-	.065-	.041-	.963-	1.556	2.809-	.133-	.704	.096	.223-
.125-	1.915	.584	1.014	.322-	.194-	.151-	.696-	2.138	1.367
1.516-	1.813-	.723-	2.594-	1.090-	1.922-	.457-	.738	1.961	1.639-
.977	1.303	1.591-	.105	.894	.147-	1.504-	.206-	1.908-	.569
.954-	.906-	.653	1.186-	2.092-	.251	.226-	.256	.185-	.622-
.316-	2.072-	.239-	.552-	.108	.372	2.643	.543-	.750-	.647
.918-	.260	1.342	.370-	1.036	.246	1.840-	1.860-	1.424-	.914
.403-	.382-	.899-	1.817	.512	.779	.311-	.518	.475	1.302-
.498	.174-	1.272-	.495-	.916-	1.172	.396	.217	.047	1.214-
.965-	.874-	.338	.003-	1.079-	1.367-	.268-	.729	.087-	1.193-
1.931	1.406	.350	.347	1.272-	.342	1.586-	1.608	.332	1.703-
2.036-	.704-	.109	.193-	.636-	.201	.212	.019-	.258-	.187
.157	.858-	2.423	.321-	.110	.127	.737	.756-	1.081-	.440-
1.776	.068-	1.556	1.221-	1.527-	.363-	.798	.029-	.322	.812-
.442	.464-	.288	1.620-	.716-	2.555-	.743-	.138	.145-	.720-
1.649	.134	1.006-	.841	1.462	.347-	.721-	.169	.245-	1.308-

Normal Deviates (*continued*)

1.181	.209	.237	.539-	1.219	.293-	.853-	.194	.410-	.317
.581-	.445	.572	.795	1.158-	1.177-	.551	.279-	.405-	1.464
.442-	.834	.799	.825-	.880-	.186	1.392	.501	.334	.722-
.827	1.371-	.189-	.947-	1.626-	.612-	.952-	.828	.269-	1.023
.440	.298-	.847	.335	.659	.526	1.155-	1.143	.361	.598-
.099	.167-	1.405-	1.456-	.202-	1.620-	1.142-	.702-	.004	.644
.013-	.410-	1.225	1.189-	.937-	1.185-	.478	.360	.120	.499-
.878-	.634-	.109-	2.105	.588	2.093	.329-	.687	.337	.941-
.159	.258	.189	.418	.462	.744-	.166	.833	.182-	1.024-
1.289	.831-	1.292-	.122	.662-	.070	.008	.006-	1.508	2.496
.094	.047	.646-	.095	.166-	.556	.903-	1.321	1.397-	.355
.699-	.092	.482	2.194	.193	.610-	1.229	.244-	.778-	.221
1.176-	.559	.890	1.174	1.130	.417	1.077	.897-	1.142	.595-
.588	.456	.806-	2.433-	.435	.579-	2.106	.814-	1.094	1.155-
.134	.727	1.118-	1.494	.202-	.168	.893	2.202	1.366-	1.179
.192	.292-	.654-	.079-	.215-	.503-	.906	.567-	.658-	.314-
.515	.372-	.897	.173	.877-	.755-	.679	.148	1.766-	1.472
.409-	.332	.981-	.813-	.905	2.039	.876	.193-	1.220-	.405
1.068	.690	.924	.315-	1.323-	2.514-	.166-	.038-	.853	.212
.235-	.984	.621	.737	1.193	2.555	.723-	2.185-	.436	.640-
.560-	.609-	.630	.859-	.257-	.427-	.932	1.043-	.456	305-
2.173-	.274-	.427	.817-	.895	.535	.641-	.112-	.102	1.447-
1.695	.298-	.680	1.338	.862	1.264	1.251	.002-	.644-	1.502-
.731-	.819	1.478	.710	.004-	.042	.468	.618	.083-	2.189
.110	.782-	.573	.736-	2.111	.260-	1.821-	.082	1.297	.159-
.017	.692-	.452	.523	1.173-	.714-	1.553	.580-	1.000	1.134
.279-	.421-	1.299-	.138	.762-	2.263-	.128	1.126	.100-	.250
.500-	.939-	2.296	1.951-	1.011	.331	.499	.901	.318	.954-
1.764-	.018-	.744-	.487	.157-	.332	.660	.497	.683	.879-
1.589	.970-	1.614-	.345-	.113-	.101	.717-	.498	.020-	.509
.473	1.422-	.434	.718-	2.286-	.217	.880	.206	1.865-	.680-
.108	1.962-	1.277-	1.319-	.047-	.312	.436	.340-	.239-	.286-
.029-	.353	.226	1.222	.732	1.282-	1.183-	.802	.562	.041
.703-	.008-	.597-	.432-	.338-	1.210	.064	.696	1.832-	1.439-
.208	.940-	1.315-	.225	.002-	1.933-	.629	.752-	1.576-	.798-
.096	.778	.684-	.206-	1.318	.955	.129-	.727	.524-	1.330
.443	1.829	.456-	.833	.775	.037	2.082-	.354-	1.512-	2.135
1.790	.132-	1.381-	.610-	2.211-	.287-	.883	1.466	.714	.416-
1.787	.605-	.937-	.964	.637-	1.115	.012	.025-	.251-	1.855-
1.066-	.895	.544-	1.681	.369-	.046-	.451	.331-	1.214-	1.113
.934-	1.728-	.433	.218	1.451	.158-	.656-	1.897	.028-	.097-
.765	.938	.720	1.720	.874-	.477-	1.191-	.837-	1.141	2.258-
.774	1.809-	1.876	1.319-	1.488-	.468	2.273	.916-	.938-	.001
.656	.796	.528	2.544	1.316	.588-	.427-	1.035	.464-	1.730
1.600-	.266	.576	1.980	.405-	1.210-	.628	.144	.949-	.828
1.705	.092-	.535	.391-	.176	2.151-	1.641	1.542	.614	.791-
1.095	.320-	.731	.140	.435	1.296	1.236-	1.063-	.125	2.499
.575	.054	.446-	2.720	.709	1.259-	.484	.612	.305-	.108-
.254-	.084	1.273-	1.341-	.589-	1.002	1.636	.621-	.431-	.534-
1.011-	.408	.264	.854	.186-	1.346	1.849	1.007-	1.227	.500

Normal Deviates (*continued*)

.551	.506-	1.077-	2.834	1.318	.660	.034	.222-	1.566-	.488
.298-	.241-	1.959-	.489	1.086	.409-	.078-	.614-	.782-	.115-
.036	.932	2.513-	2.090-	.342-	.044-	.268	.885-	.456-	.069
.420-	.444-	.399	.531-	1.334-	1.097-	1.127	.482-	.152-	.562-
2.191	.513	.965-	1.623-	.087	.010	2.107	.952	.469	.863
2.063	.047-	.379-	.862	.397	.162	1.886	1.922	1.322	.182
1.225-	.175-	.592-	1.102-	.242-	.730-	.725-	.208	.512-	1.334-
1.522-	1.188-	2.236-	.359-	.241-	.530	.198-	.240	.730	1.580
.439	.486	2.578	.940	2.686	1.544	.460-	.531-	.894	.884
.000	.236	.224-	.594	.391	.141-	.206	.079	.427-	1.165-
.740-	.289-	1.152	1.295	.000	.236	.965	1.969-	.141	.722
.160	.867-	.096-	.161	.148	.701-	.305-	.251	.560-	.410-
.667-	.190-	.049	.045	.469	.471	.689	.394	.275-	.063
.251	.992-	.037-	2.125	2.712-	.385-	.102-	.054-	.864-	.264
1.443-	.399-	.171-	.027-	1.209-	.027	1.008	.427	.689-	1.653
.071	.279-	.344-	.743-	1.102	.425	2.174-	.127	.608-	.156-
1.697-	.406-	.274-	.813-	1.053-	.191-	2.004-	.123-	.394	1.865
1.045	.113-	.321-	.200	.371-	.057	.237	.036	1.063	1.105-
.348-	.979	.792	1.160	1.612-	1.777	.662	.205	.639-	.344-
.136	.693	.258-	.566	.515-	.441-	.257	.599-	.934-	.689-
.092-	.602-	1.017	.509-	.664	.917	.310-	.168	.555	.548
1.655-	.376	.738	.112-	.223	1.042-	1.537-	.568	.363-	.485
.433	1.437-	1.085-	.240-	.793	1.840-	.411-	.168	.417-	.083-
.171-	1.148-	.083-	1.849	.217-	1.617	.607-	.868-	.970	.931-
.425	.258	1.313	.475-	.444	.702-	.981-	.365-	.059-	.631
1.020-	1.553-	.346-	2.919-	.138	.679	.308-	1.099-	1.211	1.429-
.094-	.450	.609	.037	.638-	.949	.807	1.159-	1.017	.728
.142	.190	1.034-	1.214-	.413-	.652-	.047	.190-	1.249	.865
1.962-	1.075	.324-	.306	.934	.141	.415	.056	.056	1.444
.186-	.641-	.507-	1.741	.167-	.246	.311-	1.604	.650	2.399-
.712	1.026	.610-	.482	2.242	.110	.926-	.012	.700	.506-
.699-	1.173-	.245	.126	.729-	1.253	.229-	.554-	.669	1.015
.878	.064-	1.341-	.457	.394-	1.392	1.810-	1.307	.146-	.604
.173	1.282-	.326-	.157	1.974	.666-	.930	.021-	.632-	1.231
.434-	.934	.903	.778-	.161-	1.525-	.822-	2.779	.841	1.251
.156-	.765-	1.692	.460-	1.483	1.208-	1.918	.937-	.490-	1.726
.261-	1.457	2.202-	.269-	.747-	.492	1.525	1.308	.257	1.389
.319	.730	.722	.242	1.004	.094-	1.533-	.381	.027	1.422-
.124-	.207	.686-	1.985	.517-	.235	.255-	.792	.077	.829
.515	.036	.932	.480	1.026	.654-	.503-	1.031	.283-	1.976-
.970	2.162	.254	.240	2.859	1.053	.473	.638-	.176-	1.024-
1.700-	1.337-	.008	.226	1.097-	1.474	.247	.265-	.066	1.361-
.365-	1.297-	.090-	.248	.236-	.786-	1.391	.242-	1.260-	2.679-
1.272-	.946	.429	1.315	3.074-	2.082	.953-	1.035-	1.069	.804-
1.272	1.133-	1.164	.383	.801	1.488-	1.168	.292-	.657-	1.393
.143	.882	.749	.601	.193-	1.087	.497	2.348	.148-	1.639-
.059	1.238-	1.465-	.462-	.563-	1.218	.508	.308	.135	.108-
.422-	1.033-	.516-	1.986-	.096-	1.657-	.055-	.170-	.878	.866-
.297-	.800	1.079-	.375-	.075	.016-	1.679-	1.005	1.485	1.354-
1.130-	.396	.978-	.992	.391	.282	.096-	.439-	.068	1.680-

Index